Collecting Experiments

COLLECTING EXPERIMENTS

Making Big Data Biology

Bruno J. Strasser

The University of Chicago Press
Chicago and London

The University of Chicago Press, Chicago 60637
The University of Chicago Press, Ltd., London
© 2019 by The University of Chicago
All rights reserved. No part of this book may be used or reproduced in any manner whatsoever without written permission, except in the case of brief quotations in critical articles and reviews. For more information, contact the University of Chicago Press, 1427 E. 60th St., Chicago, IL 60637.
Published 2019
Printed in the United States of America

28 27 26 25 24 23 22 21 20 19   1 2 3 4 5

ISBN-13: 978-0-226-63499-9 (cloth)
ISBN-13: 978-0-226-63504-0 (paper)
ISBN-13: 978-0-226-63518-7 (e-book)
DOI: https://doi.org/10.7208/chicago/9780226635187.001.0001

Library of Congress Cataloging-in-Publication Data

Names: Strasser, Bruno J., author.
Title: Collecting experiments : making big data biology / Bruno J. Strasser.
Description: Chicago : The University of Chicago Press, 2019. |
Includes bibliographical references and index.
Identifiers: LCCN 2018052450 | ISBN 9780226634999 (cloth : alk. paper) |
ISBN 9780226635040 (pbk. : alk. paper) | ISBN 9780226635187 (e-book)
Subjects: LCSH: Biology, Experimental—Data processing. | Biology, Experimental—Databases. | Biological models—Data processing. | Biological specimens—Collection and preservation—Technological innovations. | Big data.
Classification: LCC QH324.2 .S728 2019 | DDC 610.72/4—dc23
LC record available at https://lccn.loc.gov/2018052450

♾ This paper meets the requirements of ANSI/NISO Z39.48-1992 (Permanence of Paper).

To Eole and Eloi

Contents

Acknowledgments

This book took shape over many years and places, and I owe a great debt to the many people who helped me refine the argument presented here. The book started as a history of bioinformatics, but I came to realize that the introduction of computers, although important, could capture only a small part of the profound historical transformation of contemporary biological and biomedical research. I thus broadened the historical framework, focusing on what seemed to be one of biology's enduring epistemic tools: the collection.

Initially, the argument was, wrongly, conceived as a story pitting experimental biology against natural history, with contemporary biology growing out of natural history. It was first presented, in inchoate form, to historians of biology at the 2005 Ischia Summer School, "Gathering Things, Collecting Data, Producing Knowledge: The Use of Collections in Biological and Medical Knowledge Production from Early Modern Natural History to Genome Databases," organized by Janet Browne, Bernardino Fantini, and Hans-Jörg Rheinberger. The organizers, participants, and presenters, especially Nick Hopwood, Soraya de Chadarevian, Gordon McOuat, Anke te Heesen, Lisa Gannett, Manfred Laubichler, Staffan Müller-Wille, Andrew Mendelsohn, and Garland Allen, provided constructive feedback. They helped me realize that what was at stake was not disciplines or fields, but "ways of knowing," to take John Pickstone's very useful expression. The story I wanted to tell was no longer one of mutually exclusive approaches to biology battling for academic supremacy, but a story of different "ways of knowing" becoming intertwined and hybridizing—and here my intellectual debt to Rob Kohler should be obvious—resulting in today's complex research practices. These ways of knowing aligned with different moral economies, which helped explain the tensions and frictions that could be felt when they interacted through the twentieth and twenty-first centuries. This shift in my perspective arose from the mix of encouragement and criticism that senior scholars such as Gar Allen provided. Over the years, Gar has been a model of kindness and generosity; I have tried to keep his example in mind whenever I am approached by a young and enthusiastic scholar with half-baked ideas like myself.

The research for the book started in earnest when, thanks to Angela Creager's kind invitation, I was a visiting fellow at Princeton University's Program in History of Science in 2005–6. Discussions with Angela, Michael Gordin, D. Graham Burnett, Robert Darnton, Michael Mahoney, and especially Joe November—whose insights and friendship I have enjoyed ever since—as well as graduate students were most enriching and helped get the argument on the right foot.

Most of the research for the book was accomplished while I was on the faculty in the Yale University School of Medicine's Section for the History of Medicine and the History Department's Program in the History of Science and Medicine. The five years I spent there were intellectually and humanly the most enjoyable of my career. Dan Kevles and John Warner taught me, in very distinct ways, far more than they could possibly imagine, and their examples remain for me models of how to be an academic scholar, mentor, and administrator. My colleagues in the section, Naomi Rogers, Sue Lederer, Mariola Espinoza, Bill Summers, Toby Appel, and the late Joe Fruton, as well as in the program, Frank Snowden, Paola Bertucci, Bill Summers, Ole Molvig, Bill Rankin, Bettyann Kevles, and a number of stellar graduate students and colleagues at the Whitney Humanities Center and across the university, including Mark Gerstein, Alondra Nelson, Paul Sabin, and Jennifer Klein, nurtured my work and made for an exceptionally collegial working environment.

A large part of this manuscript was written at a somewhat less distinguished institution that I visited during a sabbatical year at Yale—the Ashbox café in Greenpoint, Brooklyn. At the end of a dead-end street, it provided an ideally quiet environment to elaborate on the arguments of this book and chats over coffee with a few other academics on sabbatical leave.

I set the book aside for a few years after I undertook new teaching, administrative, and familial duties upon returning to Switzerland as a member of the faculty of the University of Geneva's Section of Biology. As president of the Section of Biology, Didier Picard did his best to make my odd institutional environment compatible with my intellectual interests. I benefited greatly from discussions with him and other biologists who read part of this manuscript, including Denis Duboule, Amos Bairoch, and Graham Robinson. Other colleagues at the University of Geneva, notably Marcel Weber and Marc Ratcliff, have been valuable resources on the history of biology, while colleagues including Jean-Dominique Vassali, Jean-Marc Triscone, and Jérôme Lacour have supported my work institutionally. In Zurich, Jakob Tanner, Michael Hagner, and David Gugerli have been great sources of inspiration on the history of knowledge.

The arguments of this book have been presented in numerous places where they received helpful comments and criticism from faculty and students, including at the University of California–Berkeley, University of Pennsylvania, Johns Hopkins University, University of Wisconsin, Massachusetts Institute of Technology, University of South Carolina, National Institutes of Health, University of Exeter, University of Manchester, University of Lancaster, University of Copenhagen, University of Milan, University of Naples, Max Planck Institute for the History of Science, University of Munich, University of Zurich, University of Lausanne, University of Bern, ETH Zurich, University of Paris-Sorbonne, University of Strasbourg, École des hautes Études en sciences sociales, and École normale supérieure. The 2010 MBL-ASU History of Biology seminar, at Woods Hole, was particularly stimulating, especially thanks to the input from Jane Maienschein, Paul Farber, Michael Dietrich, Lynn Nyhart, John Beatty, and Kristin Johnson. Lynn Nyhart and Rob Kohler have been exceptionally generous in helping me rethink what was going on in biology around 1900. Lorraine Daston's scholarship and workshops, especially the recent Science of the Archives organized at the Max Planck Institute for the History of Science in Berlin, have also been a constant source of inspiration for me. Dan Kevles helped me think harder about ownership in the late twentieth-century life sciences, and I have felt privileged to enjoy his support and friendship.

The main argument of this book was also deeply influenced by the late John Pickstone, who invited me as a visiting professor to the Center for the

History of Science, Technology and Medicine at the University of Manchester. Between two art exhibitions, John and I debated enthusiastically about how his "ways of knowing" could be put to work historiographically. I miss deeply our discussions and friendship. My stay at Manchester was also the occasion to share ideas with Jon Harwood, Robert Kirk, Sam J. M. M. Alberti, Steve Sturdy, and others in the UK.

A number of other colleagues have provided valuable feedback including Nathaniel Comfort, Marianne Sommer, Betty Smocovitis, Dave Kaiser, Janet Browne, Edna Suárez-Díaz, Jean-Paul Gaudillière, Dominique Pestre, Michel Morange, John Krige, Hans-Jörg Rheinberger, and Jérôme Baudry, as well as Joel Hagen, Mary Sunderland, David Sepkoski, and Sabina Leonelli. All have accompanied and stimulated this project longer than I can remember.

I have been lucky to access numerous unprocessed archival resources: those of the National Biomedical Research Foundation, the Protein Data Bank, the National Institutes of Health, the European Bioinformatics Institute, the European Molecular Biology Organization, and the European Molecular Biology Laboratory, as well as the personal archives of Judith Dayhoff, Ruth Dayhoff, Temple Smith, Christian Burks, Norton Zinder, Helen Berman, Ed Meyer, Margareta Blombäck, and Helen Boyden. I thank their proprietors for their trust. I am also most grateful for the indispensable help of the archivists at the American Philosophical Society, Massachusetts Institute of Technology, Harvard University, John Hopkins University, the Bancroft Library at Berkeley, Rockefeller Archives Center, Rutgers University, Yale Peabody Archives, Caltech Archives, National Archives, and the American Society for Microbiology Archives.

During the course of this book, I interviewed and corresponded with numerous scientists who shared the kind of important recollections that often leave no trace in written archives. I would like to especially thank for their generosity and time Frank H. Allen, Carl W. Anderson, Winona C. Barker, Helen Berman, Frances C. Bernstein, Dennis Benson, Joe Bertani, Howard S. Bilofsky, Frederick Blattner, Margareta Blombäck, Mark S. Boguski, Douglas L. Brutlag, Christian Burks, Graham N. Cameron, Christine K. Carrico, Judith E. Dayhoff, Ruth E. Dayhoff, Scott Federhen, Joe Felsenstein, Peter Friedland, Greg Hamm, Maximilian Haussler, Elke Jordan, Patricia Kahn, Laurence H. Kedes, Olga Kennard, Ruth L. Kirschstein, Thomas F. Koetzle, David Lipman, Robert S. Ledley, David J. Lipman, Edgar F. Meyer Jr., Ken Murray, Daniel Normak, Jane Richardson, Richard J. Roberts, Temple F.

Smith, Dieter Söll, C. Frank Starmer, Hans Tuppy, Michael S. Waterman, and Norton Zinder.

Karen Darling, at the University of Chicago Press, has nurtured this project in many ways, and her kind encouragements kept it going. Her patience and understanding have been beyond belief, and I am deeply grateful that she has kept faith in this book. My copyeditors, Margaret Hyre and Russ Hodges, polished the first draft before I dared send it to the Press, where it was polished once again by Susan Tarcov. The reviewers for the press offered a number of constructive suggestions; the book would have many more flaws without them.

The Swiss National Science Foundation generously supported this project.

Finally, I would like to thank my wife, Muriel, and my children, Eole and Eloi, for keeping my mind away, at times, from the history of collections.

Introduction

The "data deluge."[1] This metaphor, pointing toward an event of biblical (or at least historical) dimensions, has taken a firm hold in current discourses about science and society. The "data deluge," and the associated notion of "big data," are increasingly used to characterize the present era, so concerned about collecting, comparing, and classifying data of all kinds, stored in data collections hosted by companies like Facebook and Google and in scientific databases in fields from genomics to high energy physics. Coping with this deluge is not just a matter of building larger and faster computers. As the amount of information in databases explodes, we are being forced to reassess our models about what knowledge is, how it is produced, to whom it belongs, and who can be credited for producing it. These questions have significant epistemological, social, and moral dimensions, and apply just as much to everyday life as to scientific inquiry. Consider the passionate debate about the trustworthiness and legitimacy of Wikipedia, the collectively and anonymously produced online encyclopedia, compared with the classic *Encyclopedia Britannica,* whose entries are composed by identifiable expert authors with "credentials."[2] Such controversies reflect a broad uneasiness about standards for (and the quality of) knowledge in the "information society" and illustrate a current destabilization of many long-held assumptions about the relationships between knowledge and people.

The American rock composer and performer Frank Zappa was not the first—but was certainly the most vocal—to point out that "information is not knowledge" (and that "knowledge is not wisdom . . . and music is the best").[3] Since then, the data—information—knowledge—wisdom (DIKW) hierarchy has become a standard way of thinking about our representations of the world and their relationships. But whereas early authors focused on the relationships between information, knowledge, and wisdom (and music), since the late 1980s "data" has come center stage as the foundation of everything we know. Data stands at the far end of the long continuum of representations going from data to information to knowledge that humans produce to transform nature into understanding.[4] Data provides understanding, meaning, and power. The central place given to data motivated *The Economist* to devote an issue in 2010 to the "data deluge" and its "vast potential."[5] Contributors presented data and modes of analyzing it as crucial assets in an "information economy." Data is a key commodity in this new market, and a growing number of companies depend on it. In this picture, data alone has little significance without this extra dimension of analysis, "distilling meaning from data."[6] Two spectacular examples illustrate the importance currently attributed to data analysis: The authors of the *9/11 Commission Report* pointed to the failure of US governmental agencies to process intelligence data that might have prevented the terrorist attacks on September 11, 2001, and numerous financial analysts claimed that the meltdown of global financial markets in 2008 might have been prevented through a better analysis of readily accessible economic data.[7] Beyond the capacity merely to collect data, the ability to compare, classify, and interpret information has become a strategic asset in the modern world.

Databases provide the foundation for these capacities. Today they have become a mainstay of our lives and capture nearly everything, as reflected in the diverse formats of their contents: numbers, words, sounds, images (still and moving). Without databases, we could neither store nor analyze the vast amounts of data we produce. They have become a sort of self-fulfilling prophecy in which the act of accessing data creates new information and value in its own right. Any Google query both retrieves information and creates it, through tools geared toward improving the efficiency of search algorithms and understanding the behavior of users. That meta-analysis generates information about strategies and people—as users and consumers—that a lot of companies are willing to pay for.

Databases have revolutionized most aspects of our lives, but the best example of their power and importance can be found in the practice of science. There, they have become more common than microscopes, voltmeters, and test tubes. Today every scientist—whether in the laboratory, field, museum, or observatory—draws on them to produce scientific knowledge. The increasing amount of data produced by disciplines from astronomy to zoology has led to deep changes in research practices. It has also led to profound reflections on the role of data and databases in science, and the proper professional roles of data producers, collectors, curators, and analysts.

In 2008, *Nature* devoted an issue to these themes, with a cover simply entitled "Big Data."[8] That same year, the technology magazine *Wired* bluntly announced "The End of Science," explaining that the "quest for knowledge used to begin with grand theories [but] now it begins with massive amounts of data."[9] According to the article, old ways of doing science based on the experimental testing of theories were on the verge of being replaced by a "data-driven" approach: the comparison of large amounts of data in search of patterns.[10]

Do "data-driven" approaches constitute a turning point in the history of science? Such claims have become widespread since the 1990s, in the context of whole-genome sequencing and especially of the Human Genome Project.[11] Data-driven science was presented as a logical (and thus necessary) consequence of the scaling up of genome sequencing efforts, which were producing more data than any individual could analyze. At the same time, "data-driven" science was put forward as a philosophical justification for these massive endeavors in genomics, which had sometimes been criticized by experimental scientists as being intellectually shallow.[12] Instead of fighting these claims directly, proponents of genomics (and the later "-omics") enterprises argued that their scientific value should be measured by a different standard. They distinguished the standard deductive and "hypothesis-driven" research from the new inductive and "data-driven" research. In "data-driven" science, new knowledge would be produced by the collection, comparison, and classification of large amounts of data.[13] In 1999, molecular biologist David Botstein of Stanford, for example, called for the "collection" of DNA microarray data, which would then be systematically compared in order "to discover things we neither knew or expected" through a process that did not involve "testing theories and models" and that was "not driven by hypothesis."[14] These attempts to legitimize a new way of doing science were all the more important given that science funding agencies, especially in the United States, explicitly relied on a "hypothesis-driven" model of scientific research in their evaluation of research grant proposals.[15]

The computer industry, particularly Microsoft, has been quick to embrace—and promote—"data-driven" research as the future of scientific research (and, incidentally, as a selling point for the software that "data driven" science will require). In 2009, Microsoft published *The Fourth Paradigm: Data-Intensive Scientific Discovery*, available free of charge under a Creative Commons license (a rather unusual move for the company).[16] The book was a tribute to computer engineer and "silicon valley legend" Jim Gray, who had introduced the notion of the "fourth paradigm" as a successor to the empirical, theoretical, and computational paradigms in a talk two years earlier.[17] Just three weeks after the pronouncement, Gray, an avid sailor, was lost at sea in the Pacific.[18] In the volume honoring his memory, contributors from Microsoft and from academic institutions described the mounting level of data in the environmental, health, and life sciences and the vision of a new science relying on new tools to store, curate, and analyze this massive amount of data. "This dream must be actively encouraged and funded," concluded the Microsoft computer scientist Gordon Bell, who had made a similar call in a piece published in *Science* that same year.[19]

Critics of "data-driven" science have questioned the epistemological underpinnings of this new way of producing knowledge. A cell biologist argued that without a hypothesis, trying to derive knowledge from data is "asking for the epistemological equivalent of a perpetual motion machine." Because of the illusion that induction could lead to universal knowledge, he added, biology was "now threatened with a new dark age of positivism."[20] Others have argued that even "data-driven" science necessarily relies on some sort of hypothesis, or scientists would be testing "the effect of Italian opera on yeast."[21] The point is that scientists' imagination shouldn't be completely unrestrained by hypothesis and theory. The discussion goes on, but all participants agree that whether for testing hypotheses or for generating them, large amounts of data have become indispensable.[22]

There is more than epistemology at stake in these debates. There is also a defense of bench experimentation as a "way of life," which is perceived as being threatened by the computational approaches inherent in "data-driven" science (figure I.1). More important, databases have changed not only how we produce knowledge, but also who produces knowledge. New professional roles and research communities have emerged and are transforming the traditional social and moral order in the sciences. Instead of individual researchers

Fig. I.1 Caricature of two experimental scientists turned bioinformaticians, reluctantly returning to the bench to produce experimental data, a critique of the excessive focus of researchers on data analysis at the expense of experimentation and the devaluation of "the hard work of the protein chemists." Hodgson, "A Certain Lack of Coordination." Printed with permission of Elsevier.

gaining authorship and credit for results drawn solely from data they produce themselves, researchers consume data produced by others and made accessible through databases to produce new knowledge. The American writer Alvin Toffler coined the word "prosumer" ("producing consumer") to designate the blurring of these traditionally distinct economic roles.[23] In the sciences, this community of "prosumers" is far from homogenous and is rife with tension. Databases draw on the work of large numbers of individual researchers who contribute data they have produced to answer their own intellectual questions, but also of researchers specializing in the production of data alone. Open access to databases has also led to the emergence of professional communities of individuals specializing in data analysis, challenging the authority of those who produced it (and leading to some name-calling: "parasites").[24]

Furthermore, open access has also allowed amateurs to participate in the analysis and sometimes contribute to the production of new knowledge. Millions of individuals have examined and analyzed data about genetic ancestry or extraterrestrial life on their home computers.[25] Thus databases reflect not only the coming of age of modern computer technology to deal with the data deluge but, more important, the creation of a new social and moral order with distinct communities that collectively contribute to the production of knowledge.

This book is about the development and use of data collections in the experimental life sciences from the early twentieth century to the present: their emergence, their development, their meaning, and their effects on the production of knowledge and on scientific life. Data collections, or databases as they are more commonly known, play a particularly important role for experimental research carried out in laboratories around the world. At first sight, they might be thought of as the equivalent of books, journals, and other means of communication. But more significantly, they are instruments for the production of knowledge. Studies of genes and genomes, for example, rely crucially not only on the substances and instruments traditionally found in laboratories, but also on computerized databases that are now indispensable in the experimental exploration of nature. The early introduction of databases in the field provides a good opportunity to examine the emergence of this particular "way of knowing" (see below), to explore the challenges that it presents, and to understand how the data deluge is changing the relations between knowledge and people in the sciences and beyond.

Biology, Computers, Data

Recently, scholars have begun to address these issues. In *Biomedical Computing: Digitizing Life in the United States*, Joseph November offered the first scholarly account of the introduction of computers in the life sciences. But instead of proposing a technologically deterministic argument, he shows how visions of a computerized biology and biomedicine in the 1960s and 1970s, such as analog computing or automated diagnostics, never became mainstream. November argues that today's alliance between computing and the life sciences, with its massive use of databases, required other contingent historical factors (a sobering counterweight to today's hype about a computer revolution in the life sciences). But computers did nevertheless, as November shows, deeply change biology (and biology changed computing, but that's another story). Computer technologies were the vectors of profound transformations: epistemic (mathematization of biological research practices), political (federal support for computer infrastructures), and social (a community of experts on biomedical computing).[26]

Following the work of Joseph November, Soraya de Chadarevian, and my own, Miguel García-Sancho has also focused on the period from the 1950s to the 1970s to understand how the intimate embrace of computing and biology, so visible in the genomics projects of the 1990s, came to be. Instead of focusing on computers, García-Sancho examined the production of sequence data, first from proteins, then nucleic acids (that is, DNA and RNA), and rather than telling a story of technological innovation in instrumentation, he described the practices of sequencing as a "form of work" (building on John Pickstone's "working knowledges") that emerged in the laboratories of academic (bio)chemists and molecular biologists and was sustained by the development of commercial sequencing instruments and data analysis software. He showed how computational practices were developed to assist the experimental determination of sequences, and not just as a tool for data analysis.[27]

Hallam Stevens's ethnographic work on contemporary bioinformatics continued this line of inquiry by looking at how computational practices have transformed biology, especially since the 1990s. His account, based on his "conversations with those . . . working in bioinformatics" and written sources, offers a vivid insider's view of computational biology and bioinformatics research practices. Most illuminating, Stevens has argued that the successive formats of computational infrastructures, specifically sequence databases (from flat-file to relational), reflected changing views about biology (from gene-centric to multi-element). For Stevens, "computers imported statistical approaches from physics" and transformed biology to the extent that "a large proportion of contemporary biology . . . turns on the production of a product—namely, data."[28]

Such generalizations from bioinformatics (or genomics) to biology as a whole are at odds with other accounts showing that the use of computers was (and is) far more diverse than data analysis (or even production), including data recording, simulation, and expert-systems, as November made clear. The field of bioinformatics (or computational biology) thus cannot credibly stand for all of biology, even laboratory biology. While keeping a focus on the role of data, computers, and databases, Sabina Leonelli's rich "empirical philosophy" of what she calls "data-centric biology" showed that the transformation of biology has affected a much broader range of experimental, theoretical, and computational research practices. She argued that databases have been successful only when they attended to the needs of "multiple epistemic communities" and that "data-centric" practices cannot be reduced to a single epistemology but include diverse epistemic and material practices in which data are "a central scientific resource and commodity." Most important, she highlighted, as in the present book, the importance of data curators—the "invisible technicians" of laboratory work—who make it possible for data to "travel" in widely different contexts, by packaging it with the relevant metadata, thus providing data with its evidential value. Taking a more plausible view of the impact of computers and data, she concluded that "while data mining does enable scientists to spot potentially significant patterns, biologists rarely consider such correlations as discoveries in themselves and rather use them as heuristics that shape the future directions of their work." Indeed, today, the vast majority of scientific papers published in *Science, Nature,* or *Cell* do not just report the production of data but use data as a resource to support claims about the mechanisms producing biological form and function.[29]

In popular accounts, these transformations have mainly been explained through changes in technology: the broader internet, faster computers, bigger servers, and instruments producing data at an ever increasing pace and decreasing cost ("high-throughput").[30] All of these have certainly made possible the emergence of electronic databases, but more profound historical forces were at work and need to be taken into account to explain this deep historical transformation. Why did life scientists start collecting, comparing, and classifying large amounts of experimental data in the first place? Finding an answer requires examining databases in a much longer historical perspective.

Biology Transformed

The present book shares similar intellectual concerns with the other scholarly work discussed here on the deep transformations of the life sciences in the twentieth century that made computers, data, and databases essential to contemporary research practices. The central argument of this book is that this transformation is best understood as part of a much longer tradition of collecting, comparing, and classifying objects in nature.[31] The tradition of collecting has been most closely associated with the endeavor called "natural history," including taxonomy, paleontology, and geology, but it is also connected with comparative anatomy and embryology.[32] Here I hope to show that the practices of collecting were essential to the development of the experimental life sciences, especially when they focus on the level of molecules, such as DNA and proteins. In a nutshell, today's databases are to the contemporary experimental life sciences what museums have been to natural history: repositories of things and knowledge, as well as key tools for their further production. This perspective lends a new sense to many of the issues that have been raised concerning contemporary databases. To whom does the knowledge that they store belong? Who should have access to it? What is the status of the data collector? Who is responsible for the integrity of the data? How should databases be organized? These questions are nothing new to naturalists dealing with their own "deluge" of specimens, bones, skins, and fossils for several centuries. The answers they found were appropriate for their time and context, and now can provide guidance and inspiration for current attempts to understand the role of databases in the production of knowledge.

A historical approach to databases might seem odd because we intuitively feel that today's "information overload" is unique in quality and quantity. The feeling of being overwhelmed with information, however, has a long history. As historian Robert Darnton has argued, thinking of the French Enlightenment, "every age was an age of information, each in its own way,"[33] and every age has devised its own means of coping.[34] Even in the Renaissance, as historian Ann Blair has demonstrated, there was "too much to know."[35] The technologies that scholars developed were primitive according to present standards but were, in their own context, very effective in dealing with the amount of information and making the best use of it. Libraries and museums, encyclopedias and card collections arose long ago, yet they served the same purpose of storing, organizing, and making sense of overwhelming amounts of information as today's databases,[36] which are merely the most recent addition to this long list of modes of dealing with data and knowledge.

These technologies have often been thought of as repositories of knowledge whose main function is preservation. All, in fact, have equally served as tools for the production of new knowledge. Natural history museums, for example, preserved rare or even unique artifacts, including specimens of extinguished species, but the point was to allow their study in the broader context of a collection. In the late nineteenth century, natural history museums emphasized their role in education but continued to expand their research activities as well. However, their spatial reorganization according to the principles of the "dual arrangement"—separate rooms for public display and for research activities—has often hidden extensive research activities carried out on the collections, even when those activities took place in the same building, contributing to the perception that museum collections were merely for display.[37] Thus the analogy between databases and museums as research technologies can serve a heuristic role to help us understand the role of databases in the production of new knowledge.

These technologies are part of a specific "way of knowing" the natural world based on collecting, comparing, and classifying, which is epistemically, socially, and culturally distinct from the "way of knowing" typically associated with experimentation and laboratories. These two ways of knowing have often been opposed, at least since the late nineteenth century and throughout the twentieth. They were rhetorical weapons in the many disciplinary battles fought among field naturalists, museum naturalists, comparative anatomists, and many kinds of experimentalists since the expansion, in the mid-nineteenth century, of experimentalism in the life sciences and medicine. The standard story, as recounted by early historians of biology and scientists alike, holds that after many centuries spent "merely" collecting and describing nature, the life sciences finally began to benefit from the laboratory revolution.[38] In the early twentieth century, the rise of genetics, microbiology, and biochemistry illustrated the growing power of the experimental approach in unlocking the secrets of nature, alongside the inexorable marginalization of natural history. The successes of molecular biology, at the mid-century, testified to the ultimate triumph of the experimental approach, confirmed by the current (post)genomics era.[39]

This narrative was crafted almost half a century ago and has shaped subsequent scholarship that has often, implicitly at least, adopted this framework opposing the "old" natural history (including morphology) and the "new" experimental biology. Yet there are a number of problems with this picture, as pointed out by historians of biology. First, as Lynn K. Nyhart and others have argued, one should be suspicious of this narrative that was designed by the proponents of the "new biology" themselves, as there were many continuities in research practices. In the late nineteenth century, naturalists conducting life-history studies, for example, "saw nothing contradictory in conducting experiments to answer questions that interested them." In the same period, even morphologists, in the comparative anatomy tradition, could claim to be both comparative and experimental.[40]

Second, studies of natural history in the twentieth century have questioned the assumption of its decline. Natural history might have been declining relative to biology as a whole, but it was growing absolutely around 1900, owing to the general expansion of biology's territory,[41] and natural history remained "alive and well," if only or "primarily within museums."[42] On American campuses in the late nineteenth century, far from being irreconcilable, laboratories and museums actually grew hand in had.[43] Similarly, in northern England, laboratories and museums were thought of as being "equal though different" in civic colleges, and the "biology laboratory supplemented, rather than eclipsed, the museum."[44]

Third, natural history transformed itself deeply, even incorporating a variety of experimental approaches.[45] Similarly, practices in the field, as Robert E. Kohler has shown in his *Landscapes and Labscapes,* also drew from the experimental ideal—quantification, isolation, purity—to the extent that the twentieth-century practitioner of the "new natural history" no longer looked like the "butterfly collector" experimentalists loved to ridicule (and the same is true of morphology, which became largely experimental).[46]

Fourth, this narrative artificially isolates biology from medicine, which had a long tradition of comparative studies performed on anatomy collections. Anatomy was comparative in several distinct ways in Enlightenment Europe, long before Cuvier theorized its epistemic groundings, turning these practices into an academic discipline, and Darwin provided its current scientific justification. As private anatomical collections grew into medical museums in the late eighteenth and especially nineteenth centuries, they became essential pedagogical tools for medical schools, as well as places for the production of anatomical knowledge.[47] Anatomical collections and museums would deserve a much longer treatment, but the focus of this book is on neither anatomy nor taxonomy or their transformations. It is on how a wide range of experimental life sciences in the twentieth century adopted and adapted the comparative ways of knowing that had been so emblematic of these other traditions.

Naturalists vs. Experimentalists?

The original narrative of a frontal opposition between natural history (taxonomy, paleontology, morphology, anatomy, or field research) and experimentalism has mainly been debated in the American context, but similar elements can be found in France or Germany. Claude Bernard famously distinguished the experimental and the observational sciences, arguing that only the former could provide causal explanations (to the great dismay of naturalists). By 1900, the scientific reputation of the *Muséum National d'Histoire Naturelle* was at a low point, while experimentation was flourishing at the University of Paris and the Pasteur Institute and enjoying wide support.[48]

Although a simplistic narrative based on the opposition between experimentalists and naturalists has largely been put to rest by subsequent historical scholarship, a few of its points have remained valid. Beginning in the late nineteenth century, the laboratory became an increasingly central place for the production of biological knowledge, and researchers in physiology, microbiology, embryology, and heredity were setting the agenda of the "new biology." In 1886, an anonymous contributor to *Science* put it bluntly: "A good museum is valuable, but a good laboratory is immensely more valuable."[49] In the first decades of the twentieth century, the public figures of (male) experimentalists like Robert Koch, Paul Ehrlich, Jacques Loeb, or Thomas Hunt Morgan were defining the modern scientific *persona* in the life sciences. Popular movies such as Sidney Howard's *Arrowsmith* or James Whale's *Frankenstein*, both released in 1931, reflected the hopes and fears about the modern laboratory sciences.[50] Collecting, on the other hand, remained gendered as a female pursuit, and male naturalists around 1900 resisted the idea that collecting botanical specimens was "suitable enough for young ladies and effeminate youths, but not adapted for able-bodied and vigorous-brained young men who wish to make the best use of their powers."[51] At the same time, the once prominent disciplines of (idealistic) morphology and taxonomy had "fallen somewhat into a state of desuetude" and "lower repute in the mind of the general biological public," according to the American biologist Raymond Pearl in 1922.[52]

Unsurprisingly, those who had made a career outside the laboratory, in the museum, the garden, or the field, resented this changing landscape. A litany of speeches from retiring presidents of naturalist societies grew into a literary genre through which they voiced their fears and sometimes desperation about the current state and possible future of biology. Their discourses might not have accurately reflected the state of biology, as historians have subsequently shown, but their resentment was very real. The American naturalist C. Hart Merriam, first president of the American Society of Mammology, first secretary of the American Ornithologists' Union, and first head of the Division of Economic Ornithology and Mammalogy of the United States Department of Agriculture (it helped to have been a founder of each of these organizations), was a leading figure of American biology at the turn of the century and a leading critic of the focus on laboratory experimentation. In 1893, at the height of his scientific and political career, he wrote in *Science* about "the perversion of the science of biology." His wrath was directed toward those who spent their lives "peering through the tube of a compound microscope and in preparing chemical mixtures for coloring and hardening tissues; devising machines for slicing these tissues to infinitesimal thinness." For Merriam, "modern teachers of biology . . . while deluding themselves with an exaggerated notion of the supreme importance of their methods, . . . have advanced no further than the architect who rests content with his analysis of brick, mortar, and nails without aspiring to erect the edifice for which these materials are necessary." Merriam's greatest concern, however, was the effect of these "section-cutters and physiologists" on work in natural history, especially the "resulting neglect of systematic natural history," which had "disappeared from the college curriculum," and the "race of naturalists" that had become "nearly extinct." Yet only the naturalist who looked beyond just "a few types" could understand "the principal facts and harmonies of nature."[53]

After 1900, genetics became a prime target for those like C. Hart Merriam who were concerned about the excessive focus on laboratory research. The head of the Department of Entomology at Cornell University, James G. Needham, complained in 1919 about the fact that some "laboratories resemble up-to-date shops for quantity production of fabricated genetic hypotheses" and that the "prodigious effort to translate everything biological into terms of physiology and mechanism" was "as labored as it is unnecessary and unprofitable." A better approach, for Needham, was to adhere to a strict empiricism: "Why not let the facts speak for themselves?" he asked rhetorically.[54] William Morton Wheeler, a researcher of much greater standing, echoed similar concerns. Curator of invertebrate zoology at the American Museum of Natural History and later professor of (economic) entomology at Harvard University's Bussey Institution, Wheeler lamented in 1923 "the present depauperate glacial fauna of the laboratory, the perpetual rat-guinea-pig-frog-Drosophila repertoire." He found "genetics, so promising, so self-conscious, but, alas, so constricted at the base" because it focused on just a few organisms. Overall, Wheeler argued, biologists were divided "more or less completely into two camps—on the one hand those who make it their aim to investigate the actions of the organism and its parts by the accepted methods of physics and chemistry . . . ; on the other, those who interest themselves rather in considering the place which each organism occupies, and the part which it plays in the economy of nature." For Wheeler comparing biological phenomena in a great variety of organisms was the only way to properly understand "the economy of nature."[55]

The opposition between "naturalist" (or "morphologist") and "experimentalist" thus does not really capture what was at stake in these tensions among researchers in biology.[56] There were also important fault lines *within* the communities subsumed under these categories. Naturalists, for example, were divided with regard to the importance of live organisms, and in this respect some field naturalists criticized both the museum and the laboratory as places where only dead organisms were studied. Some experimental biologists criticized the artificial conditions prevailing in the laboratory but valued experimentation in the field.

More important, even the most zealous naturalists (or morphologists) often recognized the general importance of experimentation for attaining their own intellectual goals. Experimentation has been part and parcel of natural history from at least the eighteenth century.[57] In the nineteenth century, the French naturalist Cuvier, who became an icon of the comparative approach, recognized the similarities between laboratory and "natural" experiments. In his 1817 introduction to his *Règne animal* he noted that the diverse bodies compared by the anatomist were "kinds of experiments ready made by nature . . . as we might wish to do in our laboratories."[58] Later in the century, the American naturalist Merriam acknowledged that experiments "fulfill an important and necessary part in our understanding of the phenomena of life" but added, "they should not be allowed to obscure the objects they were intended to explain."[59]

But the bigger problem is that "experimentation" could cover a wide range of practices, from preparation of tissues for microscopic observations in embryology, histology, and cytology, to physiological investigations of live organisms.[60] The epistemic goals pursued in the name of experimentation have been widely different too. Some scholars prefer to reserve the term "experimental" to designate the results of manipulations intended to uncover *causal* mechanisms in nature. This definition might be more satisfying philosophically, but the historical actors of this study did not adopt it and used "experimental" in a much broader sense to designate results as different as microscopic observations of cells and DNA sequence data, all produced through the manipulation of nature, usually in the laboratory, with specialized instruments. "Experimental" will be used here as an analytical category to designate, at least since the early nineteenth century, this broad range of research practices, including both experimentation intended to control and experimentation intended to analyze. At the same time, "experimental" will be recognized in the historical actor's discourses as a rhetorical tool that, although it rarely did justice to the complexities of their actual practices, served as a powerful political weapon in positioning their own discipline in the professional landscape.[61]

The opposition between "laboratory" on the one hand and "museum" (or "garden" or "field") on the other also doesn't capture what deeply opposed these researchers. Around 1900, no less than in previous centuries, the term "laboratory" referred to a space devoted to a great variety of practices, including the preparation of specimens for museum collections, the instruction of students in microscopic observation, and the experimental explorations of the mechanisms at work in biological systems.[62] To make matters worse (analytically), in the early twentieth century a number of laboratories were set up in natural history museums (chapter 1), as well as collections (and even museums) in laboratories (chapter 2), making it even more doubtful that these spatial categories fully capture what was at stake in the oppositions and tensions voiced by so many biologists throughout the long twentieth century. The development of marine biology stations since the late nineteenth century—the Stazione Zoologica in Naples, the Marine Biological Laboratory in Woods Hole, the marine station of Concarneau and Roscoff in France—blurred even further these spatial distinctions, as Robert E. Kohler has argued, creating places that bridged the laboratory and the field (and, I would add, the museum). Marine stations typically included laboratories set up for experimental work, especially in embryology and physiology, as well as collections of alcohol-preserved and live (or at least fresh) animals in fish tanks. The leaders of such marine stations, such as Anton Dohrn in Naples, positioning themselves against both the descriptive natural history of museum taxonomists *and the* laboratory work of "stain-and-slice morphologists," prided themselves on working with a wide range of live organisms. The institutional landscape thus does not follow precisely the fault lines that divided discourses of turn-of-the-century biologists.[63]

Behind polarizing and antagonistic discourses, biologists often shared common research methods, and all practiced careful observation and often experimented in some way. But something fundamental still opposed them: the value of biological diversity and comparisons among species as a key strategy for unlocking the secrets of nature. Those seasoned in comparative approaches, experimental or not, resented what they perceived as the narrow focus of so many experimentalists on just a few species ("the perpetual rat-guinea-pig-frog-Drosophila repertoire") and the lack of comparison to a broader range of species. They were all the more resentful because they perceived this as a recent change in biology. Indeed, the range of species studied experimentally became increasingly constricted in the twentieth century. Thomas Hunt Morgan is a case in point. Before becoming the leading geneticist in the United States, his research focused on development and, specifically, regeneration. His 1901 book, *Regeneration,* reported experimental results from a very wide range of species, including protozoa, worms, sea urchins, starfish, fish, salamanders, frogs, lizards, and even plants. But his 1915 book, *The Mechanism of Mendelian Heredity,* almost exclusively reported on his experiments on a single species of flies (although a few other species were mentioned in passing).[64] Comparative studies implied comparisons among different *species* existing in nature, not differences in conditions created by the experimenter.[65] And by that time, experimentalists often believed that single species, which came to be called "model organisms," could stand as exemplary models for all living creatures. They were content to generalize from one exemplary species to all species, without engaging in comparative work.

This book builds on this opposition between two "ways of knowing": the comparative and the experimental, the former centered on collections and the latter on exemplary systems. Most important, it looks at how these ways of knowing interacted, conflicted, and hybridized *within* different fields of biological inquiry. This story is not about the clash of scientific disciplines or research fields (natural history against molecular biology), but about the historical dynamics of their epistemic components (comparing and experimenting). During the period covered by this book—over a century—disciplines have come and gone (remember postwar "cybernetics"?), and their content has evolved deeply and rapidly. For example, within just two decades, between 1870 and 1890, the practices subsumed under the heading of "embryology" changed profoundly. For this very reason, the narrative of this book is structured around the deep continuities in research practices, or "ways of knowing," John Pickstone's immensely helpful analytical category.

The crucial point about Pickstone's "ways of knowing" is that unlike many earlier historiographical categories, they are not taxonomic but analytic. Their goal is not to put people, practices, and places into *unique* conceptual boxes in order to write a history of *successive* historical periods—"natural history" followed by "experimentation," "museum science" by "laboratory science"—but to analyze knowledge practices into analytically distinct *components.* Ways of knowing are the ingredients, coexisting in different proportions, of scientific practice. In almost all sciences, one will find various ways of knowing interacting. This approach offers a powerful way to highlight historical continuities and discontinuities in scientific practices without reducing any of them to pure "kinds" (like Ludwik Fleck's "Denkstil," Thomas Kuhn's "paradigms," Michel Foucault's "episteme," Gerald Holton's "themata," or Alistair Crombie's "styles"). Obviously, some ways of knowing have been more important than others in a given science at a given time. The "experimental" might not have been dominant in eighteenth-century natural history, but it was present nevertheless, along with other ways of knowing that Pickstone would call "museological," "natural historical," and "analytical." Over the years, Pickstone has redefined and renamed several ways of knowing, including the initial "museological," "natural historical," "analytical," and "experimental." Here, I depart from his terminology, while adhering to his general approach, focusing on the interplay between the two ways of knowing I designate as "experimental" and "comparative."[66]

The comparative was most prominent in taxonomy, morphology, anatomy, paleontology, embryology, and natural history more generally, while the experimental was most important in physiology, microbiology, genetics, biochemistry, and later molecular biology. But the greatest historiographical benefits of analyzing scientific practices in terms of ways of knowing come from the cases where they intersect and interact, as Pickstone originally pointed out. Instead of having to decide whether there was "a dichotomy or a continuum of approaches," "revolution" or "continuity" in the life sciences, as historians of biology debated long ago, one can fruitfully analyze the historical dynamics of these two different ways of knowing *within* different biological disciplines.[67]

The experimental and the comparative ways of knowing did not intersect for the first time with the rise of collections in twentieth-century experimental sciences. The most significant precedent is unquestionably to be found in comparative embryology, an experimental and comparative science practiced in the laboratory. As Nick Hopwood has shown in his masterful history of Ernst Haeckel's embryos, images, and models, the comparative anatomy that had flourished under Georges Cuvier and Étienne Geoffroy Saint-Hilaire in the early nineteenth century should be seen not only as part of the history of taxonomy and evolutionary thought but also in the longer history of a "comparative science," including comparative embryology of which Louis Agassiz is perhaps Haeckel's most illustrious predecessor. A student of Cuvier, Agassiz was particularly keen on comparison, aligning the developmental stages of various species on a single visual plate to highlight similarities and differences. In his *Methods of Study of Natural History*, published in 1863, he put it in no ambiguous terms: "the true method of obtaining independent knowledge is this very method of Cuvier's,—comparison" (figure I.2). But it was Haeckel in Germany who brought this comparative embryological approach its greatest fame (and disrepute). After 1945, as Hopwood points out, comparative embryology became marginal.[68]

This book explores the intersection of collecting and experimenting in the life sciences from the early twentieth century to the present, in a broad range of disciplines, including genetics, microbiology, systematics, crystallography, evolution, biochemistry, and molecular biology. To do so, it focuses on the collections of specimens, molecules, images, and data that were becoming increasingly present in laboratories and that supported the hybrid way of knowing that defines contemporary biomedical research.

The central argument of this book is that the way of knowing based on collecting, comparing, and classifying, so central for naturalists, paleontologists, anatomists, and embryologists, among others, for most of the history of the life sciences, has not been overthrown by the experimental way of knowing.[69] It is "alive, and well," as Keith Benson has rightly put it, but not only, I argue, "within museums."[70] In fact, it is precisely in the laboratory—the key site for the production of experimental knowledge—that it is having its greatest impact today in combination with the experimental way of knowing. Yet laboratory researchers have not somehow reverted to a way of knowing practiced in natural history and elsewhere. They have created a *hybrid* centered in the use of databases of experimental knowledge, a synergy of experimenting and collecting. This revised account does not challenge the importance of the experimental sciences in the twentieth century through a rehabilitation of natural history, comparative anatomy, or embryology. Rather, it attempts to demonstrate that the stunning successes of the experimental life sciences, and the string of Nobel Prizes that gave them public visibility, were not founded solely on the experimental way of knowing. Since the second half of the twentieth century, the power of the experimental life sciences in unlocking the secrets of nature has increasingly depended—and today more than ever—on collecting, comparing, and classifying.

This perspective has far-reaching historiographic consequences. First, twentieth- and twenty-first-century experimentalism can no longer be viewed as a sort of historical pinnacle of the life sciences and biomedical sciences, but rather must be viewed as an episode in a much longer history that has largely been dominated by collecting, comparing, and classifying. Second, the life sciences and biomedical sciences can no longer be considered

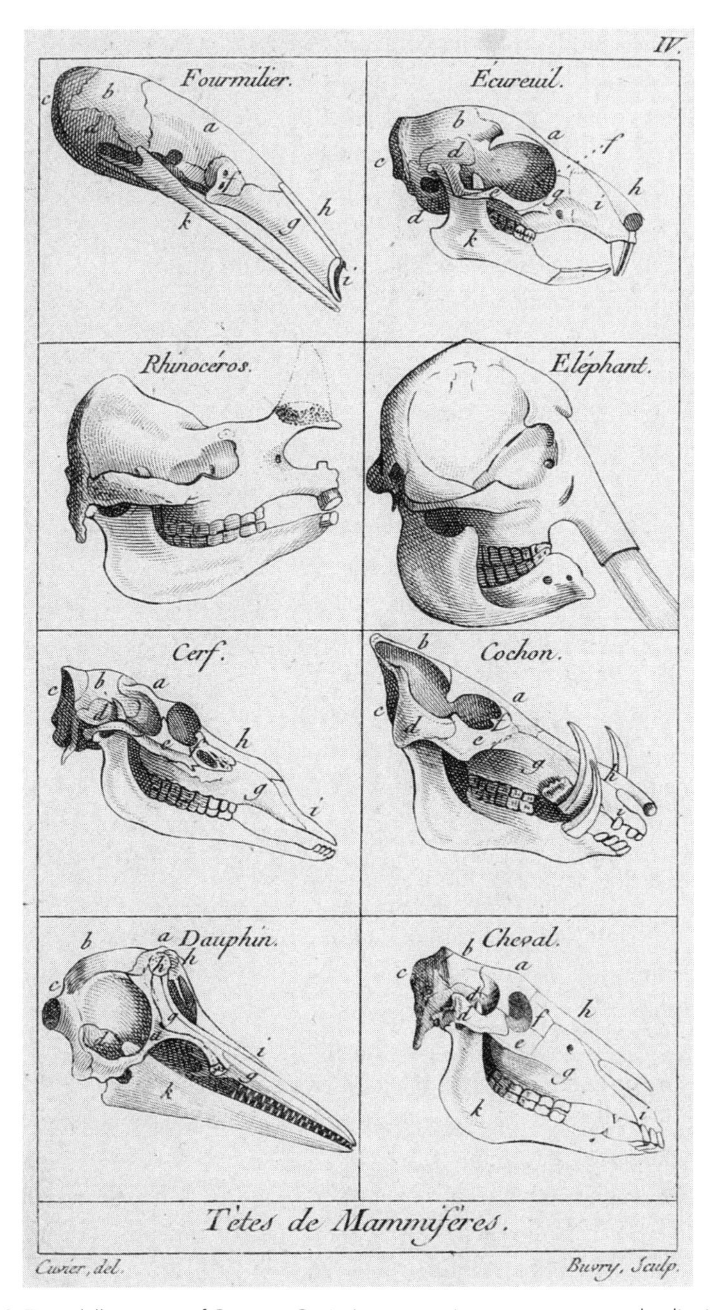

Fig. I.2 Typical illustration of Georges Cuvier's comparative anatomy approach, aligning similar subjects to facilitate comparisons among them for the reader, here in his *Tableau élémentaire de l'histoire naturelle des animaux*, published in 1797–98. Printed with permission of the Bibliothèque nationale de France.

as having followed the "typical" path toward experimentalism set by the physical sciences, but must be seen as a scientific field with a distinct history that might serve as a model for understanding the development of the other sciences.[71] Third, and most important, the recent transformations of the life sciences should not be considered unprecedented, but must be framed in a *longue durée* perspective. This will help us make better sense of many current debates in the life sciences—about the role of data collections, of course, but also about data sharing, authorship, and collaboration. We will be in a position to offer concrete solutions to these problems, because we can draw from the experiences of naturalists and others who have faced these problems over and over again throughout history.

Drawing databases into the larger history of collections, especially natural history, anatomical, and embryological collections, brings forward the two major themes of this book. First, scientists face a great challenge in turning heterogeneous things, dispersed in space, into commensurable data stored in a single location. Here, too, the problem is not unprecedented: natural history collections have all required modes of enabling "things" to travel—aboard commercial vessels, in Wardian cases, and on blotting paper. In the electronic age, this has required the development of new technologies and infrastructures to allow a smooth circulation of data. Second, diverse people, including professionals and amateurs, have to be "collected" as well. Perhaps the most distinctive social feature of scientific collecting is that it has been a collective enterprise, "a science of networks."[72] Integrating different kinds of collectors into networks has required specific social arrangements and moral economies. Both the collection of data and the collection of people turn on an ambiguous professional figure: the (female) curator. Here there is a difference: common in natural history museums, curators were an unknown figure in the experimental sciences. Experimentalists, who claimed to be the only ones with the necessary expertise to interpret the data they had produced, constantly challenged their scientific legitimacy. And the fact that many were women did not help (chapter 3).

The collecting of things and of people has gone hand in hand, resulting in a "co-production" of knowledge and communities. To understand the development of collections (of organisms, molecules, or data), these two dimensions must be considered simultaneously. This book shows that experimentalists, working within a different social and moral framework from that of naturalists, created their own solutions to this dual challenge—giving rise to the present structure of contemporary research, with centralized high-throughput data production, dispersed individual users, and tensions about access, property, and priority everywhere.

The multiplication of collections of knowledge, from scientific databases to Wikipedia, marks not just the emergence of a vaguely defined "information society" but also the rise of a culture that specifically values open access, data sharing, and collective contributions. Paradoxically, it emerged at the same time as the expansion of intellectual property rights to include the patenting of genes and organisms in the late twentieth century.[73] Resistance to the commercial appropriation of knowledge has often been ascribed to computer programmers in the "hackers' culture."[74] But experimental scientists have made similar efforts in their own arenas by guaranteeing that some of the largest data repositories in the world, such as the DNA sequence database GenBank, are freely accessible to all. Their success now serves as a model for broader changes in contemporary culture, for example, open access to the entire scientific literature. The study of the emergence of scientific databases offered by this book provides a window on the transformation of relations among people, knowledge, and the public good that is a hallmark of present times.

The Laboratory and Experimentalism

Where did these profound changes originate? Databases now live in the "cloud," that is, they are physically stored in numerous computer servers dispersed among air-conditioned rooms around the world where they can be accessed from any place connected to the internet. Striking is the fact that these collections of data are now extensively accessed from laboratories, where researchers routinely compare massive amounts of data. In 1982, the evolutionary biologist Ernst Mayr, in his monumental book *The Growth of Biological Thought*, conceptualized the "experimental method" and the "comparative method"—the first belonging to the laboratory, the second to the museum—as two mutually exclusive methods.[75] Through databases, the "comparative" method is now also widely practiced in the laboratory.[76]

To understand the significance of this shift in the location of scientific practices, we need to clarify the terms "laboratory" and "museum," as well as "experimentation" and "natural history." These notions have changed over time, have sometimes overlapped, and have often included heterogeneous elements. Yet they remain useful as analytical categories to understand the changes that have taken place in the complex world of science, for example how field research was transformed by laboratory methods. As Robert

Kohler put it: "Laboratory and field biology are distinct modes of knowledge production and have distinct political economies."[77]

What characteristics distinguish them? In the seventeenth century, the laboratory emerged as a specialized place for the conduct of experiments.[78] Perhaps its most important element was how it made it possible to control (and especially restrict) the inflow and outflow of people and things.[79] Its power derived from the fact that only select parts of nature were allowed to enter (or were produced in) it, and only select people (first gentlemen, then PhDs) with shared moral and epistemic values, and rules of scientific behavior, were admitted to witness the results of experiments.[80] The laboratory provided a much-simplified natural world to a much simplified social world. Whereas outside, very diverse crowds of people tried to agree on the description of an immensely diverse nature, the laboratory provided an environment where a (relatively) homogenous group of people tried to agree on a (relatively) homogenous nature. That goal was more modest, but more accessible.

Even though the laboratory constantly regulated the spatial circulation of things and people, it increasingly appeared as a "placeless place."[81] Initially built in the private house of the investigator,[82] in the nineteenth century the laboratory typically became a more anonymous and standardized place, often located within public institutions. Its standardization served an essential epistemic function, namely, to guarantee that knowledge produced locally (in a specific laboratory) would hold true anywhere (or at least in any laboratory).[83] An experiment conducted with a specific set of instruments, procedures, and skills should produce the same result anywhere. Thanks to standardization, laboratories allow researchers to transcend the local conditions in which they work and produce knowledge that holds true universally, at least anywhere where the same set of social and epistemic assumptions are warranted.

Laboratories have the "placelessness" of airports, for similar reasons: the same procedures (checking departure gates, getting there, waiting for boarding time) ought to result in the same outcome (such as catching a connecting flight). An airport with an entirely different system would leave too many passengers stranded. Standardization produces its universalizing effects because the laws of nature (or human behavior) are assumed to be the same everywhere, at any time, a principle that philosophers call "uniformitarianism." A belief in this epistemic norm is thus crucial to the experimental enterprise. Scientists who study organisms interpret this norm in a specific way. Living things, unlike atoms, come as individuals, each with its own singularities, so this principle does not always seem to hold true. Yet life scientists have hypothesized that certain observations made on one individual could be made on any individual of the same kind. In a laboratory, a fly could thus represent all flies of the same species. Or all insects. Or even all living creatures. In the twentieth century, as the molecular mechanisms of life were being unraveled, this belief was pushed to its extreme. As molecular biologists Jacques Monod and François Jacob put it in 1961, at a time when experimental molecular biology was triumphing: "anything found to be true of *E. coli* must also be true of Elephants."[84] That principle came in handy, as it was hard to fit elephants into the laboratory.

These epistemic norms were aligned with a material culture that rested on the study of a restricted number of "exemplary" organisms, some of which were considered "model organisms" (chapter 1).[85] Geneticists and embryologists chose specific organisms as models to investigate genetic and developmental processes, with the faith that their results would be valid more generally, if not universally. Together with experimentalists in other fields, life scientists valued precision, quantification, and a kind of objectivity resting on the mechanization of observation.[86] Experimentalists also shared a number of social norms. Authorship, for example, was attributed to individuals for the divulgation of original empirical data *they* had produced.[87] Data was considered the intellectual property of the researcher who had produced it and thus private (and often kept secret), until it was published, at which time it was considered public (but not necessarily shared).

Animals of a particular species that have been bred in the laboratory have less genetic variation than other members of their species in the wild. But the concept of standardization required that species studied by various scientists meet an even higher standard of "sameness." This led to the establishment of stock collections of organisms, usually starting in a single laboratory where they had been extensively studied, that were then shared with the scientific community as a whole. From the beginning, as chapter 1 shows, these practices had a deep effect on concepts of ownership of the strains that had been bred and of the data that had been produced by studying them.

The history of laboratories is immensely richer and more complex than this simple account, as shown by recent historical studies. They are home to a wider range of activities including experiments, manufacture, testing, and teaching and cannot be considered, therefore, simply as places of experimentation.[88] Conversely, experiments were carried on in many different places, including laboratories, fields, workshops, museums, and kitchens. Most important, laboratories have changed overtime, serving different goals and patrons.

But since the twentieth century, the focus of this book, the identity and the authority of the laboratory have derived mostly from the fact that it is a place where one could conduct experiments.

The Museum and Natural History

Museum and field researchers shared epistemic, material, and social norms that were widely different from those of the laboratory. They emphasized the epistemic value of collecting localized observations (rather than the "placeless" observations of the laboratory), engaging in extensive comparative work on a wide range of organisms (not a few "exemplary" ones), and producing classifications of their objects of study (rather than detailed descriptions of single phenomena). Most important, this way of knowing was centered on collections of specimens, images, or data.

How were these collections constituted and used? There is a rich literature on collecting practices in natural history, anatomy, embryology, and anthropology from which we can derive some of their key elements.[89] The purpose was to bring spatially dispersed objects to a central location and make them commensurable, i.e., to turn them into data. Thus collecting was (and is), above all, a spatial practice. Successfully bringing specimens from distant places requires a mastery of space that has been achieved in many ways. Colonial powers used their rule of colonies to bring specimens back to the metropole. Napoleon's 1799 military adventure in Egypt returned numerous specimens (found or pillaged), which were incorporated in the *Muséum National d'Histoire Naturelle* in Paris.[90] Similarly, the botanical collections of Kew Gardens reflected the reach of the British Empire.[91] Specimens were also commodities that followed the main commercial routes. In the nineteenth century, the busy merchant port of Canton allowed British naturalists to purchase plants and animals, and the rise of professional animal dealers in the United States supplied natural history museums and zoos with wild animals.[92] Museums such as the American Museum of Natural History also mounted expeditions to collect specimens from specific parts of the world.[93] But one of the most effective ways in which collections have been constituted, from the Renaissance to the present day, has been through the voluntary participation of "amateurs."[94] In the eighteenth century, a system of "polite indebtedness" gathered a vast network of amateur collectors around the *Jardin des Plantes* in Paris; they contributed time, expertise, and specimens to a centralized collection in exchange for patronage or the gift of other specimens.[95]

These different modes of collecting were often organized at a distance by professional "armchair collectors" who preferred to remain close to their collection rather than expose themselves to the hazards of wandering in the field. The coordination of these complex and heterogeneous networks of people including naturalists, merchants, hunters, and many others was key to success. It rested on finding the proper material or symbolic rewards to keep amateurs actively amassing specimens. As this book will make clear, in the twentieth century the creation of a system of collection among experimentalists proved to be the greatest challenge of all: there were no colonies to collect from, no established communities of amateur experimentalists to tap into, no experimental data to buy on the market. As a result, an entirely original system of collecting, adapted to the moral economy of the experimental world, was conceived.

There are several studies on the manner in which collections were displayed, including lavish dioramas, but few on how collections were used for the production of knowledge.[96] Yet some general elements can be highlighted. In brief, collections have represented a sort of "second nature" for researchers. This "second nature," like the "phenomena" or the "epistemic things" of the experimenter, was a product of the human hand and mind. Specimens—or data—were not found in nature but made by the collectors; they are the "working objects" of science, as Lorraine Daston and Peter Galison have put it.[97] The elements of a collection—specimens, skins, bones, tissues, molecules, or data—were produced by *reducing* nature to a common set of properties (usually, shape and structure). Collectors isolated a part of nature (say, a bird), deprived it of its relations to its surroundings (the forest), left behind most of its properties (such as being alive), and turned it into a specimen embedded in a new system of relations with other specimens in a collection. Birds could be found in trees, but in collections there were only specimens. One only needs to think of the indispensable role of taxidermists in preparing specimens for museum conservation to realize how much these were also human artifacts. As Nick Hopwood has put it for embryological collections, "Collecting not only brought scarce items together; it also framed nondescript or differently interpreted objects as embryos in the first place."[98] Reducing nature to artifacts possessing a common set of properties, bringing these artifacts to a single location, and organizing them in a collection were the three key operations that gave collections their epistemic power. They made the systematic comparison of diverse objects possible.[99]

This "second nature" was at the same time more subject to manipulation than the natural world (dead specimens are more easily moved around than live organisms) and simpler (specimens in a drawer have fewer properties than organisms in nature). Collections are a concentrated version of the world, turning the many things dispersed in nature into objects accessible to the limited human field of view. A researcher could compare, say, a mountain lion and a seabird, which were rarely seen together in nature. Numerous objects became accessible in a single place, in a single format, and could be arranged to make similarities, differences, and patterns more apparent to the eye of a single human investigator. Collections contained more than the sum of the individual objects created by collectors. They encompassed all the relations among these objects. As Georges-Louis Leclerc de Buffon, head of Louis XVI's natural history cabinet, put it in 1749, "At each sight, not only does one gain a real knowledge of the object considered, but furthermore one discovers the relationships it can have with those around it."[100] For this reason, the epistemic potential of collections as "relational systems" was understood to grow exponentially with their size, and collections were driven by the ideal of "completedness." Collections, archives, and "second natures" have been a prime locus for the production of scientific knowledge in the comparative tradition, making possible "Science *in* the Archives," as Lorraine Daston put it.[101]

This way of knowing was associated with a specific moral economy, which defined how authorship was granted.[102] In the natural history tradition, the production of knowledge was understood to be more collective than individual since it relied on collections of objects that could be gathered only collectively, often with the help of skilled amateurs. These objects, usually specimens, did not belong to individual researchers, but were a "common property, belonging to science rather than to an individual," as the leading naturalist Ernst Mayr and his co-authors reminded their readers in their 1953 *Method and Principles of Systematic Zoology*. Authorship could be granted to researchers who published analyses of materials and data produced by others, for example when a taxonomist published a monograph about a group of organisms on the basis of comparative work carried out on a collection assembled by a large number of mostly anonymous collectors (the fact that they were often amateurs, women, or both made it easier to appropriate their work).[103] In the experimentalist tradition, the analysis of someone else's data was generally considered illegitimate because the production of data through experiment was considered a major achievement deserving individual credit. Experimental data constituted a capital from which a researcher could draw to produce publications.

This sharp difference in what counted as a scientific author, and thus what counted as scientific knowledge, explains in part why these two ways of knowing have polarized the research enterprise and, when they met in "border zones," why tensions arose between their practitioners. Yet, as this book argues, these two ways of knowing, associated most closely with the laboratory and the museum, with the experimental sciences and natural history (and comparative embryology, morphology, and anatomy), intimately hybridized in the twenty-first century. Achieving the new hybrid form required overcoming many epistemic, material, and social differences. Those who adopted a way of knowing based on collecting, comparing, and classifying in the experimental life sciences were often misunderstood by their colleagues, who took them for second-rate scientists or outdated naturalists or discounted them as scientists altogether. Contrasting these two ways of knowing brings to light the paramount historical significance of their hybridization that is currently so central in the life sciences and beyond.

The institutions where knowledge was produced, including the library, the museum, and the laboratory, and their associated ways of knowing have flourished at different times in history. Instead of explaining these changes through some inherent superiority of the new institution or way of knowing over its predecessor, various authors have pointed to changes in the societies that supported them. The flourishing of cabinets of curiosity in the seventeenth century has been explained by a prevailing culture of curiosity and the importance of wonders and miracles,[104] and the diffusion of laboratories since the late nineteenth century by how they shared goals of modern states including mass education and the rationalization of production.[105] The fact that a way of knowing based on collecting, comparing, and classifying is once again rising to the forefront of scientific research, after being marginalized for more than a century, raises important questions about the kind of society supporting the change. The coming of age of the "information society" is too simple an answer to explain this transition, a point to which I will return at the conclusion of this book.

The main historical actors examined in this book set up collections of experimental objects and data for further laboratory work. Mostly, they did not borrow from natural history, comparative embryology, or medical anatomy. They were firmly rooted in an experimental tradition that focused on exemplary systems, not comparative work. Yet the rapid growth of the experimental sciences in the twentieth century, which accelerated even further after the Second World War, and the technological development of more powerful instrumentation led researchers to produce increasing amounts of data of all kinds offering new opportunities for comparative work. Some seized these possibilities, embracing the comparative approach while discovering the specific epistemic, social, and moral challenges of collecting, comparing, and classifying. Unlike naturalists, who had faced these problems for centuries and found workable answers, they struggled to reinvent practical solutions that would work in the specific communities of experimentalists to which they belonged. The making of this hybrid culture—experimenting and collecting—is the topic of this book.

1
LIVE MUSEUMS

Life science laboratories are usually home to just a few animal or plant species. *Homo sapiens* can be found there during the day (and sometimes at night), but the permanent residents are those species that have gained the enviable status (at least from the scientists' perspective) of "model organisms."[1] The story of the experimental life sciences from the late nineteenth century to the present can be told from the vantage point of these select species that have served as scientists' so-called "guinea pigs."[2] Guinea pigs proper (*Cavia porcellus*) are among them, and so are microbes (*Escherichia coli*), mice (*Mus musculus*), flies (*Drosophila melanogaster*), corn (*Zea mays*), and more recently worms (*Caenorhabditis elegans*), fish (*Danio rerio*), weeds (*Arabidopsis thaliana*), and many others.

For most of the twentieth century, experimental scientists' narrow focus on selected species stood in sharp contrast with the broad diversity of organisms studied by naturalists. In the first half of the twentieth century, some embryologists still worked experimentally on diverse species, but overall the range of organisms found in laboratories was narrowing. By contrast, natural history museums often housed tens or hundreds of thousands of species—up to a million in the British Museum (Natural History) in London and almost as many at the American Museum of

Natural History in New York. For historians and scientists, research in collections such as those housed in museums came to be viewed as distinctive to the naturalist enterprise, in contrast to the experimentalist's focus on laboratories and model (or "standard") organisms. Stating the opposition this way is too simplistic but is a starting place on which to build. The aim of this chapter is to capture the important role that collections of organisms have played in the development of the experimental life sciences, thus making this opposition more nuanced. This is especially true of genetics, whose practitioners were ironically keen on drawing a contrast between their experimental approach and the collection-based approach of naturalists. When we look at the histories of the two disciplines side by side, we will see that collections of living organisms, so-called "stock collections," have been indispensable in the rise of experimental genetics. These findings help to refine our understanding of research practices and to explain the historical basis of the current use of collections in experimental research.

How should we conceptualize stock collections to understand their role in the development of scientific knowledge and of scientific communities? In many ways, they have been to the experimental life sciences what museum collections were to natural history: repositories of organisms (preserved for a possible future use), centers of standardization (defining nomenclatures), centers of distribution (providing remote researchers with specific organisms), tools for research (allowing comparative studies), centers of coordination (for complex networks of researchers), and institutions defining the norms of practices (social and epistemic). Like museums, stock collections served these many roles at the same time, co-producing communities, practices, and knowledge. As with museum collections, the single most important issue for creators of stock collections was how to develop a moral economy that would support the wide participation of researchers, their contribution of organisms to the stock collection, the sharing of information, and an obedience to community norms. The solution to this problem hinged on the subtle definition and enforcement of a boundary between private and public objects and ideas, between members of the community and outsiders, between intellectual contributions that deserved individual credit and those that should remain communal.

Creators of stock collections from the early twentieth century to the present have relied on an unusual epistolary technology to achieve these aims: the newsletter.[3] Neither a private letter nor a public journal, the newsletter was a way to address a select community of individual researchers, providing information about the content of the stock collection (a catalogue), its uses in research (research results), and the members of the community (a directory). Such newsletters were invented in the late nineteenth century for internal communication within growing corporations[4] and became part of the social bond among communities of researchers working on the same organism. They helped compensate for a loss of close interpersonal relationships as research communities grew larger and helped propagate and enforce community norms and ideals. A close examination of several newsletters reveals their crucial role in the development of experimental research centered on stock collections.

The collections examined in this chapter differ from those of museums in two crucial ways. First, rather than embracing a broad range of species, they contained many variants, often mutants, of single species, which were almost always model organisms. These organisms were selected mainly for practical reasons: their anatomy, physiology, and behavior made them particularly amenable to experimental studies of some specific aspect of their biology. Thus different fields in the life sciences—embryology, physiology, or genetics—have adopted different model organisms. In the nineteenth century, the sea urchin became a favorite organism to study embryonic development because its eggs were transparent and every step following fertilization could be easily observed under a microscope. Similarly, in the mid-twentieth century, neurophysiologists focused on the Atlantic squid because electrodes could easily be inserted in its exceptionally large nerve axon, permitting experimental measurements. The qualities that made organisms well suited to genetic studies were a combination of small size, short generation time, and distinct and easily observable characteristics.

Second, unlike museum collections, those examined in this chapter are collections of *live* organisms. Curators of natural history museums were concerned about the preservation of their specimens. Curators of stock collections were concerned about keeping them alive and constant against natural mutations producing variation at each generation. The practices of maintaining collections of live specimens had been developed in zoos and botanical gardens, but also in marine stations, such as the Stazione Zoologica in Naples, which became a "clearing house for model organisms" and supported experimental, and often comparative, research, especially in embryology and physiology.[5]

Unlike embryologists and physiologists, geneticists worked with huge numbers of individuals over many generations. Typically, they performed thousands of crosses between various organisms in order to study the distribution of traits among their offspring. The probability of discovering a new trait, initially a matter of chance, could be raised by increasing the number of individual observations. Thus only small and fast-reproducing organisms could accommodate the limited space of a laboratory and the limited time span of a human researcher's career. These important constraints mean that very few organisms have actually been studied genetically, and even fewer organisms have been used in more than one field of research. Yet those species that have had the most enduring place in the history of biology are precisely those that were suited to more than one line of investigation. The famous fruit fly began its scientific career as a genetic model and only later became a model for embryonic development. The choice of model organisms in other fields has been determined by other considerations, such as their economic importance (corn) or their evolutionary proximity to humans (mice).

The science of genetics emerged after the rediscovery of Mendel's laws in 1900 and relied heavily on model organisms.[6] To obtain material for their experiments, geneticists produced and maintained huge numbers of individuals in stock collections. Thomas Hunt Morgan's "fly room" at Columbia University typically contained tens of thousands of individual *Drosophila*.[7] The small size of this organism made it possible to store them all in just a few hundred milk bottles. But more important than the number of individuals, which could be increased at will thanks to the extraordinary fertility and short generation time of the fly, was the number of distinct mutants that could be produced and collected. In 1909, Morgan found his first mutant, a fly that had white eyes in place of the usual bright red ones.[8] Within two decades, Morgan's group was caring for more than six hundred different mutant strains. This diversity was particularly important for the mapping of genes along the chromosomes, the central intellectual agenda of the Morgan school of genetics, since each strain constituted a reference point for the position of a gene.[9]

In the 1910s and 1920s, mice, corn, and flies became widely used in genetic research. This was possible because organisms and increasingly large numbers of mutants were kept alive in stock collections and made available to individual researchers. Like museum collections, these started out as private collections intended for the exclusive use of a local group of researchers and their privileged correspondents. They functioned under rules of civility established between researchers who knew each other personally and whose community was headed by respected leaders in the field. By the 1930s, the needs of the broader community prompted several of these private collections to retrace the trajectory of museums a century earlier.[10] They were converted to public "stock centers," often funded by philanthropic institutions such as the Rockefeller Foundation or the Carnegie Institution of Washington. They were thus able to foster the sharing of organisms, research data, and social norms beyond a small initial group of researchers.

Historians have sometimes considered model organisms to be akin to the physical instruments such as microscopes and spectrometers used daily by experimentalists.[11] This perspective is useful because, like instruments, model organisms are made by researchers to serve special research needs. Model organisms were produced through techniques of inbreeding, i.e., the crossing of siblings. After many generations, inbred lines consisted of highly similar individuals. These standardized organisms made experimental results more reproducible, like standard scientific instruments.

Yet at least as important as the crafting of these individual model organisms was the role of organism collections in the production of knowledge. As Robert Kohler's exemplary study of Thomas H. Morgan's fly group has shown, the members of this group shared, with some notable exceptions, a particular communal culture of intellectual and material exchange and credit attribution.[12] These norms can be linked to the very nature of *Drosophila*, specifically the extraordinary number of mutations it experienced, and to the particular research agenda of Morgan's group, mapping genes, providing more problems to solve than any individual could tackle in a lifetime. When we look systematically at stock collections of microbes, corn, mice, and flies, it becomes apparent that similar systems of norms arose in various model organism communities, which raises questions about the nature of the relationship—genealogical, functional, or some other type—between the norms (or "moral economy") in each system.[13] By focusing on stock collections and the moral economies that they sustained, one based on freedom of charge and reciprocity, and on the co-production of collections and communities, this chapter provides a revised picture of the early rise of experimental life sciences, one in which knowledge production is more comparative than analytic, where contributions are more collective than individual, and where moral economies are more collaborative than individualist.

Microbes at the American Museum of Natural History

The public stock centers for mice, corn, and flies are perhaps those whose history is best known, but they were not the first. The American Type Culture

Collection (ATCC), a collection of microbes founded in 1911 in New York as the Bacteriological Museum, forms an important precedent. In the 1930s, the ATCC and its European counterparts represented the largest collections of organisms in the world. Although it was not yet used in genetic research, the ATCC served some of the same purposes as stock collections in genetics, namely, to provide standardized organisms to researchers, especially microbiologists working in academic or industrial contexts. The history of the ATCC illustrates particularly well the roots of stock collections in natural history museums and their similar trajectories. Indeed, the ATCC began at the foremost natural history institution in the United States: the American Museum of Natural History (AMNH).[14]

In 1911, the bacteriologist and public health expert Charles-Edward Amory Winslow (1877–1957) had just accepted the position of curator of the Department of Public Health at the AMNH in New York City.[15] He had obtained his MS degree from MIT in 1899 under the direction of the bacteriologist William T. Sedgwick and taught there for over a decade.[16] He was already a respected researcher and public health figure, having published several studies on water supply contamination and microbiological issues in sewage treatment, collaborating with his former teacher Sedgwick, a leader in American bacteriology and founder of the first school of public health in the country at MIT.[17] Winslow had also made several contributions to the biochemistry and classification of bacteria of sanitary importance, especially the widely distributed *Staphylococcus*, a frequent cause of human infections.[18] The latter project was accomplished with collaborator Anne F. Rogers, soon to become Winslow's wife. Their work culminated in a 1905 book in which more than five hundred strains were described and classified. Now at the AMNH, Winslow established the first public "museum of living bacteria."[19] Within a year, this collection contained 578 strains representing 374 distinct types.[20] It was assembled thanks to forty-five laboratories in the United States and Canada, which had "contributed freely" from their own collections of bacteria after Winslow sent out a call to laboratories in both countries.[21]

The material used in this classificatory work became the basis of a collection of organisms similar to earlier natural history collections. Yet unlike the zoologist and the botanist, the bacteriologist could not examine the morphology of specimens, since most bacteria appeared as "regular spheres."[22] But bacteria could be distinguished by some of the properties they exhibited when they grew. Living cultures were therefore required for those interested in taxonomy. The Winslows thus kept their cultures alive for the entire duration of their research, and when their collection was donated to the AMNH, its intent was to echo the purpose of the zoological collections stored in other parts of the museum, except that it had to be kept continuously alive.

It might seem surprising that a bacteriological collection should be housed in a natural history museum rather than in a medical institution. But it reflects the view that microbes, like plants and animals, should be considered not just as pathogens of medical interest but as a part of nature. As Winslow explained, the AMNH was the first museum to "recognize that the relation between man and his microbic foes is fundamentally a problem in natural history."[23] An additional justification for the location of the collection was a public exhibition that Winslow planned to present "the main facts about the parasites which cause disease, their life history, the conditions which favour their spread to man [and] the means by which mankind may be protected from their attacks."[24] This initiative was part of a crusade by medical reformers and public health officials to spread the "gospel of germs" in early twentieth-century America.[25] It was also part of the efforts under Henry Fairfield Osborn's presidency of the museum's board of trustees to make the AMNH more active in education and, crucially, to please the rest of the board and potential donors.[26] Karen A. Rader and Victoria E. M. Cain have argued that the creation of Winslow's bacteriological museum was also part of the AMNH's acknowledgment of "biology's turn towards experimentalism."[27] The 1908 exhibition on tuberculosis, a major cause of death and leading concern among the general population, had attracted over one million visitors. The AMNH released photographs of the crowd lining up to see the exhibition (figure 1.1). The success of the exhibition provided the impetus for the creation of a museum department of public health,[28] first headed by Winslow. Under his leadership, the department informed the public about sanitary control, including microbial contaminations of water.[29]

Winslow was in a strong position to found the bacteriological collection, but not everyone welcomed the arrival of live bacteria at the museum. The director, who was responsible for the research agenda of the museum, wrote him a letter stating that he would not rest until the museum "had sent the bacteria elsewhere." The director and Osborn agreed that the work of the museum should focus not on microbes but on the morphology of vertebrates, which made for the most appealing public displays and supported the scientific focus of the museum on evolutionary zoology.[30]

The association between Winslow's culture collection and a natural history museum is less surprising if we consider the focus of both institutions on

Fig. 1.1 Crowd standing in line to visit the American Museum of Natural History's exhibition on tuberculosis, a major cause of death in the United States, New York, 1908. Image 32185, American Museum of Natural History Library. Printed with permission of the AMNH.

the preservation and display of natural things. Indeed, in addition to maintaining live bacterial cultures, Winslow built large models of his bacteria in order to make them visible to museum visitors.[31] Here Winslow's "Museum of Bacteria" was similar to Franz Král's bacteriological collection in Prague, the only other large bacteriological collection of this kind in the world. An entrepreneur and bacteriologist, Král developed special glass flasks for the display of bacterial cultures in the 1880s.[32] He toured hygiene congresses where his displays of various bacterial cultures created a sensation. As one commentator put it, after the International Congress of Hygiene and Demography held in London in 1891: "No clinical or theoretical institute ought to be without such a collection . . . and no physician should neglect the opportunity of making acquaintance with the dreaded producers of diseases which Dr. Král has most skillfully and securely enclosed."[33] The displays were meant to educate and could be observed with the naked eye or with the microscope. Král suggested that they be used as the basis of "bacteriological museums" like the one he had created at the Institute for Hygiene at the German University in Prague. He soon discovered that the cultures contained in his specially built flasks remained alive and could thus be grown for bacteriological experiments.[34] After setting up his own private laboratory in 1890, Král published a catalogue of his cultures in 1900 and began to sell them to microbiologists in Europe and the United States.[35] After his death in 1911, the year Winslow created his Bacteriological Museum, Král's culture collection was transferred to Vienna[36] and eventually to Chicago, although only a few cultures remained alive by that time. Some found their way into Winslow's Bacteriological Museum.[37]

Alongside display and preservation, Winslow's collection served another role typical of natural history museums: providing well-defined material to researchers. As he outlined in a pamphlet sent to bacteriological laboratories in 1913, "the opportunities offered by the bacteriological collection are unique" and the "possibilities of research . . . in the future are almost limitless."[38] These opportunities would prove essential for the rise of bacteriology in the United States.

Following the foundation of the Society of American Bacteriologists in 1899, the number of researchers working in the field grew slowly at first, then exponentially in the late 1910s. The Society of American Bacteriologists had only 100 members in 1905, 300 in 1915, and 1,200 in 1923.[39] Winslow's collection made it possible for researchers to enter the field of microbiology and develop new lines of research by obtaining specific bacterial strains from the AMNH. It was made necessary by the growth of the scientific community and supported it at the same time. In the first two years of its existence, 1911–12, Winslow distributed more than 1,700 bacterial strains; in 1914 alone, he distributed 3,283. Throughout the First World War, this number remained constant, then rose sharply in the postwar period.[40] As he put it in a note published in *Science* to promote the collection, the cultures had been distributed "in every case without charge," emphasizing the public role of the collection and its continuity with previous, informal practices of exchange of natural history collections.[41]

Before the existence of the AMNH collection, American bacteriologists had obtained their cultures from colleagues or from Král's collection in Europe. Winslow's collection had the advantage of being geographically closer, thus facilitating shipment within the United States, and also of being larger, aiming to be exhaustive. Since the cultures were given away free of charge, Winslow was in a good position to urge researchers—something he did on every possible occasion—to promptly send him cultures of all the new bacterial species they isolated.[42] He intended the collection to become a "reference center" akin to the type specimen collections in natural history museums. Plant and animal systematists who described a new species traditionally deposited their original specimen (the "type specimen") in a museum. Other naturalists who tried to identify specimens could always turn to the precise individual on which the species definition had originally been established.[43] But since bacteriologists could not derive "information of any special value from the study of stained slides which would correspond to the dead herbarium specimen of the botanist," they needed a collection of live bacteria for identification purposes.[44] The curators of the Bacteriological Museum at the AMNH thus maintained living cultures by growing them constantly and regularly transferring them to new growth media.

By the time the AMNH's Department of Public Health closed in 1921 owing to a lack of funds, the Bacteriological Museum had distributed 43,911 cultures to more than eight hundred institutions mainly in North America and Europe, where they were used for experimental research, industrial production, and education.[45] The collection was also used in-house for the identification of new bacterial strains and for taxonomic work, leading to several publications. When Winslow joined the Yale School of Medicine in 1915 to create a department of public health, his collection began a long series of institutional moves under the authority of the Society of American Bacteriologists. In 1922, it was transferred to the Army Medical Museum in Washington, apparently in a suitcase carried by the president of the Society of American Bacteriology, Lore A. Rogers.[46] In 1924 it went to the McCormick Institute in Chicago. In 1925, the committee in charge of the collection was incorporated as a nonprofit scientific institution under the name American Type Culture Collection (ATCC).[47] By that time, there were only 175 viable cultures left. This number had been restored to 1,500 by 1937, when the collection was transferred to Georgetown University Medical School in Washington, DC. Ten years later, it was moved among different locations in Washington, DC, before settling in 1964 at a "million-dollar building" in Rockville, Maryland, constructed especially to house it and where it remains to this day.[48] The postwar search for antibiotics, and the rise of microbiological research more generally, had made such a bacterial collection particularly desirable.

One of the main concerns of stock collections curators was to find a way to prevent the microorganisms from changing over time. Until the late 1930s, microorganisms were kept alive by constantly culturing them, presenting a continuous risk that the cultures would change, or by drying them to keep them alive but not reproducing, a technique that did not always allow for long-term preservation. But by 1940, the ATCC switched to a new technique of preserving cultures: freeze-drying. This had the advantage that the cultures could be stored in a freezer at -79° C without further maintenance. Starting in the 1960s, cultures were stored in liquid nitrogen at -179° C; experience had shown that at that temperature "cells are viable indefinitely."[49] Freezing proved the ideal solution for microbes and later for seeds. By 1990, the ATCC had "50,000 microbes suspended in a sleep of absolute biochemical inactivity."[50]

The other main concern of stock collections curators was to secure funding for their collections. For the first half-century of its existence, the ATCC bacteriological collection was at the mercy of uncertain or insufficient funding, yet its guardians remained deeply attached to the ideal of sharing bacterial cultures at no charge. While the ATCC was located at the AMNH, cultures were always distributed free of charge. This practice followed the common museum practice of lending specimens free to researchers. The maintenance of the collection was funded by the AMNH and special gifts from the trustees.[51] When the collection moved out of the museum, its financial position worsened.[52] Modest grants from the Rockefeller Foundation did not suffice to keep it afloat.[53] In 1927, finances had become so tight that the curators of the collection decided to charge $1.00 for each culture. The chairman of the ATCC apologized for this move and explained in *Science* that the charge was made only "to help defray the costs of maintaining the collection."[54]

The new fee represented a significant source of income, but in 1934 the Great Depression brought the ATCC once again to the brink of bankruptcy. The directors decided to double prices to $2.00 per culture, explaining contritely that it was a financial "necessity," that they had never wanted to "make the collection self-supporting," and that it was impossible "to provide this service only through the sale of cultures."[55] Other disclaimers of this kind illustrate how the commodification of the cultures seemed to require justification to researchers who were accustomed to obtaining biological material free of charge. Historical accounts of the ATCC written by its members take pains to justify why the freedom of charge policy of the early days was abandoned, explaining that the ATCC served not only academic institutions but also for-profit industries such as Coca-Cola, Abbott, and Park Davis.[56]

Yet some exceptions to the commercial arrangement were granted: researchers who contributed strains could still receive cultures free of charge,[57] and so could other collections, mimicking a standard practice of specimen exchange among natural history museums. In 1950, the price of cultures was raised to $10.00, but nonprofit institutions were entitled to a 70 percent discount. Even so, the financial situation of the ATCC remained precarious. A virologist put it in no uncertain terms: "It is my frank opinion that the ATCC cannot continue to exist in its present form or in any minor modification of its present form. It is a bastard child for whom many scientific groups are willing to read a benediction and to contribute a fathering amount but for whom no group will assume the responsibilities of parenthood."[58]

Three years later, a group of leading American bacteriologists published a plea in *Science* to gather support for the ATCC, reaffirming that the ATCC was basically a museum. The authors included Kenneth B. Raper, who had led the effort to produce penicillin during the war, polio researcher John F. Enders, and Paul R. Burkholder, the discoverer of another antibiotic, chloramphenicol. They noted that support for museums of natural history had "long been recognized as a proper responsibility of governments and universities" and that the collections housed in these museums constituted the "principal bases" upon which knowledge of "the relationships, phylogeny, and taxonomy of the higher plants and animals" rested.[59] They lamented that while the ATCC served the same purpose for microorganisms, it had never received comparable support. Furthermore, the ATCC served constituencies far beyond natural history museums, including basic researchers in biochemistry, genetics, physiology, and other fields, as well as more applied activities such as waste disposal, textile deterioration, and forest technology. Without additional support, the authors concluded, the ATCC would reach the "limits of its usefulness."[60]

By the end of the 1950s, the National Science Foundation (NSF) and the National Institutes of Health (NIH) began to support the ATCC more substantially (including the new "million-dollar" facility that opened in 1964).[61] By that time, federal funding agencies had begun to recognize that biological collections were an indispensable infrastructure for the development of science. As discussed at the end of this chapter, other collections had faced the same difficulties as the ATCC, and their mounting demands for support eventually led to this change of policy.

The Industrialization of Mice

The story of the ATCC forms an important backdrop to the rise of biological collections in the field of genetics in the first third of the twentieth century.

It illustrates how the constitutions of collections accompanied and supported the growth of a research community. In the first decades of the twentieth century, another experimental discipline, genetics, followed the same growing trend, and it too saw the foundation of collections of live organisms to support it. Around 1900, the most common laboratory animals used in experimental physiology and medicine—rabbits, rats, frogs, and guinea pigs—were joined by mice, corn, and flies. These three new organisms became the pillars of genetics research after the rediscovery of Mendel's laws. The mouse was the first organism to become widely used for genetics studies and was popularized as a model by arguably the first full-time geneticist in the United States, William E. Castle.[62] Trained in botany and zoology in Ohio and then at Harvard, Castle worked under the biologist and eugenicist Charles B. Davenport at Cold Spring Harbor's Station for Experimental Evolution. By 1901 he had begun to study the inheritance of coat color in rats, guinea pigs, rabbits, and mice.[63] At Harvard's Bussey Institution for Applied Biology, where he taught animal genetics from 1908, he developed his research into the inheritance of various traits in these four organisms. He also trained a number of students, many of whom would become key figures in genetics, including Sewall Wright, Clarence C. Little, Leslie C. Dunn, and Rollins A. Emerson.[64] Little (1888–1971) came to play a crucial role in building a community of mouse researchers that connected many of Castle's former students, especially after the Bussey Institution closed in 1936.

Beginning in 1919, at Cold Spring Harbor, Little and Edwin Carleton McDowell, another student of Castle, took several steps to bring together the "diaspora" of the Bussey Institution.[65] They invited the researchers who were otherwise dispersed around the United States to spend their summer at Cold Spring Harbor to work on problems of mouse genetics. This loosely organized community became known as the "Mouse Club of America" and included all the visitors to Cold Spring Harbor and a few researchers working elsewhere.[66] They began to correspond and distribute an informal *Mouse Club Newsletter* to share news of recent results and mouse strains available from the Cold Spring Harbor stock collection.[67] Little and McDowell began to scale up the production of inbred mice at Cold Spring Harbor to meet both their own needs and those of the growing community.

The newsletter and the stock collection did much to cement the mouse community when researchers returned to their home institutions, sustaining communication within the group. Most important, by informing other researchers about the different mouse lines and mutants available at the stock collection, the newsletter made it possible for several laboratories to work on the same strains and thus make their experimental results comparable. But the Cold Spring Harbor stock collection could provide only limited numbers of mice; each laboratory needed its own animal facility to produce the larger numbers needed for experiments. It took another initiative of Little and stress caused by the Great Depression to make mice widely available.

When he established the Jackson Memorial Laboratory in 1929 in Bar Harbor, Maine, Little founded what he conceived primarily as a research institution centered in the genetics of cancer.[68] But the facility would also provide mice without charge to other laboratories, following the model of the Cold Spring Harbor stock collection. However, when the financial situation of the laboratory became tense in 1933, Little reluctantly decided to begin selling mice, even though he was still "opposed to the vulgarities of commercialism in matters of science."[69] The Jackson Memorial Laboratory took on a dual function as a research laboratory and a factory for the production of mice. It was a significant shift, not only for the institution, but also for the values governing mouse research. As Karen Rader put it, the Great Depression "forced [Little] to modify the accepted practice of exchanging results and animals for free—now, by selling animals for cash."[70] The development of big biomedicine in the postwar period and its focus on topics such as laboratory studies of cancer, the effects of radiation, and the screening of new chemotherapeutic compounds combined to make mice an essential component of a range of research projects. Little had a strong influence on this agenda, promoting mice as an essential tool for cancer research, for example in an article that made the cover of the magazine *Life in 1937*, transforming the Jackson Laboratory into a large-scale factory that shipped more than one million mice in 1960 and welcomed visiting researchers from around the world (figure 1.2). But Little remained uncomfortable with the transformation of his research materials into commodities. As late as 1964, he apologetically explained that mouse production at the Jackson Laboratory had "been a very great service" and that he had "not gotten rich on it. . . . There has been no business. There has been no industry."[71]

The case of Jackson Laboratory was not unique. In the 1960s, guinea pigs and rats could be purchased from private institutions such as Carworth Farms in New York and the Wistar Institute at the University of Pennsylvania. Rats, which were the first organism to be inbred for scientific research on an industrial scale, constitute the most direct precedent to Little's "JAX mice." Researchers at the Wistar Institute, founded in 1892 as a museum of anatomy and pathology, set out in the first years of the twentieth century to produce

Fig. 1.2 Visiting scientist from Brazil and a laboratory assistant injecting mice at Jackson Memorial Laboratory, Bar Harbor, Maine, in 1948. In the background, boxes containing different mouse strains with water bottles sticking out. Printed with permission of the Jackson Laboratory Archives.

rats for their own brain research before scaling up production to supply other researchers. The application of the principles of Taylorism, the optimal management of the production line, to the inbreeding of rats produced, after many generations, the "Wistar rat," a genetically highly homogenous organism that was first sold in 1906 and was trademarked in 1942.[72] By 1913, the institute was selling over three thousand animals annually.[73]

The sale of small mammals such as mice, rats, and guinea pigs was, however, something of an exception in the history of material exchange practices in biology. As Little's apologies imply and his earlier practices attest, the norm among biological researchers was a free distribution of organisms. The two other major organisms used for genetics research in the first decades of the twentieth century—corn and flies—make this point abundantly clear. Corn genetics began in the same two institutions as mouse genetics, Harvard's Bussey Institution and Cold Spring Harbor's Station for Experimental Evolution, before taking hold at Cornell University.

Corn in an Agricultural Station

In 1900, corn was already a crop of major agricultural interest in the United States, and this became only more so after the development of hybrid corn by George H. Shull at Cold Spring Harbor and Edward M. East, one of William Castle's colleagues at the Bussey Institution, in 1910.[74] A year later, plant breeder Rollins A. Emerson (1873–1947) from the Nebraska Agricultural Research Station spent a year at the Bussey working with East on the inheritance of quantitative traits in maize, obtaining a PhD for the work in 1913.[75] In 1914 he moved to Cornell, which had large fields for conducting agricultural experiments and breeding programs (figure 1.3). There he established a productive research group in corn genetics that focused on physiological genetics rather than chromosomal mapping, a move echoed by his colleagues working on *Drosophila*.[76] Like Castle at the Bussey Institution or Little at Cold Spring Harbor, Emerson fostered a climate of cooperation among the circle of maize researchers at Cornell, which initially included Barbara McClintock, George W. Beadle, and Marcus M. Rhoades and soon expanded its reach. He encouraged the free exchange of materials, data, and ideas.[77] As with mouse genetics, the need to stabilize the extended community was felt as the first generation of students left the institution. So it is no surprise that in the early 1930s Emerson followed in the footsteps of mouse geneticists of a decade earlier, establishing a club, newsletter, and stock center.

Since at least 1918, Emerson had been trying to organize corn researchers by holding "cornfests" or "corn-fabs" at meetings of the American Association for the Advancement of Science and "hoping that all the men [*sic*] in this country who are working on related problems with corn may cooperate to such an extent that we can cover the field more quickly."[78] In 1928, about fifteen men met in Emerson's hotel room to discuss problems of maize genetics.[79] As another plant geneticist noted, "At all genetical meetings [corn geneticists] hold private 'corn-fabs,' [they] are working in close cooperation."[80] One facilitating element of the cooperation was Emerson's circulation of mimeographed letters to his colleagues in the United States summarizing the new results in maize genetics, beginning in 1929 or slightly before.[81] Corn had recently become yet more promising for genetics studies thanks to Barbara McClintock, another student of Emerson at Cornell, who had just published a study linking traits to specific chromosomes. This gave corn a sudden advantage over *Drosophila*, because the results of genetics experiments in the plant could now be traced to changes in the chromosomes visible under the

Fig. 1.3. Geneticist Rollins A. Emerson collecting corn kernels in the field at Cornell University, undated. Under his belt, a stack of paper bags to hold the different kernels, which will be organized in the box at his feet and brought to the corn stock collection. Undated. Printed with permission of the Cornell University Archives.

microscope.[82] McClintock's breakthrough also stimulated the mapping of genes along the ten chromosomes of corn,[83] a project that was best tackled cooperatively for Emerson, owing to the amount of work it represented.

The number of researchers grew, leading Emerson to form a "Maize Genetics Coöperation" at the International Congress of Genetics, which took place at Cornell three years later.[84] The group discussed questions of nomenclature and progress in mapping, two activities requiring broad consensus within the community. The group numbered approximately sixty researchers, most of whom were then involved in mapping.[85] They received a copy of the *Maize Genetics Coöperation News Letter*, an extension of Emerson's collective letter project, edited by Rhoades. The newsletter disseminated research results such as linkage data giving the location of genes on the chromosomes and information on available corn stock. It also attempted to promote cooperative behavior within the maize research community. At the 1932 International Congress of Genetics, Emerson provided a vivid description of his take on relations among the corn geneticists:

> I cannot refrain from noting here a very real advantage experienced by students of maize genetics, which is in no way related to the peculiar characteristics of the maize plant. I am aware of no other group of investigators who have so freely shared with each other not only their materials but even their unpublished data. The present status of maize genetics, whatever of noteworthy significance it presents, is largely to be credited to this somewhat unique, unselfish cooperative spirit of the considerable group of students of maize genetics.[86]

Yet the sharing of unpublished data was indeed in some ways "related to the peculiar characteristics of the maize plant." Given that corn could be crossed only once a year, experimental work required "several seasons of checking" after preliminary results had been obtained.[87] The newsletter thus allowed the dissemination of tentative results, too preliminary for formal publication but highly useful to researchers planning their next experiments. At least as important as the sharing of results was the sharing of seeds of the different corn lines with which investigators were working. In the first newsletter, Rhoades addressed the corn researchers forcefully: "It is requested that, as soon as convenient, you send to the undersigned a small quantity of seed of any stocks which you think may be useful to other workers now or which should be maintained for future use."[88]

The seeds that came in were kept at Cornell and replanted every three to four years to maintain their viability, just as the bacterial cultures at the AMNH were being cultured anew every month. In 1934, Emerson's group grew eight thousand plants to maintain the seed stock.[89] The collective stock center saved individual workers "a considerable amount of space and energy" in maintaining their own corn mutants.[90] Most important, the seeds were listed in subsequent newsletters and provided freely to researchers. In January of that same year, the newsletter listed over 150 mutants and 110 mutant combinations.[91] Over the next seven months, Emerson's group sent out 350 different stocks to investigators.[92] This system of collection and distribution of seeds generalized an informal practice of gift exchange that had characterized maize genetics from the beginning. In 1929, George Beadle answered a request for seeds by Milislav Demerec, promising to "pick out some good material," while asking Demerec to send seeds of other mutant combinations in return.[93] According to the editor of the newsletter, the response of corn geneticists to the call for information and seeds was "good," and the following issues acknowledged a growing number of individuals and institutions that had done so. By 1940, the collection contained over 550 corn stocks.[94] Starting in 1934, the stock center was partially funded by a grant from the Rockefeller Foundation, always eager to foster cooperation in science.[95] Furthermore, as its trustees put it, the "small sum of money" for the stock center "will yield unusually high returns to the science of genetics." Cornell was becoming a successful "clearing house and central repository" for data and material pertaining to corn genetics, as Emerson had wished.[96]

The newsletter also helped to organize the gene-mapping enterprise and thus avoid overlapping work by announcing which chromosomes had been assigned to which research groups. Everyone remained free to work on whatever chromosome he or she wished, but it was "expected, however, that when two or more are interested in the same [chromosome], they will work in close cooperation."[97] Even though, or perhaps because, research was envisioned as a cooperative project, issues of individual credit were addressed in some detail. For example, individuals were not credited for providing stock lines because their exact origin was often unknown to the Cornell workers. However, as Rhodes acknowledged, credit was "due [to] those investigators who have spent a great deal of time in building up good genetic strains."[98] He encouraged researchers to communicate unpublished results, such as data from linkage experiments that would help other researchers in their own attempts to position genes along the chromosomes. These results would be credited to their author, but the editors insisted numerous times that the newsletter did "not constitute publication."[99] Anyone who wished to use results in print was expected to contact the author for permission. The newsletter was thus conceived as extending the space of private communication, customarily made orally or through private correspondence, to all those who received it. As anthropologist Christopher Kelty has argued, the "newsletters constitute a closed community at the same time that they facilitate and even demand the unrestricted sharing of flies, techniques, results and other information within the community."[100]

The newsletter did not challenge the boundaries between private and public communication, or upset the reward system based on publication in a journal. Yet when Emerson drew, with their permission, on the work of the many authors who had contributed to the newsletter in order to publish a summary of the linkage studies in corn, he lauded this "almost unique example of unselfish cooperation" and hoped that his presentation would "prove to be sufficiently helpful to the contributors to compensate them in some measure for their aid in its preparation."[101] Researchers were to be rewarded for their work not by individual credit and authorship, but by the knowledge derived from their various contributions. The reward was collective, not individual. At least this is how Emerson, as leader of the community, saw it.

The corn community has been described by historians and participants alike as an entirely open community, not a "membership organization" or one with "boundaries of 'membership.'"[102] This might have been the perception of those who were part of the community, but it hardly represents the intent or the actual practice of the corn researchers. Boundaries were drawn and enforced. Only those who complied with the rules of cooperation set out by Emerson were allowed into the group. Rhoades put this rather strongly in 1934: "We feel that anyone who does not value these letters sufficiently to include his own data has no claim to the unpublished data of others who have generously cooperated."[103] In other words, reciprocity was the condition for receiving the newsletter and thus being a full-fledged member of the corn community. A few years later Rhoades reaffirmed that the newsletter was being sent only "to those who are now coöperating or who have furnished material in the not too distant past."[104]

This tightening of the corn community's boundaries came as a result of an incident where one "foreign geneticist and plant breeder" who was not working with corn and who had thus not contributed to the newsletter published a summary of conclusions presented in the newsletter without the author's permission. Emerson took the matter very seriously and consulted a number of active corn researchers on the problem. Opinions ranged from "disastrous" to benign, but the majority seemed to think that some kind of enforcement should be put in place to foster cooperation, which led Emerson to adopt the policy outlined above. In the end, he reaffirmed his sole authority in deciding, based on his "own judgment (good or bad),"[105] who should receive the newsletter and thus become part of the community. Earlier, Emerson and Rhoades had tried to point fingers at those who did not cooperate by marking the corn stocks that had not been contributed to the seed collection, but by 1942, after the community had grown, a more stringent policy apparently became the order. Within the community, the democratic, open, and free sharing of materials and ideas might have been the rule, but the regime had limits that were ultimately determined autocratically by the uncontested father of the community. As Emerson put it himself, the Maize Cooperation Genetics Group "is my baby."[106]

Sharing Flies

The third, and best-known, genetics community in the early twentieth century was the "fly group," which grew in parallel—and in a kind of rivalry—with the corn community. Under the leadership of the biologist Thomas Morgan (1856–1945), first at Columbia University and, after 1928, at the California Institute of Technology, the fly community was governed by a communal working regime among "the boss and the boys."[107] Efforts to organize the *Drosophila* community beyond the laboratory of the patriarchal figure of Morgan took precisely the same course as those for mice and corn, based on a newsletter and stock centers.[108] In 1934, Morgan's student Calvin B. Bridges at Caltech and Milislav Demerec (1895–1966) at Cold Spring Harbor launched the first issue of *Drosophila Information Service* (DIS). The newsletter was explicitly meant to emulate Emerson's circulars, which had "proved to be so useful" to corn workers.[109] It was also meant to compensate for the loss of the "intimate contact which existed between the *Drosophila* workers of the past" now that the community had grown and dispersed.[110] It provided a forum to communicate information that had previously been shared through personal communication, especially lists of new mutants, linkage data, stock lists, technical information, directories of researchers, and bibliographies.[111] The first newsletter mostly consisted of lists of stocks available at various institutions.

Caltech, holding by far the largest collection, listed 573 different stocks of *Drosophila melanogaster*; Cold Spring Harbor listed only 91, but the number soon grew into the hundreds.[112] Its number of mutants had begun to rise rapidly after the geneticist Hermann J. Muller discovered X-ray mutagenesis in 1927. Like the corn newsletter, the *Drosophila Information Service* (DIS) contained a disclaimer on its cover page that it was "not a publication," and thus that the public use of materials presented in the newsletter required the "specific permission of the author."[113] When Muller, who was something of an outsider to the Morgan group, criticized this clause,[114] it was agreed that the public use of the DIS communications required personal acknowledgment in print, not authorization from the authors (even so, the warning on the cover was maintained).[115]

As provisions were made to ensure individual credit to the authors, the editors of the DIS insisted on a more collective form of reward. They pointed out that the production of scientific knowledge was necessarily a collective enterprise and that usually "the discoverer did what somebody else would have done soon afterwards or what somebody else was doing at the same time." Thus "the larger share of credit" belonged not to "the discoverer himself, but to the common work done earlier." The editors requested that authors give up their personal claims to credit because "the less the claim for personal credit is stressed the better are the chances for the harmonious and productive working within the given group." They reminded their readers that "after all, the main driving force behind our efforts is the extension of knowledge," not the accumulation of personal credit. They thus asked researchers to stress the personal element "as little as possible" for the sake of a collective reward, knowledge, just as Emerson had done earlier.[116] In both cases, however, the principal promulgators of the "collective reward" ethic would benefit the most from it, since all new knowledge would advance a field in which they were the most visible leaders.

Like the corn newsletter, the DIS also specified norms that defined the community, and the selective distribution of the DIS was a means of enforcement. The first issue was mailed "rather widely," but following issues, the editors warned, would be sent only "to those who are actively cooperating in the project."[117] Furthermore, past cooperative behavior was not sufficient to entitle a researcher to receive the DIS: he or she had to answer the questions of the editors for each issue.[118] Requests for a copy of the newsletter by geneticists who had not answered were denied.[119] They were expected to share not only information about new mutants or linkage data but also their research materials. As the foreword to a DIS made clear in 1937: "The free exchange of material is the established policy of the Drosophila group," and "stocks kept in any one laboratory are available to others."[120] For Demerec, this "unwritten law" was "contributing more than any other single factor toward the usefulness of Drosophila as research material."[121] The DIS served at the same time both to enforce community boundaries and to promote controlled openness within these boundaries.

In practice, the sharing of flies could be time-consuming, and their shipment through the mail was illegal in the United States (researchers shipped them anyway).[122] Shipments to and from foreign countries were almost impossible, because when discovered, they were systematically destroyed by customs officials. In 1933, Leslie Dunn, who had succeeded Morgan at Columbia after his departure, complained that he didn't want to "turn into a supply department" because he couldn't "afford the cost or the time." He was especially burdened by requests for flies for teaching purposes in high schools and colleges, and suggested either turning them down or charging $1.00 per shipment. However, he hastily added that of course none of this "applied to requests of stocks for research purposes."[123] In order to ensure the proper supply of flies to anyone who needed them in the United States and abroad, Demerec set out to establish Cold Spring Harbor as a second stock center for *Drosophila* (the first being Caltech, where Bridges was in charge of the fly collection) to facilitate shipments to the East Coast and to Europe. For Demerec, "more could be accomplished by the use of different materials selected to suit the problem than by adjusting the problem to suit the material," and thus the broadest possible range of *Drosophila* mutants should be made available as widely as possible.[124]

The new institutional status of the *Drosophila* collection made it possible for Demerec to negotiate with the US Bureau of Entomology and Plant Quarantine for an authorization to ship live *Drosophila* across the country and abroad. At the time, Caltech and Cold Spring Harbor, the two largest *Drosophila* collections, maintained 825 and 442 stocks, respectively. In May 1934, with the support of the Rockefeller Foundation,[125] Demerec established his collection as a stock center at Cold Spring Harbor (figure 1.4) at the same time as Emerson established the corn center at Cornell. Strains were delivered free of charge for research purposes.[126] Just a year after the foundation of the stock center, a private breeding company from New York contacted Demerec to include his *Drosophila* stock in their sales catalogue, offering payment in return. Demerec flatly refused the deal because he did not want to "charge or receive compensation."[127]

In 1939, Demerec made the case that the Cold Spring Harbor stock center had become essential to the development of genetics as a whole in the United States. With the retirement of Castle and East, genetics was at a "standstill" at Harvard. Emerson was also soon to retire, having lost active interest in theoretical genetics, as had Morgan.[128] Demerec pleaded that two independent stock centers were indispensable in order to prevent the accidental loss of

Fig. 1.4. Cold Spring Harbor Drosophila laboratory, 1937. On the shelves, hundreds of jars containing Drosophila mutants to be shipped to researchers around the world. In the background, microscopes used to examine the morphology of the flies. RF 200D, box 28, folder 1776. Printed with permission of the Rockefeller Archive Center.

mutants that could be irreplaceable. Demerec argued, if "anything should happen to wipe out all these stocks, genetics research would be greatly affected for many years to come." An air-conditioning failure had almost destroyed a fly collection in Texas, showing that such hazards were a real possibility. But most important, Demerec wanted to turn Cold Spring Harbor into a genuine "organized stock-collecting center" that would attempt, like most natural history collections, to cover all existing *Drosophila* strains, not just those of interest to researchers at any given time.[129] The center would serve the community as a whole while adding prestige to the host institution and reinforcing its identity as a hub for the exchange of materials and information.

Even though the stock center served its purpose and was lauded by its users, its funding remained precarious, especially after the first Rockefeller Foundation grant awarded in 1934 ran out. The foundation and the Carnegie Institution of Washington remained hesitant to fund Demerec over the long term, especially in view of the existence of Morgan's similar stock collection at Caltech. Finally, in December 1943 the Carnegie Institution of Washington approved a modest five-year grant to support the stock center at Cold Spring Harbor, which by that time was serving not only researchers but also high schools and colleges that wanted flies for teaching purposes. That year, Demerec sent out over seven hundred cultures for educational purposes alone. The stock center was also increasing its geographical scope, sending out flies not only to the United States and Europe but also to Asia, Africa, and South America.[130] The Rockefeller Foundation was particularly pleased by this aspect of the stock center because of its desire to foster international cooperation. It estimated that its modest investment in the stock center "proved to be a strategic move in terms of international cooperation" and that the center, "with fruit flies as its ambassadors," had "established helpful scientific contacts throughout the world."[131]

While the center fostered friendly international cooperation, Vannevar Bush pointed out to Warren Weaver that "apparently all has not been harmonious among the fly geneticists."[132] Indeed, Morgan considered the plan for a stock center at Cold Spring Harbor "something of an affront" and a "strategic move on the part of Demerec to shift, to some degree, the center of importance in Drosophila work to the east coast."[133] Elsewhere, he described the center as a "competitive enterprise to the stock center at Pasadena."[134] The feeling that Demerec was challenging Morgan's authority was perhaps exacerbated by the latter's approaching retirement. Apart from the matter of personal pride, Morgan argued that a second stock center was made unnecessary by the development of "land air routes and clipper ships as quick means of transportation for flies." He also questioned Demerec's qualifications, pointing out that he had formerly been a botanist, not a zoologist. The geneticist Alfred H. Sturtevant, Morgan's favorite student, claimed that Demerec's stocks were "infested with lice" and that cultures had to be "quarantined" before they could be used. According to Morgan, Demerec underestimated the skill needed to handle a stock center. This enterprise required "a man like Bridges to keep the stocks under continuous observation to prevent contamination and gene deterioration." Demerec had offered the position only to women, a further indication to Morgan that he thought of the task as solely clerical and subaltern. The status of the stock collection curator—should it be a male researcher, as Morgan thought, or a female technician, as preferred by Demerec?—was still in flux.[135]

Morgan and Demerec's divergences obviously reflected personal rivalries. Morgan stated firmly that "Pasadena stocks remain the one and only, unique and irreplaceable material of this kind in the world."[136] Morgan's reaction to

Demerec's efforts also illustrates how stock centers could contribute to the authority of those who held them. Even though the collection's maintenance might be seen as something of a clerical job, it also brought a unique power to name research objects, such as mutants, and control their distribution among researchers. In other words, it defined the very boundaries of the research community. Stock collections thus served as tools to co-produce knowledge and communities.

Viruses, Bacteria, and the Rise of Molecular Genetics

Mice, corn, and flies dominated genetics research in the first half of the twentieth century. In all three cases, researchers established stock collections and fostered the free exchange of organisms. By the 1940s, a tiny newcomer was making its way in the restricted circle of genetic model organisms: the bacteriophage. These viruses, which preyed on bacteria and were generally nicknamed "phages," had been discovered by Félix d'Herelle in 1917 and had been primarily studied by medical bacteriologists.[137] In 1939, the German physicist Max Delbrück and biologist Emory L. Ellis published a paper showing that phages multiplied not exponentially like other microbes, but in a stepwise manner (incidentally confirming that phages were viruses and not the result of cellular biochemical processes).[138]

Delbrück, who had moved to Caltech two years earlier to work on *Drosophila* genetics in Morgan's laboratory, hoped that understanding this process would generally illuminate the problem of gene replication, as viruses were then often considered "naked genes."[139] Two years later, with the Italian émigré Salvador E. Luria, Delbrück and Ellis showed that bacteria underwent mutations just like other organisms, and that the changes were thus not environmentally induced adaptations.[140] Building on these results, Delbrück and a small number of researchers began to use phages and bacteria as model organisms to study problems in genetics, especially replication. This approach produced a number of key insights into the nature of genes and their modes of action and replication. In the 1950s, it contributed to the emergence of molecular biology and molecular genetics.[141] The growth of the phage and bacterial genetics community resembled in many ways that of the earlier genetics communities and unfolded in some of the familiar places such as Caltech and Cold Spring Harbor.

By 1940, a handful of phage researchers were disseminated across the United States and in a few places in Europe, including the Pasteur Institute in

Paris. In the United States, Max Delbrück took a leading role in building the community, as Little, Emerson, and Morgan had done before him. Starting in 1941, he encouraged phage researchers to spend the summer at Cold Spring Harbor, where they could perform experiments and discuss results collectively. Three years later, he suggested that researchers focus on a small subset of phages, the T1 through T8 phages, in order to make results more easily comparable.[142] This recommendation served the same purpose that the stock center had for other organisms, namely, to standardize research materials within the community. The small size of the subset and the ease with which the phage could be preserved meant that a centralized stock center was not required. Instead, researchers could write directly to each other when they needed particular strains. Further steps to standardize the research practices and ethos of the phage group were taken in 1945 with the establishment of an annual phage course at Cold Spring Harbor, and two years later through the limited circulation of a newsletter, the *Phage Information Service*, edited by Delbrück.[143] Like its predecessors, the newsletter contained unpublished results as well as the cautionary clause that it did "not constitute a publication" and that permission from the authors was required before any results could be referenced. In 1950, bacterial geneticists set up their own newsletter, the *Microbial Genetics Bulletin*, modeled after the *Drosophila* and phage newsletters, under the editorship of Evelyn M. Witkin, who was working at Cold Spring Harbor.[144]

Delbrück, who was becoming something of a cult figure in the phage group, promoted the free exchange of materials and ideas and provided the most open—and often harsh—criticism of research results among members of the group. The cooperative ethos did not seem to require any enforcement beyond Delbrück's charisma, authority, and role as a gatekeeper to the Phage Information Service. When Delbrück relocated permanently to Caltech in 1947, Morgan's former home became the "mecca" of phage genetics.[145]

Although the free circulation of materials and ideas seems to have been the prevailing practice among phage researchers, there were exceptions. In 1947, as researchers were attempting to map genes on the chromosome of *Escherichia coli* K12, Jacques Monod at the Pasteur Institute in Paris asked Joshua Lederberg at the University of Wisconsin for some strains. Lederberg sent them immediately to his French colleague. However, on several later occasions when Lederberg requested strains from Monod, the latter failed to reciprocate. As Jean-Paul Gaudillière has noted, the culture of material exchange that seems to have been the rule among geneticists in the United States did not necessarily extend to biochemists working in Paris.[146] Yet just a few years later, Joshua Lederberg seems to have been in the opposite position. In 1949 his wife, Esther Lederberg, who was working on her dissertation project, found that the commonly used bacteria *Escherichia coli* K12 contained a hidden phage, named "lambda." Its presence could be made visible with a proper indicator strain of K12 that was sensitive to lambda. When researchers asked Joshua Lederberg for this strain, he was unwilling to share it, at least until Esther's dissertation was completed. But even in 1955, five years after Esther Lederberg received her PhD, the ATCC's request to Lederberg to contribute this strain to the collection went unanswered.[147] Another phage researcher, Jean Weigle, managed to obtain it and distributed it to several laboratories in the United States and in Europe, spurring the growth of lambda genetics.[148]

During the 1950s, the phage and bacterial genetics community grew rapidly because these microorganisms were recognized to be ideal tools in the study of a number of biochemical and genetic problems, from the elucidation of metabolic pathways to the mechanisms of gene action and replication. The courses offered at Cold Spring Harbor after 1945 introduced increasing numbers of researchers to the field, which was receiving rising recognition. In 1958 Lederberg received the Nobel Prize in Physiology or Medicine (together with George W. Beadle and Edward L. Tatum), in 1965 Monod (together with François Jacob and André Lwoff), and in 1968 Delbrück (together with Salvador Luria and Alfred Hershey). But the growth of the community and the productive research dynamic it sustained were made possible only by the availability of phages and bacteria from personal collections and stock centers.

In the early 1960s, the bacterial geneticist Edward A. Adelberg at Yale, who had been a graduate student with Lederberg in Edward Tatum's laboratory, began to develop his personal collection of *Escherichia coli* K12 into a public stock center to meet the growing demand of researchers. The National Science Foundation began to fund the center in 1966, and Adelberg appointed yeast geneticist Barbara J. Bachmann (1924–99) to be curator of the stock center. Before turning to bacterial genetics, Bachmann had been the editor of the *Neurospora Newsletter*, yet another model-organism-based newsletter, and was thus already familiar with organizing a model organism community. To expand the stock center, she began to write to early bacterial geneticists such as Lederberg, asking them to contribute personal bacterial strains to the stock center, noting the side benefit of relieving them "of the chore of sending them out upon request." As Bachmann explained to Lederberg, bacterial researchers "have been very good about sharing their strains"; she knew of only one researcher who refused to share a particular mutant. Even Lederberg contributed his prized K12-S mutants (by that time, it had circulated so widely as to be ubiquitous). Some laboratories, however, delayed sharing a new mutant in order to protect a graduate student from the threat of "big labs which would love to take the mutant and quickly do the experiment."[149] Bachmann approved of this protective attitude and believed that "almost all people in the field" did too (Joshua Lederberg certainly did). However, there was an overall willingness to share mutants, so by 1971 Bachmann had about three thousand strains in the collections. In the previous year, she had sent out about six hundred strains without charge.[150]

The establishment of a stock center for E. coli K12 was long overdue. Indeed, whereas corn or *Drosophila* mutants were deposited in a stock center soon after they were identified, bacterial mutants kept circulating through personal contacts among individual laboratories, where they were used to derive even more new mutants. This fluid exchange and the lack of record-keeping as they traveled along complex routes through numerous laboratories meant that strains used in experiments had increasingly uncertain genetic backgrounds. Bachmann set herself the monumental task of not only collecting all these strains, but also identifying where they originated. Her method was to reconstruct the genealogy of each strain and follow it in space and time from laboratory to laboratory. To do so, she turned to the "ultimate source of data," the laboratory notebooks of the researchers who had produced the mutant strains. She traveled to Stanford to examine Lederberg's notebooks and to Paris to see those of François Jacob.[151] In most cases, researchers themselves did not know exactly the origin or the precise genetic background of their strains. Bachmann had to cross-check experimental results from several laboratories to identify them unambiguously.[152] She established, for example, the exact genealogy of the K12 strain that Weigle had distributed so widely after having received it from Lederberg, as well as its many descendants, thus clarifying not only their genetic makeup but also social relationships in the early community of phage researchers.[153]

The stock center was not only a reference and distribution center but also a research tool for Bachmann. Having at her disposal a collection of mutants, the literature describing them, and personal contacts with the scientists who had produced them put her in an ideal position to integrate all the data on the position of genes along the single bacterial chromosome. Like Emerson and Sturtevant four decades earlier, Bachmann authored a linkage map based on almost eight hundred published papers as well as on "generously communicated unpublished data."[154] Yet her work was not simply a summary of others' research. Indeed, she had to "reconcile the widely varying, and sometimes conflicting, results obtained in different laboratories."[155] The "friendly cooperation" of many researchers allowed her to solve nomenclature conflicts and establish a new standard of nomenclature.

Her linkage map and its proposed nomenclature had a strong impact: in the following five years, it was cited more than one thousand times. Yet Bachman's professional identity remained torn between that of a researcher and that of a custodian of a collection. As Joshua Lederberg put it in a confidential recommendation letter, "she functions very much like the curator of a museum," but "that by itself may not offer much insight into the level of independent research competence which the task entails."[156] Lederberg's letter illustrated contemporary assumptions among experimentalists about the limited expertise involved in museum curatorial work, an assumption that would hamper the careers of many curators of experimental organisms and data (see chapter 3).

Other communities built around model organisms in the twentieth century followed the pattern of development we have described. In 1963, at the Laboratory of Molecular Biology in Cambridge, Sydney Brenner began to use the tiny worm *Caenorhabditis elegans* to study the genetics of development. A small and "tightly knit" community began to work on the worm.[157] When the community began to develop beyond Brenner's laboratory and a few others, a newsletter, the *Worm Breeder's Gazette*, was launched (1975), and a stock center was established (1978) at the University of Missouri, collecting and distributing mutants free of charge, to sustain the growth of the field.[158] The rise of *Arabidopsis thaliana* (the late-twentieth-century equivalent of corn for genetics research), and of the zebrafish (the "vertebrate *Drosophila*"), followed similar trajectories: newsletters and stock centers played central roles in the circulation and standardization of knowledge and materials.[159]

Putting Stock Centers on the Federal Agenda

By the early 1960s, biological stock centers were being noticed by science funding bodies as essential for scientific research. Until that time, individual stock collections—for microorganisms, corn, mice, or *Drosophila*—had been funded independently of each other, and there is little evidence that they were recognized either as a common practice for experimental research in the life sciences or as a common problem for funding agencies. As early as 1928, however, Leslie Dunn had set up the National Research Council Committee on Experimental Plants and Animals, hoping to find support for stock collections, particularly for mice.[160] The committee included representatives of the main model organisms: Emerson (corn), Little (mice), King (rat), and Dunn himself (*Drosophila*). After a failed attempt to enroll the National Zoological Park in Washington, DC, to house a mammalian stock center, the committee disbanded in 1931.

In 1939, the engineer Vannevar Bush, president of the Carnegie Institution of Washington, DC, realized that funding for *Drosophila* stock centers was being provided both at Cold Spring Harbor and Pasadena and wanted to avoid unnecessary duplication. Furthermore—and this issue would plague the funding of stock centers for decades—foundations and other funding agencies operated with term-limited grants supporting research projects that had an end point and whose success or failure had to be measured at that time. Biological collections, by contrast, were intended to be permanent institutions and thus required indefinite support, which funding agencies were unwilling to commit to.

Bush had these concerns in mind when he wrote to Weaver at the Rockefeller Foundation hoping to "arrive at some comprehensive solution on the whole matter of mutant stocks for biologists."[161] Weaver agreed and thought it "would be much more sensible" if both organizations "did not nibble blindly away at the corner of a cracker of unknown dimensions."[162] He believed the matter to be sufficiently important to seek advice from the National Research Council. After holding a conference on the maintenance of pure genetic strains, a new committee was set up in 1940, chaired by chicken geneticist Walter Landauer and including some of the same members as the earlier committee (Emerson, Little, and Demerec replacing Dunn). The committee made an inventory of existing stock collections and attempted to draw attention to the importance of preserving genetic stocks. In 1945, the chairman argued that they should be rated as a "most valuable natural resource." Postwar conditions would "almost certainly create demand for new kinds and varieties of plants and animals for agricultural and industrial use," with hybrid corn being one of the most important.[163] Funding agencies still ignored these calls, and obtaining support for stock centers remained a challenge.

In 1957, yet another committee on maintenance of genetic stocks was set up, this time to advise the NSF.[164] It made a new inventory of eleven existing centers in the United States and recommended that collections of importance to public health be supported by the NIH, those of economic importance by the Department of Agriculture, and those essential for basic research by the NSF. At the same time, it noted that where funding for the maintenance of stock collections competed with that of basic research, priority should be given to the latter. In 1958, working under the auspices of the Genetics Society of America, the committee outlined once again the importance of stock collections by stressing that "mutant stocks of organisms are the only really important tools of the geneticist." Without them, the geneticist "could do nothing in the way of definitive genetic experiments." Once lost, these tools might never be recovered.[165]

The number of stock centers steadily rose during the 1960s. By 1965, the committee listed 42 different organism collections for research worldwide;[166] ten years later there were 67.[167] Almost all the collections made their stocks available free of charge—the ATCC and the Jackson Memorial Laboratory being the most notable exceptions for research organisms—and published newsletters for their respective communities. The majority curated organisms of economic importance, including cattle and sheep, rice and tomatoes, barley and wheat. Others were devoted to new model organisms used for basic research, such as the weed *Arabidopsis* and the worm *C. elegans*. The NSF played an important role in funding these collections: by 1970 it was supporting at least 15 of them. Surrounded by much better endowed institutions, the NSF had been attempting to find a niche within the American science funding landscape, where it could make a difference. This led the NSF to select areas that were typically neglected by other institutions such as the NIH.[168] That same year, its budget request to Congress included a special provision for stock centers for the first time. The agency reported that it was making "programmatic efforts" to provide an "improved base of support for resource centers such as museum collections, genetic stock centers, and controlled environment laboratories."[169] By the 1970s, stock collections had finally been recognized as an essential infrastructure for progress in the experimental life sciences, more than sixty years after they were incorporated into experimental research.

This rhetorical association between stock collections and museums was a mixed blessing, as the case of the ATCC makes clear. On the one hand, it allowed the promoters of the ATCC to argue that it was entitled to the kind of governmental support provided to other museums. As one of the curators argued in 1954, the Smithsonian "now has collections representative of most human interests from fossils to airplanes, and from engineering to art, but nothing as minute as bacteria and viruses."[170] The curator hoped that the ATCC would receive the same support as other national collections. But on the other hand, this rhetoric was a liability at a time when support for museums, as scientific institutions, was generally declining and museums were increasingly oriented toward lay public education (and entertainment). The same ATCC curator acknowledged that bacterial cultures were "difficult to display effectively even to educated laymen, almost impossible to the casual museum visitor."[171] As a later curator noted, no one had found a way that would "make collections of microbes entertaining and instructive . . . for the lay public." From this perspective, there was "no meaningful basis for appealing to the body politic of the national community for the funds necessary to establish microbiological collections as integral parts of museums."[172] Half a century earlier, in New York and several cities of Imperial Germany, museum exhibitions displaying models, preparations, and cultures of bacteria attracted hundreds of thousands of visitors.[173] But in the postwar antibiotic era, the fear of microbes could perhaps no longer bring the masses to look in awe at bacterial cultures displayed behind glass, like wild natural history specimens.

In 1966, at the inauguration of the ATCC's new building, speakers expressed the same tension between the desire to be recognized as a repository of knowledge and the desire to be considered primarily as a research institution. A new curator expressed this dilemma: "The spoken and written word of the tape recorder and the book . . . cannot satisfy optimally the preservation of knowledge. The museum is needed to supplement and to give substance to the library. Museums are places where, in spite of the passage of time, scholars can do research on the same materials."[174] The same could be said of stock collections: they "supplemented" the published literature, providing stable referents and permanent objects for researchers to investigate. Like museums, biological collections were thus considered not mere repositories of things but rather essential components for the production of scientific knowledge. Yet for most of the twentieth century, organism collections never established the connection to a broader public that museums enjoyed. As a result, stock centers struggled permanently for resources: their claim to be regarded as museums cut them off in part from federal funding for research, the NIH having no funds designated for facilities such as museums.

In the long run, the "museum strategy" proved successful when stock center curators seized on another aspect of the museum's function: its role in the preservation of valuable items perceived to be part of a national heritage. At mid-century, the argument was voiced that stock collections were not only

Fig. 1.5. Julie Feinstein, manager of the Ambrose Monell Cryo Collection, storing frozen tissue specimens in liquid nitrogen containers at the American Museum of Natural History, New York, ca. 2009. Photograph by C. Chesek. ©American Museum of Natural History. Printed with permission of the AMNH.

tools for research but also reserves of a unique, national, biological resource. The Genetics Society of America's Committee on Maintenance of Genetic Stocks, which was struggling to achieve recognition for the collections, argued that "the United States probably contains the largest and most complete collection of genetic stocks in the world" and that "these constitute an important part of the national scientific heritage and must be preserved and added for future as well as present use."[175] As another microbiologist working at the ATCC put it: "I like to think of it also as a storehouse of genetic diversity."[176] This diversity, or "biodiversity" as it came to be known in the 1980s, could be understood as a resource of national importance to be preserved in "germ plasm repositories," containing just tissues or isolated DNA, in part for its potential economic value, in addition to its importance in biological warfare. It comes then as no surprise that museums came to host such genetic collections, as they possessed the expertise to collect and curate diverse biological objects. Instead of keeping specimens on shelves or in drawers, they stored their DNA in liquid nitrogen containers. In 2001, almost a century after hosting Winslow's bacteriological museum, the American Museum of Natural

History returned to collecting biological materials produced in laboratories and storing them in "a state-of-the-art cryofacility with a capacity to house up to one million tissue and DNA samples"[177] (figure 1.5). Stock centers and germ plasm repositories came to be understood as a means of preserving a supposedly national or universal genetic heritage, just as natural history or art museums preserved a natural or cultural heritage.[178]

Biological Collections Become Mainstream

At the beginning of the twenty-first century, "living museums" of research organisms, or stock centers, are omnipresent across the biomedical research landscape, representing a profound change from a century earlier. In 1911, when C. E. A. Winslow created his collection of live bacterial cultures at the American Museum of Natural History, the number of similar bacteriological collections could be counted on one hand. If one excludes botanical gardens and zoos, which served different purposes, plant and animal stock collections to support research needs were almost unheard of.[179] In 2016, there were at least 726 culture collections of microbes alone in 75 countries, and stock collections existed for all model plant and animal organisms used in research.[180]

In the second half of the nineteenth century, when experimentation became an increasingly common practice in the life sciences, especially in physiology, individual investigators maintained limited collections of live organisms for their personal use. The French physiologist Claude Bernard had a few rabbits at his disposal for his experiments on the glycogenic function of the liver.[181] Similarly, the German bacteriologist Robert Koch nurtured a small collection of guinea pigs for his experiments on cholera.[182] These collections differed from later stock centers in several respects. They held very small numbers of individuals and species because they were never intended to be comprehensive, but simply to answer the needs of the laboratory that hosted them. Large collections of diverse organisms were mainly found in natural history museums—but these organisms were all dead. The growing importance of live organisms for biological research reflects the rise of genetics and the growing number of researchers in need of organisms for the studies. However, similar changes are apparent in other fields, such as embryology, in which, since the late nineteenth century, researchers maintained their own collections of living creatures, as the aquariums of the Stazione Zoologica in Naples or the Marine Biological Laboratory at Woods Hole so clearly attest.[183] Another main difference between the early collection of embryologists or physiologists and later stock collections is that the former were mostly for internal institutional use, not for public distribution.[184]

What prompted the transition from institutional to public collections in the first decades of the twentieth century? First, the increasing geographic reach of the research communities and the loss of personal contact among researchers. The rising number of bacteriologists and geneticists transformed an intimate community of researchers, often trained in the same places under the same mentors, into a broader, less cohesive ensemble. In the 1930s, the community of *Drosophila* workers grew far beyond Morgan's "the boss and the boys" group (including a number of women),[185] and the community of mouse researchers expanded beyond Castle's tightly knit Bussey Institution crowd. As communities grew, so did the number of biological strains and mutants they were investigating. These two factors explain in large part why and when public stock centers were established.[186] They relieved major laboratories from the burden of culturing a growing number of organisms, increased the authority and prestige of the central "clearinghouse," and provided smaller laboratories with access to essential resources for research.

As the number of researchers working on a given topic increased, so did the need to standardize the biological materials they were using, in order to make their results comparable and thus foster the validation of scientific knowledge. Standardization was particularly challenging in microbiology and genetics because their research materials were alive and in permanent flux. Within just a few days or weeks the virulence of a bacterial culture could be drastically altered, casting doubts that the bacteria in the culture were even the same as the originals. Geneticists, on their side, were continuously trying to induce mutations in their organisms, constantly producing organisms with new genetic backgrounds, speeding up their already fast natural rate of change.

Whatever the technology—careful weeding out of variants or sophisticated freezing methods—the aim of stock centers remained the same: to provide a standardized and stabilized nature for scientists to study. It is thus hard to overestimate the importance of stock centers, since the assumption of a fixed nature is one of the conditions for scientific knowledge. More pragmatically, the fact that scientific knowledge was produced collectively required that individual researchers refer to the same organisms and mutants, which only a shared stock collection could guarantee.

Stock collections also stabilize the natural world by providing an agreed-upon ontology for the collective production of knowledge. The descriptions of a collection's contents in catalogues and newsletters defined the objects of scientific thought and work. This makes the close connection between stock collections and museums, particularly for microbiological collections, less surprising, as museums of natural history have served the same purpose. Half a century later, after the creation of the first major bacteriological collection, that of Král and Winslow, the curators of the ATCC still referred to it as a "bacteriological museum."

One of the most startling aspects of stock collections is how they came to embody a specific set of values guiding behavior among researchers. In addition to stabilizing the natural world, stock centers stabilized the social world of research communities. Stock centers formalized exchange practices of model organism communities that had grown beyond the immediate proximity of their founder or leading figure. These exchange practices, particularly strong among geneticists, were based on an open sharing of knowledge and organisms and a refusal to attach a price to biological materials. In genetics communities working with model organisms, a communal moral economy seems to have prevailed, raising the question of the relationship between the objects of study (living organisms) and moral economies. One could argue, as Robert Kohler has, that within a given research project such as genetic mapping, and in an academic environment, the biology of the fly ideally lent itself to the sharing of this abundant resource. This argument seems to hold true for most model organisms used in genetics research since they were chosen precisely for qualities such as fast replication. Another argument, more historical in nature, also helps explain the similarity of these moral economies: the mobility of individual researchers among different model organism communities. The mouse community, centering on William E. Castle at the Bussey Institution, seems to have been one of the first to adopt this communal moral economy. Corn geneticist Edward East, Castle's colleague at the Bussey, trained Rollins A. Emerson when he visited the institute in 1910. Emerson later developed a similarly cooperative community for corn genetics at Cornell. Castle worked not only on mice but also on *Drosophila* and trained Frank E. Lutz, who introduced Thomas Morgan to the fly around 1906.[187] Leslie C. Dunn, another student of mouse genetics at the Bussey, came to take over the fly room at Columbia when Morgan moved to Caltech and contributed to the circulation of *Drosophila* and the promotion of stock centers at the Genetics Society of America.

Perhaps the most central figure for these cooperative efforts was Milislav Demerec, who started out as a corn geneticist before moving to mice and later to phage. As director of Cold Spring Harbor's Department of Genetics, he promoted the free exchange of ideas and materials among summer visitors, working on each of these model organisms. His work as the editor of the *Drosophila Information Service*, with Calvin Bridges, and as the founder of a *Drosophila* stock center at Cold Spring Harbor was key to the promotion of this vision. In 1934, Demerec acknowledged his debts to the maize community where he was trained when he pointed out that the *Drosophila Information Service* "required a corn man to do the job."[188] Other connections can be found between the "first generation" genetics communities (mice, maize, and fly) and later ones, such as phage, worms, or fish. Max Delbrück, who organized the phage community, began working on *Drosophila* in Morgan's group at Caltech when he arrived in the United States in 1937. Robert Edgar, who started the *C. elegans Newsletter*, was socialized in the phage group, as was George Streisinger, who started the zebrafish community. The newsletters of these communities were explicitly modeled after each other, beginning with the mouse newsletter (1920s), maize (1932), fly (1934), phage (1944), microbes (1950), molds (1962), and worm (1980).[189]

This communal moral economy was also fostered by a few institutions such as Cold Spring Harbor's Station for Experimental Evolution (which became the Department of Genetics in 1921). It served as a breeding ground for many of the genetics communities gathered there for the summer. This tradition of summer visitors, like that of the Woods Hole Marine Biology Laboratory, created a unique atmosphere that has been described as playful, relaxed, and communal and contributed to the sharing of data and material.[190] Whatever the connection between the different moral economies might have been, it is clear that they stabilized the early communal practices of small communities. Robert Kohler is most likely correct when he describes the difference between Morgan and Bridges's supply system for *Drosophila* and the more formalized stock center of Demerec at Cold Spring Harbor as that between "an exchange of professional favors" and "a not-for-profit service."[191] Yet both carried out the same epistemic, if not social, function and embodied a similar set of values, including free exchange, reciprocity, and openness. And public stock centers proved much more enduring and thus played a more important role in perpetuating communal values in the experimental sciences than private collections of organisms. These values, promoted and enforced by the community newsletters, made it possible to successfully address the key challenge of collecting both things and people. Indeed, this communal ethos and the freedom of charge of stocks were preconditions for the growing participation of experimentalists in these collective collecting effort.

2
BLOOD BANKS

The collections of living organisms described in the previous chapter played a key role in the development of classical and later molecular genetics and thus contributed to the success of the experimental sciences in the twentieth century. These collections mirrored those of natural historical museums, except that the specimens were usually alive rather than always dead. They served some of the same purposes as museum collections, too, namely, to provide a stable referent for scientific investigations and allow for the systematic comparison of specimens. Yet the experimentalists' intention was not to mimic a typical practice of museum naturalists. Early geneticists, with a few exceptions, were rather unconcerned with questions of classification. As a result, their collections contained very few different species, mainly mutants of one (or a few) species of model organisms of interest to experimentalists. The geneticists' live stock collections thus resembled more those of the embryologists working at marine stations than the natural history museum collections of dead specimens.

A few experimentalists in the early twentieth century did, however, take an interest in classification. They were hoping to pursue the taxonomist's goal through other means. Instead of bones, shells, and fossils, they sought to classify species by focusing on their most intimate substance:

blood (and its various components). Instead of collections of specimens, they established blood collections. And instead of the descriptive morphology typical of museum practices, they resorted to experiments to describe blood molecules (proteins and DNA) in order to reveal what they believed to be a more fundamental level of relationships between species. In doing so, they contributed to a deep transformation of the field. By the beginning of the twenty-first century, data produced in the laboratory was surpassing the study of gross morphology in the classification and identification of species.

This transformation of taxonomy involved many debates over which methods (experimental or morphological) or characters (blood or bones) were most appropriate to classify species. It also involved the core epistemic values defining science and especially the meaning of objectivity. Naturalist researchers valued their personal expertise, or even intuition, in making judgments that were admittedly subjective. Experimentalists, by contrast, valued instruments as a path to an ideal of objectivity based on measurement, quantification, and precision.[1] These controversies among taxonomists were not merely "turf wars" but reflections of fundamentally opposing ideas about the production of knowledge and the meaning of objectivity.

This chapter concerns researchers who contributed to the "experimentalization" of taxonomy and thereby promoted associated values among naturalists. Here my narrative complements that of Robert Kohler, whose *Landscapes and Labscapes* explores in great depth how biologists, especially ecologists, transposed the experimental ideal from the laboratory to the field. My key argument, however, is that the history of experimental taxonomy not only reveals how experimentalist values took hold in natural history but also illustrates how the comparative way of knowing came to be practiced in the laboratory. This conclusion follows Joel B. Hagen's call to go beyond the simple opposition between naturalists and experimental biologists by looking at some of the places where these two traditions coalesced, for example in experimental taxonomy or, later, in molecular evolution.[2]

This chapter focuses specifically on how the classification of species came to be studied in laboratories at the biochemical level. Long before the rise of molecular studies of evolution (examined in the next chapter) and molecular systematics in the 1960s, researchers turned to the biochemical properties of organisms to understand their systematic position and evolutionary history. The few who did so in the first half of the twentieth century, such as George H. F. Nuttall in Cambridge, England, Edward T. Reichert in Philadelphia, and Alan A. Boyden in New Brunswick, New Jersey, were never able to mount a serious challenge to more traditional taxonomists who based their work on morphological comparisons (that came later). But their stories offer a unique window into how the norms, values, and practices of the comparative way of knowing, so prevalent in museum natural history, entered the laboratory, and, conversely, how those of the experimental way of knowing transformed natural history. In the second half of the twentieth century, the career of Charles A. Sibley, who transformed the classification of birds by relying on laboratory methods, illustrates the successful merger of these two traditions.

Measuring Species, ca. 1900

George Henry Falkiner Nuttall (1862–1937) made perhaps the first systematic attempt to study the diversity of species at the biochemical level and to understand their taxonomic and phylogenetic relationships.[3] Born in San Francisco, Nuttall gained an MD at the University of California and a PhD from the University of Göttingen before working at Johns Hopkins University and at Robert Koch's Hygienic Laboratory in Berlin, two leading institutions in the rise of experimentalism. Finally in 1899, he moved to the University of Cambridge (figure 2.1), where he eventually became professor of biology and, in 1921, the first director of the Molteno Institute for Research in Parasitology.[4] His training in medical bacteriology had familiarized him with the methods of serum therapy and the identification of microorganisms through serological methods.[5] This most likely prompted him to apply the "precipitin reaction" to the study of animal taxonomy. When serum (the fluid component of blood) from an animal of one species previously injected with serum from a second species was brought, *in vitro*, into the presence of serum from the second or a third species, it formed "precipitates" (the same principle used for the identification of microorganisms in the medical context). Anti-dog serum was prepared by injecting dog serum into a rabbit. After the rabbit had produced an immunological reaction, its serum (anti-dog) was drawn and tested against serum from various species. The anti-dog serum reacted strongly with dog serum, less so with cat serum, and not at all with crab serum. The reaction thus indicated how similar the blood antigens of two species were, and thus how closely related the species were to one another, since it was believed that blood antigens were inherited.[6]

Nuttall first used this technique for forensic purposes (i.e., identifying stains as blood),[7] before engaging in a large-scale study of the relationships among species, especially vertebrates.[8] He hoped the technique would allow

Fig. 2.1 George Henry Falkiner Nuttall, University of Cambridge, 1901. On the bench, a cage for holding mosquitoes, probably used in Nuttall's studies on malaria, and a microscope, a common feature in portraits of medical researchers around 1900, symbolizing scientific medicine. Graham-Smith, "George Henry Falkiner Nuttall (5 July 1862–16 December 1937)," plate II. Printed with permission of Cambridge University Press.

him to "measure species," as he put it in 1902, and surpass the morphological comparisons that were plagued by the "subjective element" because taxonomists relied on nonquantitative judgments in evaluating "similarities of structure in existing forms."[9] For his work, Nuttall obtained the blood of hundreds of species from "seventy gentlemen," mainly naturalists working in natural history museums, public zoos, and colonial research institutions from around the world.[10] In a monograph published in 1904 and dedicated to Paul Ehrlich and Elie Metchinkoff, Nuttall described the results of 16,000 tests performed on 586 species[11] (figure 2.2). He used the results to draw qualitative, and sometimes quantitative, relationships among species. He expressed satisfaction that his results, in the case of primates, confirmed phylogenies previously established through morphological comparisons.[12] However, he also was confident enough to argue in some cases that serological results that contradicted morphological classifications provided a better description of the order of nature. For example, he claimed that the horseshoe crab was more closely related to arachnids (spiders) than to crustaceans (crabs), whereas morphologists had grouped it, as its common name indicates, with the latter.[13] This work was widely discussed in the first three decades of the twentieth century, but was often criticized by researchers who were unable to replicate his results.[14] The field of serological taxonomy was small (most serologists were interested in questions of human immunology, physical anthropology, and forensic medicine), and rife with controversies, especially among plant taxonomists in Germany in the 1920s. But Nuttall remained a favorite example in the historical introductions of scientific papers and textbooks as a founder of the field, and later of molecular systematics and molecular evolution.[15]

The methods of serological taxonomy grew out of European research on the properties of blood and immunity in humans, such as those of the Belgian physician Jules Bordet, the German bacteriologist Paul Uhlenhuth, the Polish bacteriologist Ludwik Hirszfeld, or the Austrian pathologist Karl Landsteiner.[16] Several American researchers, who had visited these European laboratories, brought the methods of serology back to the United States. They were part of a larger movement of researchers around 1900 who attempted to base taxonomic studies on some of the new experimental methods aimed at investigating the physicochemical properties of life. The physiologist Edward Tyson Reichert (1855–1931) at the University of Pennsylvania, who embarked on a very similar taxonomic project to Nuttall's but using a different technique, is a good example.[17] Instead of characterizing blood through

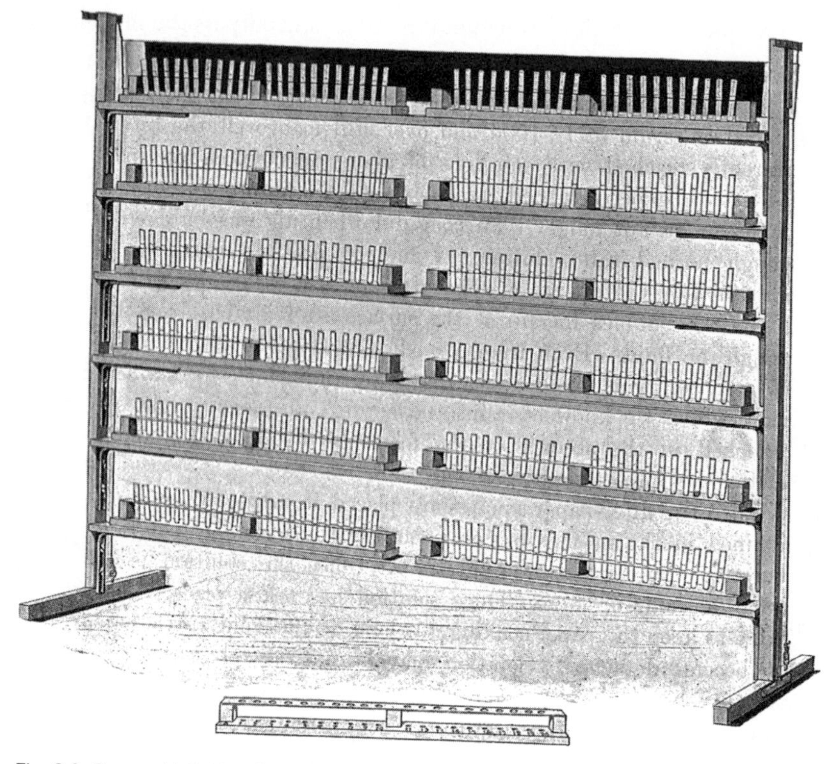

Fig. 2.2 George H. F. Nuttall's test-tube rack used for his serological studies. Each test tube contained a specific antiserum and serum from different species, and their display on a rack allowed for easy comparisons among them, against the black background. Nuttall, *Blood Immunity and Blood Relationship*, 68. Printed with permission of Cambridge University Press.

serological reactions, he grew hemoglobin crystals and compared their shapes. In 1909, he published a massive volume containing lavish photographic reproductions of blood crystals drawn from over one hundred species.[18] Reichert claimed that each species could be differentiated according to the shapes of their crystals. Like Nuttall and early naturalist collectors, Reichert met the challenge of gathering samples from numerous specimens by relying on a broad network of institutions and individuals. The Philadelphia zoo, the oldest in the United States, was one of his main sources, but he also obtained specimens, sometimes just blood samples, from other zoos (Washington and New York), public aquariums, slaughterhouses, food markets, fishermen, dealers, and individual collectors.[19] A wide diversity of animals entered his laboratory, "usually in various states of putrefaction."[20] In Reichert's physiology laboratory, rabbits, dog, and humans were common subjects, but there were also more unusual sights: Venezuelan deer, Indian pythons, or Tasmanian wolves.

Reichert's work was very well received among laboratory scientists, including the iconic experimentalist Jacques Loeb (and his less famous brother Leo Loeb).[21] In the following years, researchers including Karl Landsteiner and Michael Heidelberger would confirm the species specificity of proteins.[22] Naturalists, on the other hand, did not seem to pay much attention to the revised taxonomies proposed by Reichert.[23]

It was in newspapers and magazines that Reichert's results enjoyed the widest popularity. Journalists rapidly highlighted the potential practical applications of Reichert's methods, especially for legal medicine, presenting them in articles with titles such as "Blood Crystals Aid Detectives" or "Blood Crystals vs. Crime." Newspaper articles presented Reichert's method as being even more sensitive than "the famous Bertillon finger-print method"[24] and thus an invaluable tool in fighting crime. A preoccupation with the biological basis of race also framed the reception of Reichert's work. One author hoped Reichert's work would help "fix race relationships more scientifically"; another noted that Reichert's discovery of a "difference between the blood of the white man and that of the Negro" represented a fact of "immense medico-legal importance in crime cases in countries where the Negro flourishes."[25] Neither were the consequences of the blood crystal studies on natural history overlooked, at least by journalists. Reichert's work was cast as the "most epochal [discovery] since the time of Linnaeus," affecting the "knowledge of natural history" by showing that bears should be placed in the family not of dogs, foxes, and wolves, as was customary, but of sea lions and seals.[26]

Reichert and Nuttall's attempts to base taxonomy on physico-chemical principles rested on the creation of blood collections drawn from hundreds of different species. This was perhaps the first time that such a broad range of organisms had been brought into a single laboratory. It comes as no surprise that these species entered in the context of a study in natural history, where the embrace of biological diversity was much more common than in physiology, for example, and that they came through the usual channels of natural history collecting. However, these collections of materials, unlike those of natural history museums, were not put to further use after the initial work of their founder was completed. Although samples from Nuttall's collection were still available half a century later, they were treated as relics more than scientific objects.[27] The work of Reichert and Nuttall reflected a bold, but temporary, exploration of natural history's territory by researchers carrying the growing authority of the experimental science.

Alan A. Boyden's Serological Systematics

The writings of Nuttall influenced a number of researchers to use serological methods to solve problems in systematics (or taxonomy), evolution, and anthropology, but also in physiology and biochemistry. Starting in the 1910s, the methods became crucial for studies of the distribution of human blood groups and population genetics, human ancestry and racial types in physical anthropology.[28] Although they have sometimes been superseded by other experimental methods, such as protein or DNA sequencing, they still enjoyed wide use at the beginning of the twenty-first century as one of the most practical ways to measure genetic properties.

In the field of systematics, the zoologist Alan A. Boyden (1897–1986), inspired by the work of Nuttall and Reichert, did more than anybody else to establish the classification of species based on experimental evidence. Unlike his two predecessors, who made only brief forays into experimental systematics, Boyden would devote his entire career to the field. He became the leading figure in serological systematics and evolution in the mid-twentieth century, further developing Nuttall's techniques and applying them to various problems of classification and phylogeny. Perhaps most important, he believed that by bringing laboratory techniques to bear on classical problems of natural history, he could reform natural history by making it more quantitative, precise, and objective—the main epistemic values that experimentalists used to define their science. His quest for an "objective" method to understand the relationships among species, his creation of a serological museum where he could apply his comparative perspective, and his continued negotiations between the comparative and the experimental ways of knowing illustrate the rise of a new hybrid research culture in the twentieth century. They also provide another example of how an experimentalist's interest in taxonomy led to the adoption of comparative methods and the constitution of collections in the laboratory.

Boyden's attempt to develop a "systematic serology" paralleled other ventures to classify organisms on an experimental basis. From the 1930s on, a number of researchers tried to illuminate relationships between species and develop an "experimental taxonomy"[29] using methods from cytology, ecology, and genetics. Boyden, however, interacted more often with traditional systematists and serologists than with other experimental systematists. This is not so surprising given that experimental taxonomy never became an autonomous discipline but grew, in the United States, essentially as an informal network around the San Francisco Bay area,[30] on the opposite side of the country from Boyden's New Jersey laboratory (and museum). Additionally, most experimental taxonomists worked in botany, whereas Boyden focused on zoology.[31] This mutual isolation reflects the fact that experimental taxonomists rarely perceived the intellectual unity of their field, identifying themselves instead with the particular experimental methods they applied to taxonomic problems—reciprocal transplants, genetic crossings, cytological observations, and serological tests. Boyden's "systematic serology" was often regarded as one kind of "biochemical systematics," an attempt to use molecules as features in the classification of species.

Boyden obtained a PhD in zoology at the University of Wisconsin in 1924 for work on Nuttall's "precipitin reaction" and its application to the study of animal differences. He then joined the Zoological Laboratory at Rutgers University, where he remained until his retirement in 1962.[32] At the outset of his research he was confident that the precipitin reaction could be much improved in order to assess relationships among species and draw conclusions about phylogenies. Like Nuttall, he assumed that the degree of reaction between the antiserum and various (blood) proteins was "in proportion to the degree of relationship of these proteins to each other."[33] Boyden was likely inspired by his supervisor at Wisconsin, zoologist Michael F. Guyer, who had applied this technique in his research on experimental evolution.[34] Guyer found the work of Nuttall and Reichert of fundamental importance not only "for the detection of horse-flesh or dog flesh in sausages," but also for understanding basic principles of heredity and evolution. Indeed, the fact that the blood of different species reacted with the same antiserum showed that they possessed both similarities and differences. These facts were gradually providing "the biologist a rational biochemic basis for the study of the fundamental processes operative in metabolism, heredity and evolution."[35] The serologist Karl Landsteiner at the Rockefeller Institute also found Nuttall and Reichert's conclusions "unquestionable"[36] and published a number of serological studies in 1925 on the relationships between humans and apes.[37] He was so confident in the serological method that he claimed "one could roughly construct the zoological tree merely on the basis of precipitin reactions."[38] Landsteiner thereby joined the growing group of researchers applying the approach in the decade following the publication of Nuttall's seminal work.[39]

Boyden first attempted to improve the reliability of the precipitin reaction, which had been challenged by numerous authors especially in Germany. He drew on earlier studies of the mechanisms of immunity, carried out by researchers such as Karl Landsteiner and Jules Bordet, and also on the use of the precipitin reaction for forensic purposes, by Paul Uhlenhuth and others, essentially to identify human blood stains or as the basis of the Wasserman test to detect syphilis. He began by using the "ring test" to find the minimum concentration of antiserum that would elicit a reaction (a visible ring) in the test tube, providing a measurement of the sensitivity of the antiserum to a given antigen.[40] He used the value of the homologous reaction (for example, anti-rabbit serum reacting with rabbit serum) as a reference value. The more different two species were, the less reactive the antiserum would be. Boyden thus generated numerical values that he could use to indicate distances between pairs of species. Even though the "ring test" was clearly quantitative, the determination of the lowest concentration producing a precipitate was visual, based on personal judgment.

Boyden used variations of the test throughout the 1920s and 1930s, until physicist Raymond L. Libby, a colleague at Rutgers University, developed an instrument to measure the turbidity of solutions in 1938 (figure 2.3). Boyden then began using that instrument to assess the extent of the precipitin reaction more precisely and, he believed, more objectively.[41] Instead of visually inspecting the solution for the presence or absence of reaction, he could now estimate automatically the amount of the precipitate for a given concentration of antiserum. The rapidity of this test allowed him to make measurements over a whole range of concentrations and, after some calculations, to obtain a single value representing the strength of the antiserum reaction to serum of another species. This indicated how closely the proteins of two organisms were related.[42] He spent many years trying to identify all possible sources of experimental error in order to improve the reproducibility of his method.[43]

Boyden's dream was not simply to develop a powerful experimental method, but also to apply it to the revision of existing classifications of animals, which were essentially based on the comparison of morphological features of specimens in natural history museums. This lifelong quest was driven by his desire to make taxonomy (and eventually phylogeny) "entirely objective and independent of the interpretation of the observer."[44] Quantification was a means to achieve this goal, and precision was a necessary corollary. Serological methods would "yield measurements" of the degree of relationships among species, making the "study of relationships more exact [and] more

Fig. 2.3 Dr. Elizabeth C. Paulson, Alan A. Boyden's assistant, using the automatic photoreflectometer to measure quantitatively the turbidity of species pairings at Rutgers University, early 1950s. *The Serological Museum* 19 (Dec 1957): 8. Printed with permission of the Special Collections and University Archives, Rutgers University Libraries.

scientific."[45] In contrast, Boyden paraphrased his colleague G. Kingsley Noble, head of the Department of Experimental Biology at the American Museum of Natural History, by arguing that "phylogenists and systematists sometimes appear to work on an instinctive basis, to 'feel' their way to their systematic groupings."[46] They relied primarily on morphological data—skins, bones, and fossils. Boyden's critique thus focused primarily on the problems associated with this approach. Because naturalists had failed to develop a quantitative measure of morphological features, he argued that they necessarily depended on "interpretation as to what various structures may mean in descent." The problem, for Boyden, was that the interpretation "differs with interpreters," resulting in "an endless difference of opinion as to the relationships of certain groups of animals necessitating countless 'revisions' of them."[47]

Boyden was sometimes quite dismissive of morphology, as when he claimed that his method had "succeeded in giving us what a century or more of intensive morphological investigation has failed to provide, namely, a basis for a quantitative phylogeny."[48] He went so far as to ridicule taxonomists who were working exclusively with morphological characteristics when he wrote, referring to the rabbits he was using to produce serum, that "so far, the rabbit has actually made fewer mistakes than man in the attempt to construct a natural system of classification."[49]

Boyden hoped to delegate the task of animal classification entirely to antibodies, which would measure similarities and differences objectively. Boyden acknowledged that in its present state serology was far from a perfect method, but he argued that "biochemical evidence regarding the natures of organisms may ultimately outweigh all other bases of classification," even if at present "some 'dyed-in-the-wool' morphologists" still belittled its importance.[50]

In equating quantification with objectivity, and criticizing morphologists for their qualitative and therefore subjective approach, Boyden ignored the fact that some systematists claimed that quantification did not exclude subjectivity.[51] In *Quantitative Zoology*, published in 1939, the leading American paleontologist, George Gaylord Simpson, and clinical psychologist Anne Roe emphasized numerical methods and even the use of calculating machines in the classification of organisms.[52] They presented simple statistical methods to analyze measurements made of different specimens, mainly the size of various skeletal parts. They defined species by averaging the measurements of one or more features present in a collection of specimens. At the same time, they recognized that systematics would always remain an art because the choice and weighting of the features to be measured were somewhat subjective.[53] "Personality can no more be eliminated from classification than from any other art," Simpson declared in1961.[54] Like most systematists, he did not think this diminished the rigor and scientific character of taxonomy. In a series of letters between Simpson and Boyden, the latter criticized the term "homology" (similarity due to common ancestry) because there was often no objective way to distinguish it from mere similarity. Simpson was unmoved, recognizing that "homology expresses an opinion as to the origins of similarity" but insisting that it was "perfectly legitimate" to have opinions in science.[55] The definition of "homology" might be based on "opinion rather than on objective fact," but this did not "invalidate it or make it less useful,"[56] he claimed unapologetically.

In 1953, the leading systematist, Ernst Mayr, and his co-authors took the same view in explaining why taxonomy was both a science and an art: the "good doctor and the good taxonomist make their diagnoses by a skillful evaluation of symptoms in the one case and of taxonomic characters in the other."[57] Like some of the physicians who resisted the idea that scientific medicine was the proper foundation of clinical practice in the late nineteenth century (and later) by appealing to the irreducible "clinical art," taxonomists were trying to ground their authority in the skills that were acquired through apprenticeship to distinguished researchers in their fields and that could never be replaced by objective scientific methods.[58] Similarly, as late as 1957 botanist William B. Turrill from Kew Gardens, an experimental taxonomist like Boyden, claimed that "classifying is never entirely objective since the peculiarities and particularities of the human mind and of the individual taxonomist impose subjective elements on the result." Turrill was unsure whether the lack of objectivity was a problem for taxonomy: "How far the subjective element can be eliminated or controlled, or even how far it is desirable to attempt so to treat it, is debatable."[59] The claim that subjectivity played a role in the production of scientific knowledge was common among naturalists well into the mid-twentieth century. Experimentalists, on the other hand, had long rejected subjective judgments, believing that they were antithetical to the formation of objective knowledge, the only kind they would regard as truly scientific.

Boyden, having firmly adopted the experimentalists' ideal of objectivity, sought to eliminate the subjectivity involved in the choice and weighting of characters by relying on a single trait: the immunological affinity of blood. He justified this decision based on the fact that unlike morphological traits, an organism's biochemistry was unaffected by environmental and developmental influences and was thus better suited to measuring the true genetic relationships between species.[60]

Even though Boyden had been trained as an experimentalist, used experimental methods, worked in a laboratory, and embraced the epistemic values of experimentalism, his professional identity was that of a natural historian. In a letter to the president of his university, he claimed unambiguously: "Thank Heaven I belong to 'Natural History.'"[61] He published regularly in *American Naturalist* and *Systematic Zoology*, was invited to lecture at the American Museum of Natural History, and spent a sabbatical at the British Museum (Natural History) in London. He defended the importance of taxonomy as essential for all experimental biology. The existence of so many terrestrial species (about one million, he estimated) made it impossible to perform "the same experiments on all of them." But this would be unnecessary once a natural classification was established that would make it possible to "generalize effectively about related or essentially similar organisms and eliminate the need for countless repetition of the same experiments."[62] Boyden's background as an experimentalist positioned him ideally to make this argument, which resonated with systematists who were trying to regain prestige they were losing to experimentalists. Systematists such as Ernst Mayr, George Gaylord Simpson, and Richard E. Blackwelder took a number of intellectual and institutional steps between the 1940s and 1960s toward "upgrading, improving, scientizing" their discipline.[63] They reevaluated the links between systematics, evolutionary theory, and population genetics (the "new systematics"), introduced new kinds of empirical data (especially from experimental cytology, genetics, ecology, and serology), created the Society of Systematic Zoology in 1947, and launched the journal *Systematic Zoology* in 1952.[64]

Boyden understood his work as a contribution to these reforms, not as an attempt to downgrade natural history. Even while working with experimental data, he understood his methods as part of a renewed tradition of morphology, rather than of physiology or immunology: "Biochemical comparisons fall within the province of 'morphology,'" he wrote in 1943.[65] He was continuing the "comparative morphology" enterprise but at the biochemical level.

Boyden's attempt to build a discipline of serological zoology was largely unsuccessful until after the Second World War. He remained the leader of the field, but a lonely one.[66] Particularly in Germany, researchers were using serological techniques for taxonomic purposes, but most for only a limited period of time. Boyden's aggressive review of their work, especially when it did not live up to his high technical standards, might have discouraged some from pursuing the matter further. Unsurprisingly, morphologists remained largely unmoved by the serological approach, and serology never became a serious challenger for morphological methods. After 1945, Simpson and Mayr and other systematists were mentioning Boyden's work and the serological approach generally approvingly, if only in passing.[67] They acknowledged the great potential value of the serological method, even if they found Boyden to be a mediocre scientist who had not produced many taxonomic results in decades of serological investigations but kept making the same promises that serology would revolutionize systematics.[68] More generally, they praised the use of data from laboratory experiments involving serum, proteins, and other biochemical components as useful complements to the analysis of morphological characters, such as measurements of skeleton length, but never as a replacement.

A Museum in a Laboratory

In 1948, Boyden embarked on a more ambitious plan to develop the field by creating the Serological Museum at Rutgers University and publishing *The Serological Museum Bulletin*.[69] The move was unusual, to say the least, for an institution devoted to experimental research. True, the boundaries between natural history and experimental biology (and between the museum and the laboratory) were becoming increasingly blurred.[70] However, this was primarily a result of the burgeoning practice of field experimentation, or the creation of laboratories in natural history museums, rather than the creation of "museums" in laboratories. In 1928, the AMNH had founded the Department of Experimental Biology headed by G. Kinsley Noble,[71] an initiative that had been supported by "the younger members" of the AMNH board of directors.[72] Naturalists were recognizing the value of experimental methods, for intellectual as well as social reasons, but experimentalists rarely acknowledged the value of the comparative way of knowing, so common in natural history.

The creation of Boyden's Serological Museum is a revealing case of the way in which comparative practices made their way into the laboratory. As a reporter for *Science* put it, the new institution represented a "unique kind of museum."[73] Its purpose was similar to that of most natural history museums, which Boyden emphasized. He insisted that proteins were just as worthy of preservation and conservation as the "skins and skeletons" of organisms because they were just as characteristic.[74] Other natural history museums typically kept only the "innermost insides and outermost outsides" of animals; Boyden's collection would complement those efforts by preserving the "chemical compounds that keep their life-processes going."[75] He did not need "great halls and showcases" to exhibit his objects of study, since "bottles of sera look much like each other"; what he needed were "adequate cold rooms for the preservation of these sera."[76] A Rockefeller Foundation grant made the construction of such a room possible in 1951, in a former coal bin (figure 2.4).[77]

Fig. 2.4 Alan A. Boyden (left) and his collaborator Charles A. Leone (right) showing blood samples to a university official in the serological museum, 1951. Rutgers University Archives, Faculty Bio, Alan A. Boyden. Printed with permission of the Special Collections and University Archives, Rutgers University Libraries.

Like most natural history museums, Boyden's new institution aimed to collect, classify, preserve, share, and study serum samples. And it was driven by the ideal of completeness: "build up by collection and exchange samples of as many kinds of protein as are obtainable from all kinds of organisms, young and old, healthy and diseased."[78] By 1950, the collection held blood samples from more than four hundred species,[79] and Boyden was hoping to receive and collect many more. Completeness mattered more to Boyden than sheer quantity. The most precious specimens were those that completed some aspect of the collection. In 1950, as Rutgers University's press service proudly announced, the museum had received the last specimen that Boyden needed to complete his collection of sera from all the eighteen orders of mammals in a "securely-wrapped package, flown in by air express."[80] Boyden had just received blood from a flying lemur (which doesn't fly and isn't a lemur) from Madagascar. Now he had a new goal: to "sample all of the 118 families which divide the eighteen orders."[81]

The blood samples looked alike and were not for display, but were to be used in serological experiments. Boyden envisioned his museum as a place for study, reflecting the early modern usage of the word as equivalent to the "studio." [82] Every sample added to the collection made numerous new experiments possible, because it could be tested against all previously collected samples. Blood from a new species thus generated an entirely new set of relationships, potentially challenged those established between existing samples, and increased the number of possible comparisons that could be made. The collection was a tool to produce experimental knowledge about taxonomy through the systematic comparison of organisms. Researchers at the Serological Museum worked in the same way as naturalists in natural history museums, except that their systematic comparisons were based on experimental results rather than on bones, fossils, and preserved specimens.

Boyden's Serological Museum shared another characteristic of natural history museums: it was embedded in a complex network of collecting institutions.[83] Boyden had begun his collection with serum from domestic animals—sheep, pig, cow, horse, and dog—easily found in rural New Jersey near Rutgers University. Because these species posed little challenge for the taxonomist or the evolutionary biologist, he quickly began diversifying the collection. He collected blood of numerous organisms from zoos and field research laboratories from across the United States, from the New York Zoological Park, the San Diego Zoo, the Mount Desert Island Biological Laboratories in Maine.[84] The San Diego Zoo provided blood from aardvarks, anoas, giant pangolins, and warthogs.[85] Federal agencies such as the US Fish and Wildlife Service and the US Biological Survey provided samples of wild species living in the United States, including bison and elk. The American Museum of Natural History in New York and other museums provided even more samples. Boyden asked Simpson at the AMNH that "whenever expeditions are sent out . . . due consideration . . . be given to the matter of blood samples."[86] Marine stations in the United States and across Europe were also major suppliers: the Marine Biological Laboratory in Woods Hole, Massachusetts, the Marine Biological Laboratory of the Carnegie Institution in Florida, the Laboratoire Arago in Banyuls-sur-Mer, France, and the Stazione Zoologica in Naples, Italy.[87] Unlike natural history museums, marine stations generally kept collections of live animals and were thus particularly suited to collect fresh blood.

Boyden's efforts also resembled those of natural history museums because of his reliance on collectors in the field. At the museum's inauguration he invited the cooperation of "all naturalists, wherever they might be [who] may be in a position to collect and contribute or exchange samples." He emphasized that the success of the museum depended "on the extent of the cooperative effort."[88] A few years after the opening, Boyden renewed his call in the journal *Science*, appealing specifically to zoological collectors planning expeditions in remote regions. Boyden was aware that most would be inexperienced and "unequipped for refined serological collecting."[89] He thus described a simple method involving soaking paper with blood from wounds or from the "animal's carcass as it is being skinned." Additional amounts of blood could be obtained through "cutting open the heart and major vessels." For small animals, a blood spot "no larger than a matchhead" would suffice, but for larger ones, "a square foot or more" of blood-soaked paper was desirable. The samples should then be "hung up in the shade" to dry, and "carefully shielded from visitations by insects." True to the practices of naturalist collectors,[90] Boyden specified that each sample should come with the scientific name of the organism from which it was collected, a date, a locality, and the name of the collector.

Convincing collectors on hunting expeditions to sample blood from their game was no simple task. Even when their collaboration could be secured, the requirements of serum collecting could conflict with safe hunting practices in the wild. In 1950, *The Serological Museum Bulletin* described in detail how serum collecting would affect hunting practices in East Africa. The author began by noting that, obviously, animals almost always had to be killed before they could be bled. Unfortunately for scientists, hunters usually aimed for the shoulder, leading to massive hemorrhages and blood clots in the chest cavity. This meant that cutting the throat would not produce a flow of fresh blood. The author thus encouraged hunters to aim for the brain or "the neck across the cervical vertebrae," but recognized that this might be both challenging and dangerous in the case of big game.[91] Less dangerous, but no less challenging, was the collection of blood from marine animals, such as lobsters, fishes,

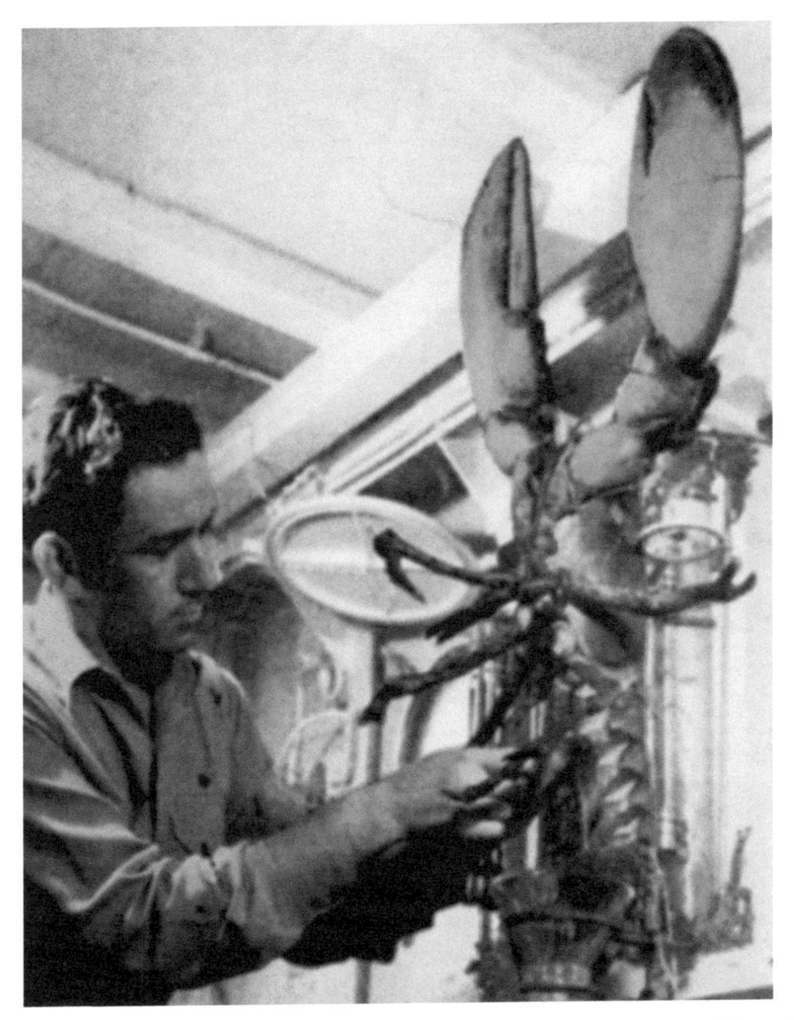

Fig. 2.5 Image used to illustrate an article giving instructions on "how to bleed a lobster," highlighting the challenges of laboratory studies on wild animals in the field. "Recent Advances in the Collection of Serum from Marine Animals," 3. Printed with permission of the Special Collections and University Archives, Rutgers University Libraries.

mollusks, or turtles (figure 2.5). *The Serological Museum Bulletin* published a detailed discussion of different techniques such as "cardiac puncture," hoping to educate and enroll naturalists embarked on sea expeditions.[92] Boyden also promoted new technologies such as the "mobile laboratory" (a truck transformed into a fully equipped laboratory) to facilitate the collection of blood in remote places (figure 2.6).[93]

Fig. 2.6 A "mobile laboratory" for collecting blood in the field, ca. 1950. The necessity of keeping blood cold and free of contamination required researchers to bring laboratory equipment into the field, rather than sending specimens from the field to the laboratory. Weitz, "Mobile Laboratory," 5. Printed with permission of the Special Collections and University Archives, Rutgers University Libraries.

Boyden's call was successful enough to ensure steady growth for the Serological Museum's collection. A collector from the New York Zoological Society provided samples of timber wolf, puma, black bear, Himalayan bear, and giant panda. Another, from the Zoological Society of London, provided samples of polar bear, hyena, and golden palm civet. The popular press got involved as well. In 1950, the *New York Times* ran a story entitled "Woman Scientist Brings Rare Blood."[94] Boyden was interviewed while he was waiting for an Australian serologist at the airport in Newark. When she arrived, she was carrying "precious samples in a brown bag under her arm." The package contained blood samples from "twenty little-known animals" including penguins, seals, sea lions, and sea elephants, native to an island in the Pacific Ocean near the South Pole. The story offered an unusual mix of narrative tropes, describing the sanitized world of the laboratory and the adventurous world of the field, the high technology of experimental science and the paper technology of natural history. This hybrid perspective became a distinctive trait of many subsequent stories about the Serological Museum (figure 2.7).[95]

To be sure, these stories belonged more to the adventure genre of natural history expeditions than to fictional accounts of white-coated laboratory scientists, such as Sinclair Lewis's *Arrowsmith*.[96]

The moral economy on which Boyden's collecting enterprise rested was a "gift relationship" by which individual collectors contributed samples to a

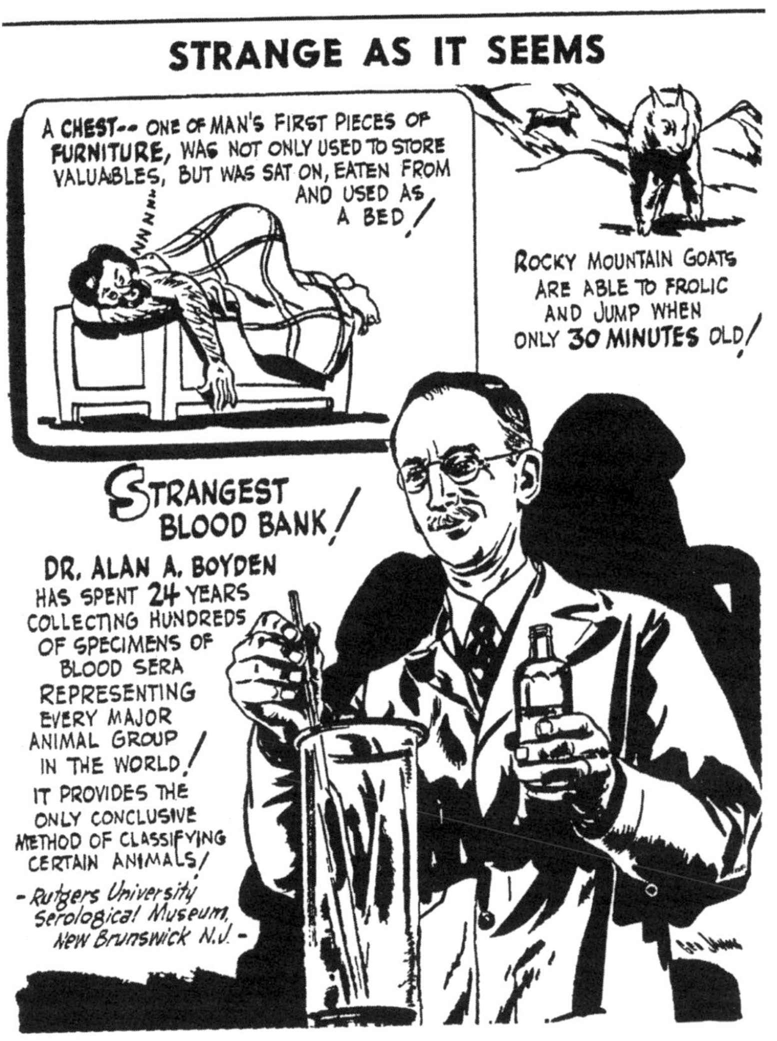

Fig. 2.7 Newspaper cartoon highlighting the "strange" nature of Boyden's experimental-naturalist endeavor, ca. 1950s. Rutgers University Archives, Faculty Bio, Alan A. Boyden. Printed with permission of the Special Collections and University Archives, Rutgers University Libraries.

public facility that, in return, offered services and assistance to collectors and institutions in many parts of the world.[97] Increasingly the contributions of individual collectors were acknowledged in publications resulting from their use, perhaps to encourage others, a common strategy adopted by authors of taxonomic publications, for example. Yet Boyden, who analyzed the material and data provided by field collectors, always remained the sole author of these publications. Even before the museum was created, Boyden often acknowledged his debts to individuals in print for "the loan or gift of specimens,"[98] and the museum simply institutionalized this practice. To those who collaborated with the museum, he distributed *The Serological Museum Bulletin* free of charge, and he acknowledged collectors in all scientific reports making use of their contributions.[99] The museum's dependence on an extended network of individuals and institutions meant that it also had to work as a service institution. Boyden insisted that the museum could not "receive only and not give in return."[100] It would make "a virtue of necessity" and serve others so that the museum would "be served in turn."[101] One service was the shipment of blood to institutions and researchers; however, the shipments were mainly domestic and rather infrequent. Boyden donated a small amount of sera from two species of turtles to researchers in New York who used electrophoresis to show that it did not contain albumin, a constituent of human blood.[102] Yet the museum received blood from all around the world. "Cooperating institutions," spread throughout the five continents, included zoological museums and universities in Europe and North America and colonial medical research institutions in Africa, Asia, and Oceania.[103] A large world map hung on the wall of the museum, with radiating lines from its location in New Brunswick, New Jersey, to all the sample collection sites, showing the extent of Boyden's serological empire.

The cooperative and service ethos that Boyden emphasized for his museum—and for science in general—contrasted sharply with the individualist and increasingly competitive ethos prevalent in most of the experimental sciences at that time. A good example of the latter was a bitter priority dispute between Boyden's colleague at Rutgers University, microbiologist Selman Waksman, and his graduate student, Albert Schatz. Their rivalry over the discovery of streptomycin went as far as a lawsuit.[104] True, the experimental life sciences had also known their cooperative moments, for example in Thomas Morgan's "fly group" in the 1920s and 1930s and Max Delbrück's "phage group" in the 1940s and 1950s (see previous chapter), but these were becoming increasingly rare in the postwar economy of science. The growing competitive environment had many causes, including the dissolution of personal relationships among researchers that arose from the growing size of the scientific communities and the decreasing authority of leading figures who had policed cooperative behavior.[105]

For Boyden, cooperation was not simply a practical necessity for the management of his museum but a commitment to certain altruistic values. In 1943, he complained to the university's president that he was "deeply concerned . . . with the character of [the] students." What bothered him most was their "complete indifference to the alarming international situation" and their "supreme selfishness." They seemed "to feel no debt to society," Boyden remarked. He had thus tried to teach his students "the ideal of cooperation [and] the biological fallacy of isolationism." These views on the social and moral orders found institutional expression in the Serological Museum. It was true, as Boyden insisted, that the museum relied for both its existence and its work upon the principle of cooperation, but he also believed cooperative scientific practices to have a broader moral and political meaning in the Cold War world. The willingness of individual scientists worldwide to donate sera was "proof that, among biologists of many lands, there can be wholehearted and friendly cooperation" and that in "these times—in any times—it is well to have occasion to reaffirm our faith in men and their capacity for mutual aid."[106] The collective dimension of museum work, as opposed to the individual dimension of laboratory work, embedded the Serological Museum both epistemically and politically within the Cold War economy of science. Museums, which were often instruments in promoting national identities, could also be put to the service of internationalist values.[107] Boyden retired from Rutgers University in 1962, but upon the premature death in 1971 of his successor, Ralph J. DeFalco, he returned from retirement to once again head the museum he had founded.[108]

Between Field and Laboratory: Charles G. Sibley

In the hands of the anthropologist Morris Goodman, the serological approach produced its most striking results the year Boyden retired (for the first time). At a 1962 conference on human evolution, Goodman announced that humans, gorillas, and chimpanzees were indistinguishable from a serological point of view.[109] Like Boyden, Goodman had measured the immunological similarity of blood proteins, using a somewhat different technique called the Ouchterlony plate. Goodman also assembled his own collection of primate blood. A few years after the presentation of these initial results, he compared blood from 118 different species.[110] The similarity between the two researcher's approaches is less surprising given that Goodman had been trained by one of Boyden's students at the University of Wisconsin, zoologist Harold R. Wolfe, who also applied serological methods to taxonomic problems.[111]

Goodman's views were perceived not only as radical but as wrong by primate systematists such as Simpson, whose vehement opposition resulted in flamboyant controversies that have lasted to the present day.[112] Yet the battle lines weren't cleanly drawn between new serological systematists and old morphological systematists. Strong disagreements erupted even among serological systematists. Boyden opposed Goodman in part because the Ouchterlony plate, which provided only qualitative visual evidence, constituted a step backward from his own quantitative methods.[113] Other researchers, such as biochemist Allan C. Wilson and anthropologist Vincent M. Sarich at Berkeley, used their own immunological technique, microcomplement fixation, to provide quantitative data that confirmed Goodman's results, yet they disagreed about their implications for rates of primate evolution.[114]

Throughout the 1960s, a new set of techniques to probe the biochemical differences between organisms were becoming increasingly popular: protein electrophoresis, protein sequencing, and DNA hybridization. However, serological methods remained in wide use, especially because they were easy, quick, and cheap to perform and thus were particularly well suited to systematic studies that often required the examination of large numbers of specimens, or to field research where only minimal laboratory equipment was available.[115] These advantages also proved crucial for studies on the distribution of human populations that were carried out through blood group research.[116]

The same advantages led to the widespread use of protein electrophoresis in the study of taxonomic relationships. It was cruder than protein sequencing, since it analyzed proteins (or protein fragments) only in terms of their electric charge and size, but it was also much faster and thus allowed large-scale studies. From the late 1950s, as Boyden was professionally fading away, the ornithologist Charles G. Sibley (1917–98) became the leading advocate of using biochemical methods for taxonomic purposes, first focusing on the electrophoretic study of blood and egg-white proteins and later on DNA hybridization. Like Boyden, Goodman, Sarich, and Wilson, he claimed to be continuing the enterprise Nuttall and Reichert had begun more than half a century earlier.

Fig. 2.8 Charles G. Sibley in a shed taking ornithological field notes while deployed on Emira Island, Papua, New Guinea, 1944. MVZ Historic Photo Collection. Printed with permission of the Museum of Vertebrate Zoology, University of California, Berkeley.

Charles Sibley was trained in biology at the University of California-Berkeley, where he obtained a PhD in zoology in 1948 for work on the hybridization of natural bird populations (unrelated to DNA "hybridization").[117] During the war, he was deployed with the Navy in New Guinea, the Philippines, and the Solomon Islands (figure 2.8).[118] In his free time (which he seems to have had quite a bit of) he collected birds for the Museum of Vertebrate Zoology and assembled observations about bird behavior that were published after the war. He was a field naturalist who had developed a taste for observation since childhood and later at the Museum of Comparative Zoology at Berkeley. His dissertation was based on the fieldwork carried out during his

Navy stint in the Pacific islands and later in Mexico and New Guinea.[119] In 1953 he moved to Cornell University, where he served as curator of birds and professor of zoology.

Sibley's methods of studying bird hybridization were based on plumage characteristics that changed with the bird's age and showed considerable individual variability. In 1956, he began to explore the use of paper electrophoresis to examine bird proteins, which could potentially offer more clear-cut results. In the 1930s, just a few years after the technique had been invented by Arne Tiselius, electrophoresis was used by Karl Landsteiner to compare the egg-white and blood proteins of ten different animal species.[120] Although Landsteiner was primarily interested in understanding the general nature of proteins, he also showed that the molecules varied between species, a crucial fact for those interested in taxonomy. A number of other researchers continued this line of investigation. In 1945, a comparison between twenty different species allowed two researchers to claim that they could produce "electrophoretic patterns characteristic of [each] species."[121] When a simpler method of paper electrophoresis was developed, investigators began to make more extensive studies of species diversity. In 1956, the biochemists Herbert C. Dessauer and Wade Fox at Louisiana State University examined 87 species of amphibians and reptiles by paper electrophoresis and published "typical patterns" of each taxonomic order.[122]

So by the time Sibley began to use paper electrophoresis in the study of birds, he was traveling along a well-established path. Unfortunately, his first extensive study of serum proteins in local bird species revealed that they were too similar for taxonomic purposes.[123] In 1958 his graduate student Paul A. Johnsgard, who was performing the experiments on the serum proteins, decided to explore the possibility that egg-white proteins might display more specific differences.[124] This proved to be the case and Sibley swiftly adopted the approach.[125] Unfortunately, collecting eggs constituted an even greater challenge than collecting the blood of live animals, as nests were often more difficult to spot than birds (and tended to be out of easy reach). Here Sibley's experience as a field naturalist proved essential.

Collecting in the Field

In 1958, Sibley set out to "assemble a collection of egg-white samples from as many species of birds as possible."[126] He followed the long tradition of trying to enlist individual naturalists in the project and managed to create a wide network of helpers, thus assembling a collection of several thousand specimens from over a thousand species.[127] Collectors included "willing students [who] acquired permits, risked their necks climbing trees and cliff faces, and combed forests, prairies, and tundras, all in search of samples from both common and rare species." He also drew hosts of "professional ornithologists and amateur birders" into the project.[128] In 1960, he published his first survey of egg-white proteins, comprising over 5,000 electrophoretic profiles from nearly 700 species—provided by "more than 150 gentlemen" (an explicit allusion to Nuttall's "seventy gentlemen") and "a dozen ladies as well."[129] Sibley personally thanked each of these "generous persons" before issuing a call to readers having "access to egg-white specimens [to] collect and send them" to him. He outlined the best procedure to collect and ship egg white (reminding his readers, "never attempt to send whole eggs"). After "supporting the egg with cotton or paper in a teacup," the collector was to "snip off a cap of shell" with sharp-pointed scissors, extract the egg white with a pipette that had previously been cleaned and dried in an oven, and place it in a sealed container. Ideally, the samples should be kept refrigerated or frozen until they were mailed. Customarily, naturalists in the field might have carried scissors (and teacups), and might even have been persuaded to bring along pipettes and small containers, but the prospect of carrying ovens and refrigerators was probably not very appealing.

Naturalists had developed a number of methods for collecting specimens in the field that reflected both practical constraints (such as portability) and the requirements of the botanical gardens and museums for which they were destined. The Wardian case was developed in the nineteenth century to transport live plants on long ship journeys from British colonies back to the metropole.[130] Naturalist-hunters were careful not to damage animal skins, which needed to be intact for display in a museum. The collecting efforts of investigators such as Boyden and Sibley needed to be adapted to the special requirements of laboratory research. Experimental taxonomists invented new tools to bridge these two worlds, including "field laboratories." Early-twentieth-century biologists had briefly used floating laboratories to study the ecology of rivers and marshes, experimental farms to investigate the physiology of crops, and marine stations to explore different aspects of marine life.[131] The collection, preservation, and shipment of proteins and DNA, from eggs or blood respectively, required "very special conditions, equipment, supplies and refrigerated shipment."[132] These infrastructures were common in laboratories, but much harder to come by in the field.

Their absence compromised the taxonomic results that could be obtained from protein studies. As Sibley complained, in 1963, to Ernst Mayr, the leading evolutionary biologist, ornithologist, director of the Museum of Comparative Zoology, and professor at Harvard:

> The main problem with all the protein work is in the field—getting good, FRESH, material. So much of the stuff arrives rotten or in poor shape that it is clear that to make really sound progress all along the line we must do our own collecting with good equipment and under proper conditions. The good-hearted volunteers have done a wonderful job but the next step is to send properly equipped and trained people into the field to obtain perfect material that can always be trusted. It will be expensive.[133]

Few naturalists enjoyed the type of training Sibley had received in the methods of field and laboratory science. He thus decided to "get into the business of finding ways and means for the collection, care, shipment, etc. of blood from field stations."[134] In December 1963, Sibley wrote to Mayr from Costa Rica: "[I am] camped in a rain forest—and it's raining." In other words, the conditions were atrocious for collecting samples. Fortunately, he had brought along a mobile laboratory in a heavy truck that included a propane gas refrigerator, making it possible to collect and store samples despite the conditions. This experience led him to conclude that the "collection of DNA specimens in the field seems to be entirely feasible."[135] Yet Sibley acknowledged that the methods had to be manageable in the hands of naturalists with little if any laboratory training, writing that the "problem has been to find an optimum combination of simplicity and effectiveness for the field collector."[136]

Sibley personally continued to travel the world to collect eggs and blood from birds. Between 1963 and 1966, he explored Costa Rica, Australia, New Zealand, South Africa, Rhodesia, Kenya, Uganda, Ghana, Norway, England, France, Germany, and Switzerland.[137] In 1969 he organized an expedition on the *Alpha Helix*, a research vessel sponsored by the National Science Foundation, to Papua New Guinea, home of "birds of paradise, men of the Stone Age, vast rain forests and snow capped peaks."[138] The *Alpha Helix* sported three well-equipped, air-conditioned laboratories (figure 2.9). It also held "a walk in freezer, a dark room, machine shop and other facilities." He stated that the "unique advantage of the Alpha Helix is that its research facilities bring areas of exceptional biological interest into direct contact with modern laboratories."[139] As Joanna Radin has argued, this "floating freezer" was also

Fig. 2.9 The *Alpha Helix* ship, 1966. The NSF-sponsored vessel was most likely named after the alpha helix protein structure, for which Linus Pauling received the Nobel Prize in 1954, as a symbol of triumphant experimentalism. Photograph by Robert Glasheen. Printed with permission of the Scripps Institution of Oceanography Archives, UC San Diego Library.

an "instrument for scientific imperialism," bringing back biological material from (post)colonial areas to the United States.[140] The shipboard environment also proved fruitful for social interactions between Sibley and researcher-shipmates including Alan C. Wilson, Vincent Sarich, and Herbert C. Dessauer, who were also seeking to collect specimens for laboratory studies.[141] During the three-month trip, Sibley collected about 3,500 birds representing 217 different species.[142] Overall, from 1963 to 1978, Sibley collected nearly 30,000 specimens representing over 2,500 species. "Can you conceive of the time, correspondence, work and cost involved" in assembling this "remarkable collection?" he asked Mayr (who certainly could).[143]

The success of Sibley's enterprise rested on several factors: his experience as a naturalist field collector, an extensive network of collaborating professionals and amateurs, and the development of specific technologies to mediate between the field and the laboratory. Techniques for the collection and preservation of specimens for museums display, such as pickling or taxidermy preparation, were of little use for experimental studies. Mobile laboratories mounted in trucks or on ships, carrying the instruments necessary for the extraction of biochemical samples such as proteins or DNA and their preservation during long journeys, provided indispensable transition zones. They offered a space that was not quite the field and not quite the fully equipped laboratory but allowed the flow of biochemical materials from one to the other. As Joanna Radin has shown, when experiments could not be performed in the field, the freezer became the key technology to preserve biological material.[144]

Sibley had joined Yale University as professor of biology and curator of birds at the Peabody Museum of Natural History in 1965, and he became its director in 1970. By this time Sibley had become a collector entrepreneur, owning the largest bird egg-white and blood collection in the world. He described it as "a catalogued museum collection" that was "available for further analysis."[145] It served the same function for him as the specimen collections in museums for the taxonomists comparing morphologies. Sibley referred explicitly to the morphological tradition, claiming that his work constituted a "comparative morphology of proteins" instead of bones.[146] When new research questions emerged, he could turn to his collection to answer them. When Sibley decided to study whether mutations occurred at the same rate in all lineages, he noted, "The nice thing is—we already have all the material needed to carry out this set of experiments."[147]

Sibley had become a leading—if often controversial—figure in American ornithology, first for his traditional studies of bird population mixing and then for his application of new biochemical techniques to their taxonomy. He used his extraordinary collection of eggs whites and blood to perform electrophoresis analysis of albumin and hemoglobin on a larger scale than anyone before (or since) and propose new classifications of bird families. By 1963 he had eight laboratory technicians producing electrophoretic patterns, and data were "piling up fast."[148] The protein studies, he told Ernst Mayr enthusiastically, were "going very well—better than ever." By that time, the laboratory had already produced more than seven thousand electrophoretic gels of egg-white proteins and begun studying hemoglobin from a collection of seven hundred species.[149]

Sibley's constant effort to gather eggs from as many bird species as possible for his electrophoretic studies was limited by the Lacey Act, a US law passed in 1900 to prevent the importation of illegally acquired animals. In 1973, the US Fish and Wildlife Service charged Sibley with violating the act for having imported eggs from Britain without appropriate permits.[150] In a context of increased public awareness regarding species extinction and of public criticism of science, newspaper commentaries about the episode dealt Sibley a blow. A *New York Times* editorial concluded that Sibley's "temptation to circumvent bureaucratic red tape" must have been strong, but that the "arrogance of science is no more appealing than the arrogance of commerce—or of government."[151] The Fish and Wildlife Service defended its action by noting: "If we're going to bust Indians for selling eagle feathers in Oklahoma, what are we going to do with this fellow? Why not go after the biggies?"[152] Sibley was eventually fined $3,000 in a civil court, where he declined to dispute the claim.[153] He attributed his troubles to collecting permits with excessive restrictions (which were not followed scrupulously by his field collectors) and to the fact that "England has an exceptional concentration of uninformed but vocal 'bird lovers' who are violently and emotionally opposed to any and all killing of birds for sports or collecting of specimens for any purposes."[154] The rise of the environmental movement had probably lent more weight to the concerns of amateur naturalists than in previous decades. Sibley had already been the target of "bird lovers" during his early collecting expeditions in the American West, but without facing any legal consequences.[155] The most curious aspect of his eventual indictment was the fact that one of the species that he supposedly collected illegally, "Torpis oocleptica," did not exist at all; it can be translated as "torpid egg stealer."[156]

In 1970, he published his first monograph on the classification of the passerine birds; two years later came another monograph, with ornithologist Jon E. Ahlquist, on nonpasserine birds, both based on the electrophoretic comparison of egg-white proteins. These classifications were generally similar to those established earlier based on morphology but also made a few new claims about relationships between bird families. Sibley might have worked with radically different data than his colleagues, but his approach to classification followed standard taxonomic practices. He regarded electrophoretic patterns as traditional morphological characters, compared their overall resemblance, and used similarities in electrophoretic patterns as the basis of establishing taxa. As he explained in 1962, "since protein molecules are the principal morphological units of the animal body at the molecular level of organization it follows that their form and structure are as relevant as sources of genetic and phylogenetic information as are the muscles, bones, organs, skins, hair, feathers and other structures."[157] A reviewer of Sibley's work noted that "the basic methodology of biochemical systematics thus does not differ significantly from that utilized by most ornithologists, past and present, with most other forms of data."[158] This did not prevent ornithologists from being skeptical about his claims. Those familiar with biochemistry doubted the reliability of his data; those familiar with bird classification took issue with particular taxonomic assignments. Some, including Ernst Mayr, warned Sibley about "over-selling" his results,[159] while others criticized them for being too tentative. As another reviewer put it, Sibley's "stated objectives are modest, which is fortunate, for the results are modest indeed."[160] Yet his standing as a "traditional" ornithologist ensured that his biochemical results were not ignored or easily brushed away.

Sibley acknowledged that his results were uncertain and that the "ultimate of ultimates" would be to determine the "complete DNA code" of each species. Until that could be achieved, however, serology and electrophoresis could provide "useful data when correctly interpreted."[161] He believed in the superiority of electrophoretic data over morphological data for many of the same reasons that Boyden believed in the superiority of serological data. Sibley bluntly told a newspaper reporter that "the white of an egg offers more accurate information on the evolution of a bird species than a bird's skeleton."[162] One reason was that electrophoretic patterns, unlike traditional morphological features, could provide an "objective" basis for species comparisons.[163] As Sibley commented to Mayr years later, morphological changes "are only perceived by human vision and a judgment, or opinion, is formed as to their magnitude."[164] Morphology had the problem that it was "simply too complex to interpret with the eye alone. Morphology sometimes gives a correct picture, but it is not consistent, not objective, and not quantitative. It leads us astray as often as it provides good answers."[165]

For Sibley, as well as most other biochemical systematists, the basic flaw of morphology-based taxonomies was that "genetic evolution and morphological evolution are only loosely coupled."[166] Indeed, superficially similar morphological features often reflected an independent process of adaptation in two species (convergent evolution) rather than common descent. Dolphins and sharks, for example, owed their overall similarity in form to selective pressures governing the common environment in which they lived. Thus, Sibley concluded, "morphology must be a poor index to genetic evolution, not a fairly good one, as we have long assumed."[167] In his mind, by basing taxonomy on biochemical methods, he and his colleagues were "fomenting a revolution . . . called SCIENCE, i.e. objective, quantitative measurements, not the Art of subjective, qualitative opinions."[168]

Sibley was quick to adopt new experimental methods into his enterprise as soon as they became practicable. Electrophoretic profiling of blood and egg-white proteins (figure 2.10) was followed by a brief excursion into hemoglobin fingerprinting and electrophoresis of eye lenses, before Sibley finally settled on the DNA hybridization technique. This method, developed in 1960, came closer to determining the true genetic identity of species for Sibley.[169] Its basic principle is fairly simple: When a molecule of DNA encounters another molecule with a complementary sequence, it forms the well-known double helix. The helix is held together by links between the bases on one DNA strand and the complementary bases on the other. Even if the sequences are not perfectly complementary, DNA may form pairs, but they don't stick together as strongly. DNA hybridization is based on an evaluation of the strength of the bonds, made by measuring the temperature at which they separate, and this gives an approximation of the similarity of their sequences.

Sibley learned the technique from the biologist Ellis T. Bolton, a contributor to its development and the first to apply it in a crude attempt at classifying animals.[170] Bolton had obtained his PhD in 1950 for a study on the use of serological methods in systematics under Alan A. Boyden. The initially enthusiastic Sibley's early attempts at DNA hybridization were fraught with technical difficulties, leading him to abandon the method in 1965 and return to his focus on egg-white protein electrophoresis.[171] After some of the problems with the method had been solved, he picked it up again in 1972 and began what would become the largest study ever carried out of the molecular basis of biological diversity, surpassed only by DNA barcoding three decades later.

Sibley had a unique advantage in revisiting classical taxonomic problems by examining species' DNA: his "huge collection of DNA's from over 2500 species of birds."[172] Actually, most of his collection was composed of blood samples, but luckily bird red blood cells, unlike human ones, contain DNA that could be extracted and used in the hybridization experiments. One of the main difficulties, however, was the sheer scale of the experiments that had to be carried out. Ideally, the DNA of every species would have to be tested against every other. Thus an examination of the relationships within his entire

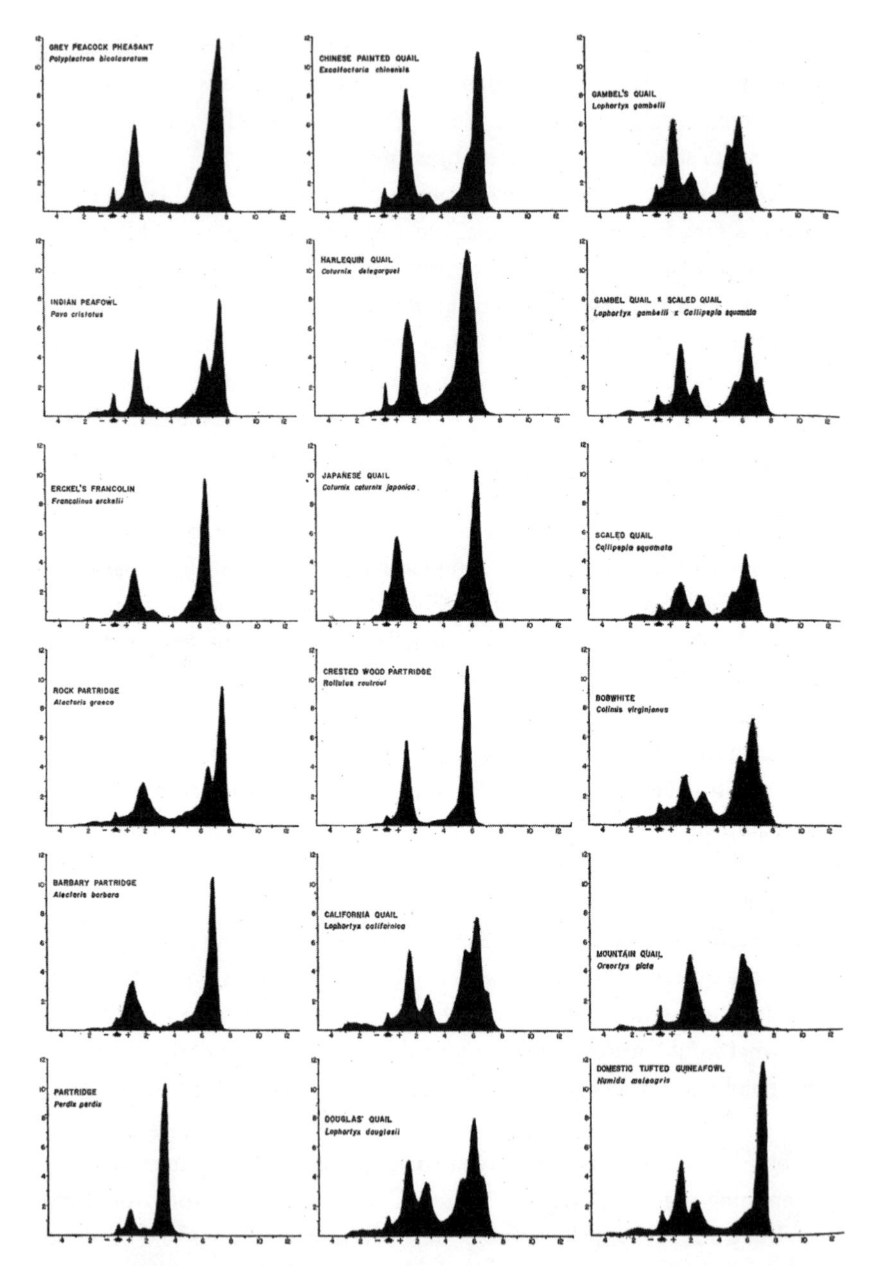

Fig. 2.10 Paper electrophoresis patterns of various birds presented side by side to allow for easy comparison of patterns. Sibley, "Electrophoretic Patterns of Avian Egg-White Proteins," 276. Printed with permission of John Wiley & Sons, Inc.

collections would require more than half a million experiments, far beyond the capacity of a single laboratory. Sibley and his collaborator Jon Ahlquist spent much time and effort on rationalizing the experimental process, developing an automated DNA hybridization machine (the "DNAlyzer"), finding ways to limit the number of experiments, and using computers in the data analysis. Thanks to the DNAlyzer (and a dozen technicians), data began to pour out of the laboratory, yielding most promising results. As Sibley enthusiastically told Mayr, "the DNA technique continues to be a joy to work with. It imparts such a sense of confidence when a set of experiments fall into place. There are many complexities, but the basic fact is that the DNA data reveal the phylogeny with clarity."[173] He was still "surprised by the extent and frequency of the discrepancies" he was "finding between morphological and genetic distances." But the data also showed "internal consistency which is the best proof that DNA hybridization is, finally, giving us the truth about phylogeny."[174] By 1979 Sibley and his collaborators had performed about five thousand comparisons and were "getting a reasonably clear idea of the overall pattern."[175] The accumulating results were stored in a "data bank" that could be scanned by a computer.[176]

During the following years, the laboratory began publishing taxonomic revisions of the main groups of birds. From 1986 on, Sibley toured ornithological congresses displaying a fifty-foot poster, nicknamed "The Tapestry," representing a phylogeny of all the bird families established by DNA hybridization. This complete representation of the biological history of birds formed the basis of *Phylogeny and Classification of Birds: A Study in Molecular Evolution,* a massive volume published in 1990.[177] Based on 26,000 comparisons, the work proposed a new classification for 1,700 bird species, representing all the orders and 168 out of 171 of the families according to the traditional classification made by Alexander Wetmore (1960). Sibley and Ahlquist revised the classification of the "pelican-looking" birds, one of "the most complex and controversial questions" in ornithology.[178] Traditional classifications had associated pelicans, boobies, and cormorants on the basis of similar morphological features, such as four toes connected by a web and a prominent gular pouch. DNA hybridization confirmed that species of pelicans were closely related, but they were more closely related to the Shoebill (an African stork) than to other pelican-looking birds such as boobies and cormorants. Overall, three-quarters of the DNA hybridization results agreed with traditional classification and one quarter did not, leading Sibley and Ahlquist to make original proposals about bird evolution and thus bird classification, which should reflect species phylogeny.[179] This was the first time that an entire class of higher organisms had been reclassified on an experimental basis. It was made possible by the application of the newest laboratory technologies as well as the field collecting technologies that brought samples to the laboratory. Understandably, the book was dedicated "To Those Who Have Helped."[180]

The work was reviewed at least thirty times. Some criticisms were harsh, to say the least. Perhaps concerned about rejections by traditional ornithologists, the authors had opened their volume with: "This is a book about birds. It may seem to be mostly about DNA, but Jon Ahlquist and I used DNA hybridization only as the means to an end."[181] Yet the most vocal critics were not morphologists but molecular systematists, who took issue with the experimental methods and data analysis. One reviewer made a shattering critique of their methodological assumptions: "The authors' phylogeny and classification are ultimately held together by a few key threads, all of which pertain to a direct correspondence between [DNA hybridization results] and genetic divergence between species. As the underlying assumptions for such a correspondence are broken, those key threads are cut, and the authors' Tapestry inevitably unravels."[182] The molecular evolutionists and systematists who had struggled for decades to establish the credibility of their method (protein sequence comparison, chapter 3) were now in the mainstream and seemed reluctant to accept another revolution.

Reviewers pointed out two other bothersome problems: data had not always been analyzed "objectively," and it was not available for reanalysis, in either the book or a public database. The charge of lack of objectivity was particularly serious because of the authors' own, quite vocal, opinion that experimental taxonomy could claim superiority over morphological taxonomy precisely because experimental data could be analyzed quantitatively without subjective interpretation. In a review of the taxonomic literature, Sibley and Ahlquist had stated: "intuition is not a substitute for measurement, and the failures of the [morphological] school litter the historical landscape."[183] Reviewers of the book in *Science* drew attention to these apparent inconsistencies: "[In the 1960s] molecular approaches showed unlimited promise by virtue of their objectivity. [However] Sibley and Ahlquist have modified an unspecified number of their [data] and the effects of these changes are unknown. [The] alterations are a posteriori and subjective. . . . As a result this work is a paradigm of how the idealized promises of molecular systematics of the '60s has been compromised by the predictable discoveries of its limitations."[184] Pressed by a reporter from *Science* a few years earlier, Sibley and Ahlquist confessed that "on occasion, it becomes necessary to correct values," but that the practice could be justified based on their "long experience in looking at [DNA hybridization curves]."[185] Similarly, in one of his earliest studies with protein electrophoresis, Sibley had also acknowledged that he relied on personal judgment to select a representative result: "All available curves for a species were spread out and examined until the characteristics of the species profile were understood. One was chosen which best illustrated the characters of the species."[186] Traditional taxonomists, derided by Sibley (and Ahlquist) for their lack of objectivity, followed the exact same approach when they evaluated morphological data.

The criticism leveled at Sibley and Ahlquist's work was colored by the bitter controversy that had erupted just two years prior to the publication of their book, when they applied DNA hybridization to the question of human origins. In 1984, they had published a paper that had stunned anthropologists because it claimed to have solved the "trichotomy" problem: the fact that the times of divergence between humans, gorillas, and chimpanzees were so close as to be indistinguishable (a point that had been made by Morris Goodman).[187] Sibley and Ahlquist claimed that DNA hybridization revealed that humans and chimpanzees were cousins, whereas gorillas had diverged earlier. What might have been just another controversy on human origins grew into a scientific storm because critics accused Sibley and Ahlquist of handling their data inappropriately—even of fraud. The controversy spilled into the pages of *Science* and other scientific journals, where questions were raised concerning the objectivity of their analysis of the data, particularly since they had not made it publicly available.[188] As a result, the accuracy of Sibley and Ahlquist's claims about human origins—and later bird phylogeny—were clouded by debates over the authors' scientific integrity.

Sibley may have been a controversial figure, but his work was steadily gaining recognition. In 1986, he was elected to the National Academy of Sciences; two years later they awarded him and Ahlquist the prestigious Daniel Giraud Elliot Medal for their book on the use of DNA hybridization to establish a new phylogeny of birds (Mayr and Simpson had been recipients of the medal two decades earlier). The jury considered that Sibley and Ahlquist's work had "revolutionized taxonomy by showing at last how to distinguish evolutionary relationships from convergent similarities";[189] in other words, they had evaded the main pitfall of all the classifications based solely on morphological characters. In 1991, the American Ornithologists Union adopted Sibley and Ahlquist's classification based on DNA hybridization for their Checklist of Birds.[190]

The polarization of the public debate around Sibley's work makes it difficult to distinguish questions regarding the role of experiments in systematics from turf battles or personality clashes. A better view of the tensions between experimentalist and natural history perspectives at the time can be found in the private correspondence between Mayr and Sibley. By the mid-twentieth century, Mayr was one of the most distinguished evolutionary biologists, taxonomists, and ornithologists. He had become a figurehead for naturalists across the world, remaining a staunch defender of naturalists' pursuits in the face of the growing popularity of molecular biology. He has often been portrayed as one of the last statesmen of a quickly disappearing generation of naturalists, who believed in the power of collecting specimens and comparing morphologies while resisting the growing power of the experimental sciences.[191]

But Mayr's letters to Sibley and others offer quite a different picture. Mayr was enthusiastic about new experimental methods and their applications to systematics; in some cases, he was even ready to recognize their superiority over morphological methods. In 1963, Mayr congratulated Sibley about the results of his electrophoretic study of egg-white proteins that were "piling up at a splendid rate." They were particularly important for the passerines, he felt, "because the anatomical clues are so few and so ambiguous."[192] His letters discuss a range of specific taxonomic assignments gleaned from Sibley's experimental methods. A decade later, Mayr acknowledged the advantages of protein sequences in establishing phylogenies: "I would think that the sequencing method should give the least ambiguous results, as far as the branching points of the phylogeny are concerned."[193] He embraced DNA hybridization even more enthusiastically, telling Sibley that his data "really sounds convincing" and that the results were "very wonderful indeed and will go a long way in telling us which the nearest relatives of the various groups are."[194] For Mayr, experimentation was a valuable research approach, at least as long as it served the taxonomic agenda and did not aim to replace comparative studies with a narrow focus on the study of single model organisms.

Mayr had been supportive from the beginning; even the taxonomic analysis of Sibley's first DNA hybridization experiments brought praise in a letter: "So far I have found no reason to disagree with any of your taxonomic conclusions."[195] When more results came out, Mayr told Sibley: "Needless to say, I am immensely interested in your latest results. Some of it is welcome confirmation, some of it is new but not shocking, and some of it is virtually shocking."[196] Apparently being shocked was something good: Mayr was receptive to the evidence.

He was not, however, completely uncritical of Sibley's data: "to base all ones' interpretation, as you seem to be doing, on the single coding DNA strikes me as skating on thin ice." Yet what Mayr asked Sibley to consider as a corrective to "coding DNA" was not morphology but "repetitive DNA" in view of the regulatory role that molecular biologists had recently attributed to these sequences.[197] Mayr also took issue with some of Sibley and Ahlquist's assumptions, methods, and conclusions when classifying birds. Even so, he concluded that "no other individual in the last 100 years has made as great a contribution to our knowledge of the relationships of birds as Sibley."[198]

Mayr's reservations did not represent an indiscriminate rejection of experimental methods, or even molecular methods, but a disagreement about theoretical issues, such as the constancy of mutation rates (the "molecular clock").[199] He was also concerned about the way new experimental methods, requiring laboratories, enormous computers, and unwieldy data storage devices, might affect the practical needs of field taxonomists. A field-worker who had to identify "hundreds of specimens each day" wrote to Mayr : "If such identifications are to be based on information stored on computer tapes, how is a poor boob like me, working in a hatch-roofed shack, with only a microscope and a Coleman kerosene lantern, going to proceed with this kind of study?"[200] Mayr concurred, stating that the new methods were perhaps appropriate in the revision of taxonomies but would be of little help in the identification of species in the field.[201]

Hybridization, Not Invasion

Reactions to Sibley's work must be understood in the wider context of interactions between the experimental and the comparative ways of knowing. This chapter's focus has been restricted to just one area: how biological diversity came to be studied at the biochemical level in the laboratory. The chapter began with practical field methods to collect blood samples from both domestic and wild animals and ship them to laboratories. The geneticists examined in the previous chapter had very different collecting practices, shipping live animals between urban centers (from New York to Pasadena, for example) and breeding them for generations to create "standardized" (inbred) strains of species. Yet both developed similar strategies to organize communities of collectors and negotiate delicate issues of credit among them. They also emphasized similar epistemic practices, centered on comparison of diverse organisms. Yet most historiographies of twentieth-century experimental life sciences place an emphasis on the growing importance of "model organisms," as they came to be known in the postwar period (chapter 1), and thus emphasize the genetic tradition. But downplaying an area of laboratory science that more closely paralleled natural history creates a starker contrast between the two cultures and emphasizes their divergence. I have tried to make this image more complex by exploring the rise of stock collections and their connection to the practices and community culture of museums (chapter 1).

But to understand more broadly the recent transformations in the life sciences, our focus needs to expand. The researchers discussed here valued experimentation *and* collection, model organisms *and* biodiversity, the laboratory *and* the museum. They represented some of the first examples, together with experimental embryologists and morphologists, of researchers embracing the experimental *and* the comparative ways of knowing. The stock collections described in the previous chapter exemplified many aspects of laboratory collection practices, especially the importance of communal values and the free exchange of materials. Some of these tasks, such as the establishment of microbiological collections, were closely modeled on existing museum practices; others arose from the habits of small communities of experimentalists. What is new with the collections examined here is the wide diversity of organisms collected, the broad comparative perspective, and the direct links with naturalist practices.

In the early twentieth century, it was rare to find researchers like Nuttall and Reichert who took an experimental approach to taxonomy. By the end of the century, researchers like Sibley were commonly represented, even in such naturalist "citadels" as the American Museum of Natural History. Spelling out the origins of this hybrid culture in terms of comparative and experimental ways of knowing gives new insights into the transformation of the life sciences in the twentieth century. Although both scientists and historians have often subscribed to the narrative of a clash between molecular and organismic biologists, or between the experimental sciences and natural history, this represents an unsatisfying oversimplification.[202] The rise of experimentation touched off debates with more than two sides. It did not simply or cleanly pit laboratory scientists against museum naturalists. Instead, the discussion spread to touch nearly everyone interested in understanding the history and diversity of life.[203]

At first glance, the stories of Nuttall, Reichert, Boyden, and Sibley might simply seem to epitomize an invasion of naturalist territory by experimentalists. They were certainly staking a powerful claim to territory traditionally inhabited by natural history—taxonomy and phylogeny. The naturalists had been studying the results of evolution on species, usually at the level of the visible body; now laboratories were developing the experimental tools to look directly at its causes: the subtle changes of genes and cell biochemistry. As they applied new methods they brought along values from the experimentalist tradition such as objectivity, measurement, quantification, and precision. Even so, their epistemic practices remained largely those of naturalists: they were still collecting and comparing, although in the form of experimental data instead of bones and fossils.

The epistemic authority of their results rested on the value assigned both to experimental data and to the comparative methods that were being applied to large collections. Thus, these developments are better understood as the emergence of a "hybrid culture" rather than as the domination of one culture over the other. But at the time, many of the participants perceived the changes in terms of a power struggle. In the early 1960s, molecular biologists and biochemists boasted about their taking over the study of evolution from supposedly antiquated morphology-minded naturalists; problems of phylogeny, they claimed, were better addressed by methods such as protein sequencing and serology. And there was something deeper going on: the influx of methods was accompanied by a more subtle transformation in the norms and values of scientific research. By the 1960s, taxonomy had already been progressively transformed through the adoption of experimental methods such as serological and biochemical taxonomy and by values from the experimental sciences such as objectivity, measurement, quantification, and precision.[204]

The next chapter explores another reason why the simple "invasion narrative" is inadequate. The very same biochemists and molecular biologists who boasted of the cultural and epistemic superiority of experimentalism over natural history were ignoring an important component of their own research practices. Experimental embryologists had often relied on the comparative way of knowing since the late nineteenth century, and a large amount of work was produced in that tradition. But in the mid-twentieth century, as experimentalism was triumphing everywhere, this integration was true even of biochemistry and molecular biology, the flagships of experimentalism in the life sciences.[205] Studies of the relationships between protein structure and function relied heavily on the collection and comparison of data across species (chapter 3).[206] At the same time, studies of taxonomy in the comparative way of knowing grew increasingly experimental, as this chapter has made clear.

This two-way process is thus better described as the progressive creation of epistemic and cultural hybrids, vigorous and exceptionally fertile.

The professional identity of Boyden and Sibley remained that of naturalists, even if traditional systematists tended to consider them as experimentalists. During the closing discussion of a conference on systematic biology in 1969, where traditional systematists had sometimes voiced their opposition to experimental methods, Sibley addressed one of them: "Perhaps you underestimate how much experience with actual organisms most of us have had. Some of the remarks indicate that you might consider us biochemists. I certainly am not. I am a birdwatcher."[207] A year later, in the midst of his electrophoretic studies, Sibley wrote to Mayr that he and his colleague Ahlquist were "birdwatchers and not ashamed of it."[208] Their naturalist identity was particularly important for the success of their collecting enterprise. As naturalists, they could make calls to other individuals sharing the same identity to help them collect materials, blood and egg white, and establish extended networks of researchers who would work without recognition beyond a simple acknowledgment of their participation. This collaborative mode was common in natural history, and experimental taxonomists simply took advantage of this established tradition for their laboratory studies. For most other collections in the experimental life sciences, examined in the previous and the following chapters, the professional identity of the curators was much more unstable because they operated among experimentalists who did not regard this function as scientifically respectable. For Boyden and Sibley, on the other hand, operating among naturalists, there was nothing wrong with being a curator, in addition to being a researchers. Their intellectual and scientific biographies suggest that it is more productive for the historian to reexamine the cognitive and material practices of these historical actors than to suppose that they fall necessarily on one side or the other of an imaginary experimentalist versus naturalist divide.

One of the aims behind Nuttall, Reichert, Boyden, Sibley, and others' attempts to use experimental methods for taxonomic purposes was to make classifications more "objective."[209] For Boyden, objectivity could be attained through quantification, automation, and reduction of systematic comparisons to a single character (serum proteins). The drive for more "objective" (according to the historical actors) methods that has pervaded the field of molecular evolution since the 1960s was thus already acutely felt in the interwar period.[210] The emphasis in the historical literature on the rise of molecular evolution since the 1960s should not obscure the longer historical trajectory: the reliance on experimentation (especially on molecules), quantification, and instruments to achieve greater objectivity in areas considered to be part of natural history is a broad trend that runs throughout the twentieth century.[211] In addition to objectivity, experimentation promised to bring additional control over the natural world, especially by allowing the creation of "unnatural" phenomena. Serological taxonomists could examine the biological reaction produced when whale and rabbit serum were brought together, but naturalists could not hope to examine the results of a crossing of these two species. The cultural and epistemic authority of the laboratory, based on objectivity and control, led not to the exclusion of natural history but to its transformation.

The growing consensus that taxonomists, experimental or not, should adopt the same standards of (methodological) objectivity as experimentalists is also what made naturalists such as Simpson, with their claim that instinct, intuition, and personal judgment were legitimate means for the production of knowledge, seem increasingly archaic within the life sciences. What had once been considered a reliable way to learn about the natural world was becoming suspect in the twentieth century. Indeed, as Lorraine Daston and Peter Galison have argued, by the late nineteenth century, objectivity and subjectivity had become opposite and mutually exclusive epistemic values. Experimentalists sided with a specific notion of objectivity (and of the self), what Daston and Galison have called "mechanical objectivity," while naturalists in the early twentieth century had a more difficult time choosing their side.[212] As late as 1961, Simpson noted in his *Principles of Animal Taxonomy* that the identification of species depended "on the personal judgment of each practitioner of the art of classification." He added that classification could not be objective: "To insist on an absolute objective criterion would be to deny the facts of life, especially the inescapable fact of evolution."[213]

How and where, exactly, objectivity could be attained were matters of great difference among experimentalists, however. Some placed their hopes in molecules, statistics, and computers—or any combination of these—while others rethought the basic tenets of classification and its relationships to phylogeny.[214]

In addition to the changing value and meanings of objectivity in natural history, the stories of experimental taxonomists examined here show how biological diversity came to be studied in laboratories. This transformation is historically particularly significant because experimentalists were increasingly focusing on one (or just a few) model organism over the course of their careers.[215] Experimental taxonomists developed new methods to collect specimens for laboratory studies and brought a broad range of species to laboratories, where they eventually became the focus of various experimental investigations.[216]

Boyden and Sibley's taxonomies based on the comparison of experimental data from a large number of species, like many other systematic endeavors based on the comparative perspective, required the constitution of a collection. Boyden's serological museum and Sibley's egg-white or DNA collection, like Linnaeus's herbarium, Cuvier's anatomical gallery, Simpson's paleontological collection, or the NIH's GenBank database (chapter 5), were the expression of the basic necessity of making various objects present in a single place if they are to be compared by a researcher. As soon as experimentalists, such as Boyden or Sibley, adopted a broad comparative perspective and engaged with the diversity of life, they established collections or databases, as so many naturalists before them had done. But in the twentieth century, when the values of experimentalism and the authority of the laboratory became dominant, these scientists were left with the delicate task of inventing institutions and practices that combined the naturalist's comparative perspective and the experimentalist's quest for objectivity. As Sibley's example demonstrates, experimental data were not inherently more objective than morphological data. Both could be considered subjective in view of more recent methods that established new standards of objectivity. But in all cases, the public access to the data collected was deemed essential to insure the objectivity of the conclusions drawn from them. Natural history museums, by storing specimens and often giving access to any trained naturalist who wanted to see them, effectively contributed to the objectivity of the taxonomies that were based on them. Sibley and Ahlquist, by withholding their experimental data, ignored this important lesson from natural history and acted like so many experimentalists who, except for the model organisms communities, never made their raw data public (for examples, see the following chapters). But in doing so, they also undermined the objectivity of their taxonomic conclusions, precisely the edge they claimed to have over traditional systematists.

Today, systematists working in museums and elsewhere rely overwhelmingly on data produced in the laboratory, such as DNA sequences. In the late 1960s, vigorous debates among systematists still opposed proponents of morphological data and molecular data. In 1967, at the final roundtable discussion of a conference on systematic biology, chaired by Sibley, these battle lines were still very clear. Biochemists, such as Emil Margoliash, and molecular biologists presented different kinds of molecular data, especially protein sequences, for systematic work. Several traditional taxonomists were perplexed: " [if] we are being quite logical in breaking these organisms down to their molecules to get closer to the truth, why not carry it further and break them down into atoms and count the atoms?"[217] These taxonomists challenged the idea that molecular data was in some sense superior to morphological data and asked their laboratory colleagues: "What is so much better about your information than any of the rest of it?"[218] Their criticism, directed at theoretical positions, was also a defense of a way of life, a life in the field rather than the laboratory. Another taxonomist, after expressing concerns that "molecular biology not only is getting a considerable number of grants but also is going into a blind alley," quoted J. H. Fabre's *Life of the Fly*: "I make my observations under the blue sky to the song of the cicadas; you subject cell and protoplasm to chemical tests . . . you pry into death, I pry into life."[219]

Overall, however, the growing consensus among participants was that this polarization was detrimental to the development of systematics. The numerical taxonomist Robert R. Sokal observed that "of all the facts and ideas presented, none is at variance with the principles and practices of systematics based on other evidence—cytological, morphological, whatever."[220] Sibley similarly made the point that molecular data was of "exactly the same caliber as the other morphological data; it is just the morphology of molecules."[221] In the following years, the debates shifted even further from the opposition between morphological and molecular data toward theoretical discussion about the most appropriate ways to analyze the data.[222] A further reason for the decline of this opposition was the observation that the rates of morphological and molecular evolution were different.[223] These data, taken alone, told different stories, and only their combination could get the systematist closer to the actual history of life.[224] Recently, however, the attempt to identify species by a unique DNA sequence ("DNA barcoding") led to a brief revival of the opposition between morphology and molecules, but this debate seemed to die out once it became clear that DNA barcoding, a simple technique for species identification, did not overlap with the attempts to produce taxonomies relying on a much broader set of data.[225] What everyone agreed upon, however, throughout the twentieth century, was that collections were indispensable for systematic work. Sure, the nature of the specimens preserved in collections was changing—frozen tissues instead of stuffed animals—but the basic fact remained that collections were indispensable for the comparative work associated with all systematic work.[226] One taxonomist put it most succinctly in 2000: "no collection, no data; no data, no knowledge."[227]

3

DATA ATLASES

In today's experimental research in the life and biomedical sciences, collections of DNA and protein sequences play a central role. They have become indispensable for all experiments attempting to understand the role of genes in the mechanisms of inheritance, development, physiology, and pathology. They are equally indispensable for all research in the taxonomy and evolution of organisms. In 2017, almost 14 million researchers accessed GenBank, the major DNA sequence data bank (chapter 5).[1]

Sequence databases originated with Margaret O. Dayhoff's *Atlas of Protein Sequence and Structure* (1965), the first public collection of sequence data. It was particularly significant in that it aimed at being comprehensive, computerized, and an instrument to understand life in informational terms. It opened up the possibility of a computerized analysis of large amounts of experimental data for the first time in the life sciences. Although the *Atlas* was based on computerized data and would eventually become an electronic database, it was initially a book, printed on paper and bound between two covers. It was a special kind of book, however, an atlas. Following the many scientific atlases that had flourished since the early nineteenth century, it contained limited narrative text and many oversized illustrations. Like its predecessors, explored in great depth by

Lorraine Daston and Peter Galison, it constituted "a systematic compilation of working objects" (in this case sequences) for researchers, it was "intrinsically collective" in its making, and it aimed "to be definitive."[2] The *Atlas* was widely used in different fields and quickly became an indispensable tool for researchers working on topics as different as biochemistry, genetics, and evolution. The creation of the *Atlas* in 1965, along with other data collection in other fields, reflects the growing preoccupation of the 1960s with the rapid increase in the overall amount of information. In the sciences, researchers believed that "data" was being produced at an unprecedented rate and that they could no long keep up with results from their own fields. Science historian Derek J. de Solla Price's 1963 book, *Little Science, Big Science,* testifies to this preoccupation with the overwhelming growth of scientific knowledge.[3] Thus looking closely at Dayhoff's *Atlas* offers a window into the strategies developed to deal with the postwar "information overload" in the sciences and the growing importance of databases in scientific research.

Understanding the crucial role and place of molecular sequence in contemporary biomedicine, genomics, and bioinformatics requires paying attention not only to the emergence of DNA sequencing and recombinant DNA technologies in the 1970s—the "biotechnology revolution"—but also to the early history of protein sequencing in the 1950s and 1960s. Although some accounts trace the origins of contemporary bioinformatics exclusively to the analysis of DNA sequences, it was in the context of protein research, decades earlier, that sequences became "epistemic objects," that they became an essential part of the new discipline of molecular biology, and that sequencing molecules became a particular "form of work."[4]

The focus of this book, however, is less on the production of data (such as sequences) than on collecting, comparing, classifying, and computing data. As it should be clear by now, the success of any scientific collection rests on its ability both to collect objects (or data) and to develop a community to sustain it. The curators of stock collections (chapter 1) and of blood collections (chapter 2) found it sometimes difficult to convince researchers to participate in their efforts. Yet their task was made easier by the fact that collectors and users were generally the same people. Individual geneticists contributed to collections of the mutants they had produced and obtained other mutants for their research. But in the case of molecular data, the community of collectors and users slowly diverged. Some researchers provided experimental data, while others specialized in its analysis, and curators of data collections were caught in between. Dayhoff faced a tremendous challenge to collect data from researchers who were neither naturalists nor geneticists and thus did not share the same communal ideals. An examination of the development of her collection of experimental knowledge is particularly interesting because it illuminates some key tensions that have arisen between collectors and producers of experimental knowledge since the 1960s. Issues of ownership, credit, and authorship loomed large over this early attempt to build a collection of experimental knowledge.

The scientific value of the collection was just as hotly debated as the legitimacy of producing knowledge through systematic comparisons of data. For naturalists, the comparative approaches adopted by Boyden and Sibley (chapter 2) were familiar, even if the items that were being compared (blood and molecules instead of bones and feathers) were unusual. Not so in the experimental sciences, where comparative approaches were often considered to lie in the domain of natural history. Yet comparative approaches were also occasionally adopted by experimentalists in fields such as "comparative biochemistry" (see below) and persisted in (comparative) embryology, even if, as Nick Hopwood notes, it remained marginal between 1945 and the 1980s in the broader biological and biomedical research landscape.[5] These earlier comparative approaches, so central in natural history, morphology, and embryology, formed a crucial source of inspiration and material practices for the researchers examined in this book and are thus essential for understanding the way systematic comparative approaches were adopted in the experimental sciences, an early hallmark of the hybrid research culture that has become, I argue, so characteristic of the life sciences in the twenty-first century.

Dayhoff's professional struggles illustrate the unstable status and questionable legitimacy of her role as a data collector and curator in the experimental life sciences. Furthermore, Dayhoff was an outsider who did not produce the kind of data contained in her collection, whereas Bridges, Sibley, and the other collectors and curators examined in the first two chapters were respected figures within the respective communities from which they gathered data. Her story reveals the slow acceptance of collectors and curators in the experimental life sciences, a role that had enjoyed scientific legitimacy in natural history for a long time.

Finally, Dayhoff's collection, which contained data from a wide variety of species, many of which lived in the wild, highlights the challenge of bringing biological samples from the field to the laboratory that has been discussed in the previous chapter. The problem was made even more acute by the fact that experimentalists working on protein sequences, unlike the naturalists, had no experience in fieldwork. Determining the sequence of the cytochrome c protein or gene from a reindeer first required getting a reindeer (or at least its blood) into a laboratory.

Dayhoff's *Atlas of Protein Sequence and Structure* was put to many uses; most important, it became the primary source for researchers anywhere in the world who wanted to reconstruct the history of life by comparing the molecules of various species. In the 1960s, the comparison of protein, and later DNA, sequences became a central practice among evolutionary biologists.[6] Up to that time, the study of evolution had primarily focused on the anatomical features of organisms, just like the taxonomic efforts examined in the previous chapter. Researchers such as Nuttall, Reichert, Boyden, and Sibley, who tried to understand the relationships between species by comparing their molecules, were the exception. It was only in the 1960s that this approach, known as "molecular evolution," gained momentum. In 1962, physical chemist Linus Pauling and biologist Emil Zuckerkandl suggested that differences in amino acid sequences between two species accumulated at a constant rate and could thus be used to measure evolutionary distances.[7] They considered sequences "documents of evolutionary history" and explained how entire phylogenies could be based on their comparison.[8] On this premise, the field of molecular evolution took shape in the 1960s, leading to occasional clashes between its advocates and the proponents of morphology-based evolutionary classifications.[9]

Most accounts of the rise of molecular evolution have focused on theoretical insights of Pauling and Zuckerkandl regarding "molecular clocks" and the subsequent debates over "neutral evolution." Scant attention has been paid to the source of the data on which molecular evolutionists based their theoretical discussions or infrastructures, such as the *Atlas* that made it possible. Who had determined the protein sequences included in Dayhoff's *Atlas*? Why did these researchers determine the sequences of the same protein in a wide range of species? And how did proteins from organisms in the wild reach the laboratory? In an experimental culture that valued research on "exemplars"—model organisms, model proteins, model systems—where did the focus on the diversity of molecules come from? Answering these questions requires shifting the focus away from the great stories of experimental and theoretical breakthroughs toward the deeper transformations in scientific practices centered on the collection and comparison of molecular data. Two examples taken from mainstream molecular biology—studies on protein functions and studies on the genetic code—highlight that practices of collecting and comparing were already essential components of molecular biology research in the 1950s.

Understanding How Proteins Work

The many protein sequences from diverse species included in Dayhoff's *Atlas* were determined by researchers with little direct interest in elucidating the history and classification of organisms. They were biochemists who wanted to understand how proteins worked. Specifically, they hoped to explain the functions of molecules—for example, how hemoglobin carries oxygen—by studying their amino acid sequences and other aspects of their structures. But instead of pushing their experimental virtuosity to its limit to learn all the structural details of molecules, they adopted a comparative approach. In the first half of the twentieth century, European researchers in particular promoted "comparative biochemistry" as an alternative to mainstream biochemistry. They believed that researchers should seize the opportunity to compare the biochemistry of many different organisms beyond humans and microbes to better understand how biochemical functions were carried out.

This was the approach taken by the British biochemist Ernest Baldwin in his popular *Introduction to Comparative Biochemistry*, first published in 1937 and reprinted in several editions through the late 1960s.[10] In line with Frederick Gowland Hopkins's programmatic vision, Baldwin's main interest was to produce generalizations about the biochemical basis of life.[11] The study of various species was a way to reach that goal, and for Baldwin "a starfish, or an earthworm, neither of which has any clinical or economic importance per se, is as important as any other living organism and fully entitled to the same consideration."[12] The Belgian biochemist Marcel Florkin also published an influential little book in 1944, *L'évolution biochimique*, which was translated into English five years later.[13] Florkin, too, reviewed the biochemistry of numerous organisms in order to stress "the unity of the biochemical plan of animal organization."[14] He also suggested that biochemical characters might also serve as the basis of phylogenies as soon as more facts about the biochemistry of different species became known.[15]

The practice of comparing sequences from various species emerged just as the first protein sequences were being determined. The method became standard among protein biochemists in the 1950s. The rationale behind the comparison of sequences of the same protein from different species was that (almost) identical regions of the protein, preserved through evolution (and thus homologous), might be "active centers" or other essential parts of the molecule. This meant that the functional importance of specific parts of the molecule might be revealed by a comparison of sequences. More variable regions, on the other hand, might be the parts of the molecule that had not been under the pressure of natural selection and that were thus probably of lesser functional importance.

The research of British biochemist Frederick Sanger, who determined the first sequence of a protein, perfectly illustrates this comparative approach. As an undergraduate at Cambridge University, Sanger had studied comparative biochemistry under Baldwin, and his father happened to be the physician who had collaborated with Nuttall in his studies of blood relationships.[16] In the mid-1940s, Sanger began to develop biochemical methods to investigate the structures of proteins. He focused on a small protein, insulin, which he purified from the blood of oxen. In parallel, he investigated insulin from other organisms. As early as 1949, he found that the insulin of oxen, pigs, and sheep had different amino acid compositions. In 1953, he succeeded in determining the complete sequence of insulin from oxen. Shortly afterward, he completed a study of pig, sheep, horse, and whale insulin. By comparing the sequences from his limited collection, he observed that the main differences were confined to a small portion of the molecule called the disulfide bridge. The result was puzzling because he believed this region to be important for the physiological role of the protein, perhaps even its "active center."[17] Yet this did not cause him to question the rationale behind sequence comparisons; on the contrary, he called for more studies of species differences.[18]

A number of other biochemists in the 1950s and 1960s were busy determining and comparing protein sequences from various species. In Vienna, the biochemist Hans Tuppy, a student of Sanger, pursued similar goals by sequencing parts of the cytochrome c protein from the horse, ox, pig, salmon, and chicken. Tuppy, like his mentor, took advantage of the first known sequence to infer the others from data on amino acid composition alone. After the sequence was known in one species, it was reasonable to assume that it would be similar in a related species, except for the few differences that the amino acid comparison would reveal. Tuppy hoped that such studies would help determine how cytochromes carried out their functions. "Those features which turn up invariably in all various cytochromes c," he argued, "are likely to be essential to the specific catalytic function, whereas structural differences will indicate points not directly concerned with catalytic activity."[19] In the United States, biochemist Emanuel Margoliash was conducting similar

Fig. 3.1 Margareta and Birger Blombäck preparing blood proteins in a laboratory at the Department of Medical Chemistry, Karolinska Institute, Stockholm, ca. 1955. Margareta Blombäck personal collection. Printed with permission of Margareta Blombäck.

studies of cytochromes c under Emil L. Smith at the University of Utah, studies he had begun in 1951 at the Molteno Institute under David Keilin, the successor of George Nuttall.

Medical researchers Margareta and Birger Blombäck at the Karolinska Institute in Stockholm extended this approach to a much broader range of species (figure 3.1). In the early 1950s, they had embarked on a lifelong study of the clotting factor fibrinopeptide. In nearby Lund, the biochemist Pehr Edman had developed a new degradation technique that made protein sequencing much easier than Sanger's method,[20] and they applied it to the study of fibrinopeptides from various mammalian species. In addition to the usual domestic species studied by Sanger and Tuppy—cat, dog, ox, horse, donkey, pig, rabbit, goat, and sheep—they investigated wild species such as the badger, bison, fox, green and rhesus monkeys, llama, mink, red deer, and reindeer. In 1965, after having compared fibrinopeptide sequences from twenty-two species, they observed that certain positions had "been stationary during mammalian evolution." They argued that these amino acids were thus likely to be "of importance for directing thrombin action."[21]

Several other researchers in Europe and in the United States were exploring the sequences of hemoglobins from various species as well as pathological variants found in humans, such as those responsible for sickle cell anemia. Biochemist Gerhard Braunitzer at the Max-Planck Institute for Biochemistry in Munich focused on hemoglobin from a wide range of species, whereas the main interest of chemist Vernon Ingram, at the University of Cambridge, lay with abnormal hemoglobins in humans.[22]

In the 1950s, sequences from insulins, cytochromes c, fibrinopeptides, and hemoglobins as well as other proteins from various species were beginning to accumulate in the scientific literature. These sequences were typically a byproduct of the attempts of individual comparative biochemists to understand the functions of their favorite molecules. As American biochemist Christian Anfinsen summarized in his popular book *The Molecular Basis of Evolution*, published in 1959, "variations from species to species may yield valuable information on the location of the site of enzymatic activity,"[23] and similarities could indicate "the minimum structure which is essential for biological function."[24]

Some biochemists turned the argument on its head, trying to draw conclusions about evolution from sequence variations. In 1956 Sanger had suggested that "more extensive studies of species differences in amino acid sequences of polypeptide chains may lead to interesting conclusions concerning evolutionary trends in protein biosynthesis."[25] Two years later, Tuppy was much more explicit: "The more proteins differ, due to the exchange of amino acids in different places of the polypeptide chain, the farther away in evolution the organisms from which they originate are. The comparative search for amino acid sequence in proteins could become an aid to discover evolutionary relationships."[26] This is one of the first published statements of the idea that the quantitative comparison of amino acid sequence changes might yield information about evolutionary distances. In 1959, in his *Molecular Basis of Evolution*, Anfinsen similarly suggested that the "rate at which successful mutations occurred throughout evolutionary time" may serve as "an additional basis for establishing phylogenetic relationships,"[27] yet he did not propose any phylogenies himself. Thus, comparisons of protein sequences between various species were commonly presented as a key to evolutionary problems in the 1950s, even if protein sequences had not yet assumed the preeminent position that they would be granted, a decade later, by molecular evolutionists.[28]

In retrospect, one might be surprised that sequence data from homologous proteins that had been acquired by the late 1950s did not lead to a more direct attempt to reconstruct phylogenies. Two explanations can be offered: First, the amount of sequence data remained limited and was often restricted to the active site of a molecule. Such sites were of most interest to biochemists investigating protein functions, but because they were also the most constant portions of a molecule, they were the least useful for evolutionary studies of closely related groups. Only when automated amino acid analyzers became available after 1958 were larger numbers of complete protein sequences, and from somewhat more exotic organisms, determined.[29] Second, during this early period the relationship between protein sequences and mutations at the level of DNA was not well understood. Until about 1960, it was unclear whether DNA sequences determined protein sequences entirely or if other components of the cell intervened.[30] In 1959, Christian B. Anfinsen noted: "Many readers will not be willing to swallow, whole, the thesis that proteins represent the *direct* translation of genetic information."[31] It was only when this question was considered unambiguously resolved in the early 1960s that protein sequences were confidently considered to directly reflect mutations that had occurred during evolutionary history and could thus safely be regarded as "documents of evolutionary history," as Zuckerkandl and Pauling put it in 1965.[32] In the meantime, however, sequence data came to be used for a very different purpose.

Cracking the Genetic Code

Biochemists often built their entire careers around a single protein: in the United States, Christian Anfinsen focused on ribonuclease; in the United Kingdom, Max F. Perutz focused on hemoglobin. When they collected sequences from many species, they usually concentrated on just one protein, or a small family of related molecules, in order to gain some insights into the structural basis of their function. But other researchers saw the potential for a more comprehensive collection of sequences. The first collection of various proteins from different species was assembled to help solve a crucial problem: the genetic code. Between 1954 and 1966, this was considered one of the most important challenges for the molecular life sciences. In 1954, George Gamow (who had proposed the Big Bang theory in cosmology) suggested that the genetic code could be solved as a cryptogram and made a proposal for a code (that was soon shown to be flawed.)[33] He then invited a number of molecular biologists and physicists, including Francis H. C. Crick, Martynas Yčas, and Sydney Brenner, to join the RNA Tie Club, which he founded to organize theoretical efforts to decipher the code. These theoretical approaches borrowed, sometimes liberally, concepts from cybernetics, cryptography, and information theory.[34] The history of the code is often told as beginning with failed attempts of theoretical approaches (1954–61) followed by the successful breakthrough resulting from experimental approaches (1961–66). But what is overlooked is that in both phases collections of protein sequences played an essential role.

Solving the "coding problem," as it was frequently called in the 1950s, consisted in finding how a text written in the four-letter nucleotide alphabet of DNA determined a text written in the twenty-letter amino acid alphabet of proteins. It required the combination of at least three nucleotides ($4^3 = 64$), a "codon," to specify uniquely the twenty amino acids. Knowing a DNA or RNA sequence and its corresponding protein sequence would have served as a sort of "Rosetta Stone" and led to a relatively trivial solution. But in the 1950s, only protein sequences were available. Nucleic acid sequences remained almost impossible to determine until the mid-1960s for RNA and the mid-1970s for DNA.[35] Thus those who wished to decipher the genetic code were stuck with examining whatever protein sequences were available, a situation analogous to that of cryptanalysis when confronting a message whose content was unknown. Members of the RNA Tie Club applied a typical cryptological strategy to the case: namely, to search for correlations between adjacent letters in the encrypted message. In human languages, some letters are more frequently followed by others (*q* often precedes *u* in English and most European languages), and similar associations within protein sequences might provide clues to the underlying nucleic acid codons. For these studies, any protein sequence, even one as short as two amino acids, could be used. So the members of the club were the first to systematically collect sequences from different proteins and different organisms, not just sequences of a single protein in various species. With this information, the biologist Sydney Brenner, a member of the club, was able to rule out certain types of codes.[36]

Collecting and comparing homologous sequences from different species also played a key role in these efforts. As early as 1956, in their review of the coding problem, Gamow, Rich, and Yčas listed and aligned all known sequences of six different sets of homologous proteins in order to test their hypothetical code. The rationale was that the change of one amino acid to another would most likely involve a change of just one nucleotide—not two or three—in the underlying DNA sequence.[37] The same year, in another paper on the code, Yčas presented twelve sets of aligned homologous proteins.[38] After August 1961, alignments of protein sequences from different natural strains and mutants of a single organism, the tobacco mosaic virus (TMV), or closely related viruses, came to play a particularly important role in the cracking of the code. The most direct approach was taken by Heinz Fraenkel-Conrat, in Wendell Stanley's laboratory at the University of California-Berkeley, and Heinz G. Wittmann, in Georg Melchers's Max Planck Institut für Biologie in Tübingen.[39] Both groups believed that this approach might be the key to solving the genetic code and published a number of sequence alignments as potential clues.

These collections of amino acid changes became crucial after August 1961, when Marshall Nirenberg and Heinrich Matthaei announced the discovery of the first codon as the result of experiments with synthetic polynucleotides.[40] Indeed, assuming that a mutation involving an exchange of one amino acid for another depended on a single nucleotide change, a collection of amino acid changes would drastically simplify the determination of the remaining codons once a few codons were known.[41] At the New York University School of Medicine, biochemist Severo Ochoa relied extensively on this reasoning and on data from TMV mutants to confirm his codon assignments and infer new ones, while biochemist Emil Smith at the University of Utah was examining the large body of sequence data of cytochromes c, insulin, hemoglobin, and other proteins that had been obtained from organisms as different as pigs and bacteria.[42] Smith also speculated on the evolution of protein functions and hoped that the approach might provide "a new tool for the study of species relationships."[43] The same year, at the American Cyanamid Company, in Princeton, New Jersey, biologist Thomas H. Jukes used known amino acid changes from an equally wide range of species to suggest new codon assignments.[44]

Biologist Richard V. Eck, later a co-author of the *Atlas of Protein Sequence and Structure*, also began to collect sequences as he worked on the genetic code. After studying chemical engineering and then plant biology at the University of Maryland, Eck joined the National Cancer Institute in 1954. There he developed mathematical models to evaluate complications from cancer surgery until, in 1960, he turned to theoretical study of the genetic code. In 1961, Eck published a paper in *Nature* in which he compared all the sequences of hemoglobin variants such as sickle cell hemoglobin and all the sequences of homologous proteins such as insulin from different species. He suggested that "the published data on amino acid sequences can be sorted, tabulated and arranged in a great variety of ways [and] any such manipulation will produce some sort of pattern."[45] Shortly thereafter he prepared a more extensive treatment of his analysis for the recently launched *Journal of Theoretical Biology*. After "compiling the published sequences," he presented sixty-one protein sequences aligned with those of their homologues in different species, the largest published collection of sequences at the time. He then proposed a complete solution to this "protein Cryptogram."[46] Eck's papers were composed before Nirenberg and Matthaei's August 1961 announcement that they had solved the first codon of the genetic code experimentally, but appeared in print just afterward—Nirenberg and Matthaei's publication providing much more compelling evidence for their solution than theoretical approaches pursued by Eck and others.

The three complete solutions to the genetic code that had been proposed by 1962—by Smith, Jukes, and Eck—as well as the later codon assignments derived by biochemist Walter M. Fitch[47] relied extensively on the comparison of homologous sequences from organisms including humans, pigs, sheep, oxen, horses, sperm whales, finback whales, humpback whales, seals, salmon, chickens, turkeys, silkworms, frogs, rabbits, bacteria, and viruses. In the following years, they became an essential part of the nascent field of molecular evolution as promoted by Linus Pauling, Emil Zuckerkandl, and others. Interestingly, after Smith, Jukes, Eck, and Fitch carried out their work on the code, they all became involved in the study of molecular evolution. The comparative way of knowing they had adopted to solve the genetic code was easily transferred to the determination of phylogenies in the context of molecular evolution.

From the Field to the Laboratory

Although the intellectual origins of these early sequence collections are clear—research on the function of proteins and on the genetic code—their material basis is harder to trace. Where did samples of deer, rattlesnakes, and camels come from? Who collected the samples in the field? How did they travel to the laboratory?

In most cases, researchers investigating proteins obtained their material from local slaughterhouses where they could get large amounts of animal tissue at a low price, often organs that were not sold for human consumption. Proteins were then carefully extracted and purified from the tissues in the laboratory. Many studies were thus conducted on cows, pigs, horses, and chickens—easily obtained from slaughterhouses. The cytochrome c proteins in beef were purified from "freshly minced heart muscle" (requiring about one kilogram for each experiment).[48]

The industrial meat-packing industry also provided scientists with material. The Chemical Research and Development Department of Armour and Company (best known in the postwar United States for its hot dogs with "open fire flavor"), purified ribonuclease, lysozyme, and other proteins from bovine pancreas and offered them for sale to researchers.[49] The whaling industry was another source, providing sperm whale meat for Max Perutz's studies of hemoglobin in Cambridge and similar studies of insulin carried out in Japan. Unlike other molecules of biological interest, very few proteins were provided by the pharmaceutical industry, because proteins were rarely used as drugs in the postwar period (insulin was a notable exception).[50]

Pathological hemoglobins in humans were provided by clinics in regions where the prevalence of disease was high. Because sickle cell anemia was most common among African Americans in the United States, Linus Pauling secured a blood supply from a clinician in New Orleans for his studies performed at Caltech. In Cambridge, UK, Vernon Ingram relied on sickle cell anemia blood brought from Kenya by medical researcher Anthony C. Allison.[51] Later, Ingram explored molecular differences in the hemoglobins of patients with a range of pathological conditions. In this case, the blood samples were taken from the blood collection that clinician Hermann Lehmann had established in Cambridge after his travels in several African countries.[52]

Supplies of biological material from wild animals posed a much greater challenge to laboratory workers. Here, too, a laboratory's local environment played a key role. Protein sequences from deer were determined in a laboratory in Stockholm, from camels in Udaipur, and from rattlesnakes in Los Angeles. Laboratory researchers, unlike naturalists such as Sibley, usually had no prior experience in field collecting. Medical researchers Margareta and Birger Blombäck, who were leading researchers on the molecular basis of blood coagulation in the 1960s, came to the field with no experience. For their studies on the mechanisms of coagulation, however, they secured blood samples from a wide range of organisms, beginning with domestic animals and later moving on to wild ones. With a visitor from the United States, the biochemist Russell F. Doolittle, they flew to northern Sweden for the annual reindeer hunt, where "a Laplander and his lasso" captured a few specimens from which blood was drawn. In Lapland the problem of storing biological samples at freezing temperature, which had stymied so many blood field collecting expeditions in Africa and Central America, was easily solved, with "nature providing excellent refrigeration."[53] Sweden's northern islands also provided a source of seals whose blood was investigated in Stockholm.[54]

In 1963, Margareta and Birger Blombäck temporarily moved to Australia to work in the biochemist Pehr Edman's new laboratory. They seized this opportunity to gather blood from various species of kangaroos and sharks that were readily accessible.[55] That same year, the Blombäcks extended their interest in fibrinopeptide variation to human populations. Margareta wrote enthusiastically that they had gathered "blood from different [human] races, as pure as they possibly can be, such as Maoris (New Zealand), New Guinea natives, East Africans and Australian Negros" and that they had started "a new field of biochemical anthropology" based on protein sequences rather than serological affinities.[56]

In addition to field collecting, the Blombäcks, like many other biochemists and naturalists, relied on gifts from colleagues around the world who had access to local species. The method had its limitations, mainly because regions hosting the most exotic species also had the fewest laboratories. As anthropologist John Buettner-Janusch complained when he was unable to obtain blood for hemoglobin studies of primates: "we have not yet been able to beg, borrow, or steal a sample of Tarsius hemoglobin."[57] Finally, most researchers adopted the strategy of Russell Doolittle, who had moved to San Diego and relied extensively on the extensive animal collection in its public zoo. He determined protein sequences of elks, camels, buffalos, and many other animals. For aquatic species, marine stations such as the Marine Biological Laboratory in Woods Hole or the station of the Collège de France in Concarneau (Brittany) were drawn in as contributors.[58]

The choice of a particular species as a source of experimental material was sometimes dictated by the properties of the molecules it contained. John Kendrew focused on sperm whale myoglobin—the protein that carries oxygen in muscles—not just because whale meat was available, but because its myoglobin produced better crystals than that of any of the other organisms he had investigated.[59] Large animals were also favored because purifying even tiny amounts of proteins required vast amounts of fresh material. A sufficient quantity of cytochrome c could be purified from a single ox heart, but the hearts of more than three hundred rattlesnakes had to be pooled together to obtain enough material.[60]

The practicalities of fieldwork often determined what animals could be studied experimentally. The distance between the field and the laboratory, the size of specimens, and their (sometimes dangerous) behavior had to be taken into account by researchers. These considerations were new to experimentalists who were used to obtaining their biological material from an animal facility often located down the hall from their laboratory. The adoption of the comparative way of knowing by experimentalists thus had deep consequences for their research practices, best understood as the hybridization of ways of knowing commonly found in the field and the laboratory.

Margaret O. Dayhoff, Computers, and Proteins

By the early 1960s, sequence data from a growing number of organisms could be found dispersed in the scientific literature and the laboratory notebooks of individual researchers. Then, Margaret O. Dayhoff (1925–83) began to collect this data systematically. More than anyone else, she shaped modern databases in the experimental life sciences. She established the first publicly available computerized collection of experimental life-science data, the *Atlas of Protein Sequence and Structure*, and developed computer tools to store, analyze, and distribute its contents. She played a crucial role in defining the role of data collections and collectors in experimental research. Whereas collections were legitimate scientific institutions and curators were legitimate professional figures in natural history, this was not the case in the experimental life sciences. Dayhoff led the way, against much resistance, to a transformation that eventually made databases and curators indispensable to the production of knowledge in these fields.

Margaret Belle Oakley was born in Philadelphia in 1925. Her father was a small business owner and her mother a high school math teacher.[61] At the age of ten, she moved to New York City, where she attended public schools, graduating first in her class in 1943. With the support of a fellowship, she then attended New York University and obtained a bachelor's degree in chemistry in 1945. Her next step was to enroll at Columbia University, supported by another fellowship, where at the age of just twenty-three she obtained a PhD in quantum chemistry in 1948 under George Kimball.[62] As a fellow at the Watson IBM Computing Laboratory from 1947 to 1948, she applied

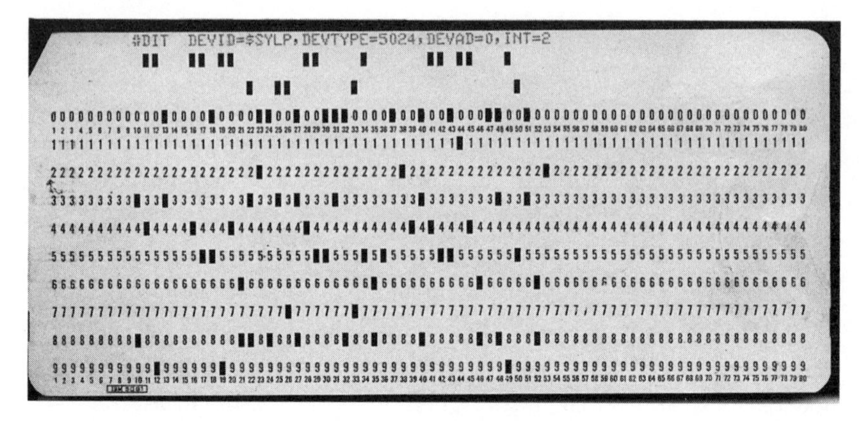

Fig. 3.2 IBM punch card, like those used by Margaret O. Dayhoff to store sequence data, 1968. Eighty numerical values, from zero to nine, can be coded on each card by punching a hole on the appropriate line. NBRF Archives.

calculation methods developed by Kimball for punch-card computers to calculate resonance energies in small molecules (figure 3.2).[63] After graduation, she married Edward S. Dayhoff, who was working toward a PhD in physics at Columbia University under Willis E. Lamb; he contributed to the research on the fine structure of the hydrogen atom for which Lamb was awarded the Nobel Prize in 1955.[64] Margaret Dayhoff worked at the Rockefeller Institute (now Rockefeller University) as a research assistant in electrochemistry (measuring the density of proteins) until 1951.[65] Once her husband obtained his degree the following year, they moved to Washington, DC, where he was offered a position at the National Bureau of Standards.[66] That same year, they had their first child (Ruth), and three years later their second (Judith). Margaret Dayhoff gave up research for eight years to raise her children, except for part-time scientific activity for two years as a postdoc at the University of Maryland, while Edward Dayhoff pursued his career at the Naval Ordnance Laboratory, working on weapons development and building electronic devices in his basement at home.

In 1960, Margaret Dayhoff began to work full-time again (figure 3.3), joining the National Biomedical Research Foundation (NBRF), which had just been established in Silver Spring, Maryland, by Robert S. Ledley, a childhood friend of Dayhoff's husband. The NBRF was a unique environment in which computers, biology, and medicine were brought into close proximity. The scientific goals of the NBRF, the professional background of its vibrant founder, and the research projects carried out there go a long way toward explaining the creation of the *Atlas of Protein Sequence and Structure* in 1965.

The NBRF was created as a private nonprofit research institution by Ledley to explore the possible uses of computers in biomedical research.[67] Funded by research grants, research contracts with industry, and later royalties from patents, it resembled the Worcester Foundation for Biomedical Research (of oral contraceptive pill fame) and other similar research institutions, which emerged with the generous postwar funding for science. It was envisioned as a place where computing and "biology or medicine could be combined intimately."[68] Born in 1926, Ledley went to dental school, where he received the minimal biomedical training, before obtaining an MA in theoretical physics from Columbia University and becoming interested in digital computers.[69] Starting in 1952, he worked at the National Bureau of Standards programming the SEAC, one of the first stored-program electronic computers in the United States. In 1960, he

Fig. 3.3 Margaret O. Dayhoff sitting outside her home in Silver Spring, Maryland, around the time she resumed a professional occupation at the newly created National Biomedical Research Foundation, ca. 1960. Dayhoff personal archives. Printed with permission of Judith Dayhoff.

finished a manuscript (published five years later as a 900-page monograph) entitled *The Use of Computers in Biology and Medicine*.[70] This constituted the first general introduction to the principles and methods of digital computing and an exploration of their possible applications in a number of fields of biology and medicine. The publication of this book was only one example of Ledley's lifelong commitment to promoting the use of digital computers in biomedicine, from the automated recognition of chromosome images to computer-assisted medical diagnostics and the analysis of molecular sequences.

Of particular significance in understanding how computers came to be applied to sequence collection and analysis by Dayhoff is the fact that Ledley was invited by George Gamow in 1954 to become one of the twenty original members of the RNA Tie Club.[71] Gamow believed that Ledley's expertise in digital computers and symbolic logic would be useful in solving the genetic code. Ledley's first and only contribution to the club was to outline a very general "system of digitalized computational methods" to be applied to practical problems in "science, industry, and government." As an example he gave the evaluation of overlapping codes by analyzing amino acid sequences.[72] Ledley noted that it "should take a computer no more than a hundred hours" to work out a solution, whereas if all possible solutions had to be tested, "a computer put to work in the days of the Roman Empire, at a rate of one million solutions per second, 24 hours a day, all year round, would not yet be close to finishing the job."[73]

This initial contribution was a failure: it produced no tangible results and was completely neglected by other researchers.[74] But Ledley envisioned another application of computers in sequence analysis. This time he suggested that computers could assist biochemists in their efforts to determine protein sequences. A standard experimental method consisted of cutting the polypeptide chain into several overlapping fragments and establishing the sequences of each. The problem was then to reassemble these partial sequences into the complete sequence of the original protein. In the 1960 draft of his book,[75] which was published only in 1965, Ledley outlined a method to solve the problem using a computer.[76] He invited Dayhoff to join the NBRF in 1960 to continue investigating this question under an NIH grant.[77] In reports published between 1962 and 1964, they described a set of FORTRAN programs they had devised for the IBM 7090, a mainframe computer located at Georgetown University, which could assemble partial sequences in the right order in less than five minutes (figure 3.4).[78] One of the programs searched the peptide sequences for particular characteristics, while another compared all peptide sequences in search of overlaps.[79] These two practices—searching

Fig. 3.4 Bead model of the amino acid sequence of the protein ribonuclease on computer listings, circa 1962, illustrating how computers could help experimental research. NBRF Archives. Printed with permission of the NBRF.

and comparing—would later become essential to the computational use of sequences in molecular evolution and other fields. Simultaneously, a similar approach to sequence analysis was being pursued by Richard Eck at the nearby National Cancer Institute in Bethesda, where he tested an algorithm in a "paper experiment" designed from published sequences.[80]

Ledley and Dayhoff made clear that their computer programs would not downgrade the role of a protein chemist to that of a simple technician, but that the computer would serve merely as an aid: "These routines may be thought of as analogous to the staff of a laboratory. Each routine has a function to perform just as a laboratory has people each with a job to perform, cleaning people, technicians, senior research workers, a librarian, a machinist, etc. The programmer and protein chemist have been upgraded to the chief of the computer staff."[81] In pressing this analogy, where the protein chemists and the programmer were in charge, Ledley perhaps wanted to avoid the outrage that physicians had expressed in response to his suggestion that computers could be used to make medical diagnoses.[82] Computers would not replace humans, he insisted, but serve as their assistants. Indeed, the computer programs he designed would print out intermediate results "for examination by the biochemist," and the process thus reflected "a close cooperative effort between the computer and the biochemist."[83]

Ledley, Dayhoff, and Eck hoped that such computer methods would be used by the increasing number of biochemists engaged in protein sequencing. Yet their methods seem to have had no visible impact on the actual practice of sequencing.[84] Many biochemists did not have access to computers in the early 1960s, and those who did often lacked the programming skills to use them.[85] More important perhaps, even when they could have secured the help of a programmer, they seem to have resisted the use of computers, which they perceived as particularly foreign to the culture of the "wet lab." In 1966, when one of Dayhoff's students began searching for a job in a laboratory where she could use her expertise in programming and in biochemistry, Dayhoff warned her to make sure "that the biochemists are sympathetic to the computer."[86]

Dayhoff and Eck's early attempts to use computers in the analytical effort led them to compile the data that had been published on amino acid sequences, which eventually became the *Atlas of Protein Sequence and Structure*. The development of computer programs also made them start thinking about the best ways to handle sequences with a computer. For example, they adopted a one-letter notation for amino acids, instead of the usual three-letter code, in order to save computer memory and to make alignments more readable on fixed-space printers. Most earlier sequence comparisons had used the three-letter code and failed to present the data in a way that lent itself to easy comparisons, because representations of three-letter amino acid sequences had different typographic lengths ("Ile" is shorter than "Asn").[87]

Other research projects carried out at the National Biomedical Research Foundation played a role in the development of computerized sequence analysis as well. For example, Ledley and Dayhoff devised computer programs to draw contour and density maps from X-ray diffraction data.[88] In this field, unlike that of sequence analysis, Ledley and Dayhoff were building on a long tradition in crystallography in which computers had been used to assist in determining protein structures (see next chapter).[89] This led them to further investigate the question of the relationship between a protein sequence and the structure of its active site, another field of protein science that would become important in the *Atlas*.[90] Once again, the attempt illustrated the belief that computers could produce meaningful results through the analysis of empirical data, the key premise of the *Atlas*.

Overall the computer tools created the need for a collection of protein sequences to which they could be applied. At the NBRF, a strong commitment prevailed that while the biological world was highly complex and irregular, systematic computational comparisons of experimental data could uncover regular patterns that explained biological functions. Neither the mind of the experimentalist nor that of the theoretician could make sense of the complexity of life. As Dayhoff put it, a "collection of strange reactions is carried on by living cells. The synthetic routes are odd and devious, almost unforeseeable from ordinary laboratory chemistry." But by comparing numerous protein sequences, one could discover the "evolutionary constraints" that had led to the development of molecular reactions. Only from "this basis of understanding" would it be possible to "more truly and readily form models of the mechanisms in living cells which can then be tested against reality."[91] For Eck, "hypothetical interpretations of [modern biochemical laboratory] data are usually not easy to find, and the data continues to accumulate in the expectation that some day they will all 'make sense.'" A better approach "takes a large mass of empirical data, and attempts to find some regular pattern in it." This approach is distinguished from the empirical approach in that it "reverses the usual methods of attempting to validate a hypothesis by finding predicted patterns in experimental data."[92]

The role of the computer in the creation of the *Atlas* was not only to analyze sequences but also to store, tabulate, and print them. The *Atlas* was an electronic data collection, but it was also a printed volume, and remained so up to the late 1970s, illustrating the persistence of paper, even in an age of computer, in the distribution of scientific information. In the 1960s, all the data of the *Atlas* (the sequences themselves and all related information) was stored on paper, "punch cards" that could be fed into a reader that would transform the data into digital electric signals. Each card constituted an entry, and the complete collection of cards was regarded as an "Amino Acid Sequence Library,"[93] which could be subjected to increasingly sophisticated computing techniques that were being developed in the field of library science.[94] Ledley took part in these developments as well, which provided another important resource for the creation of the *Atlas*. In 1958, he developed a new system for coordinating the indexing of book-format bibliographies that he called TABLEDEX. His method, another application of symbolic logic, allowed the user to search for entries containing several keywords, instead of just one as in most indexes. The National Science Foundation was actively promoting computing in American universities[95] and supported Ledley's attempt to utilize "a digital computer to assist in the automatic preparation of a bound-book form bibliographical index."[96] Similarly, in 1961, Ledley proposed to the

National Library of Medicine a method for using digital computers to publish the *Index Medicus*,[97] which would include programs to search it. The primary reasons for Ledley and others' concerns with the organization of scientific information was the perception—already—that the amount of published information was "exploding." In 1957, Ledley claimed that the "rate of doing research" had doubled since 1950 and that it was continuing its exponential growth, a product of the postwar expansion of science.[98] The same "explosion of information"[99] perceived in the field of protein sequencing would prompt the use of computers to organize and make sense of information in this field.

Ledley had long believed that computers were ideal tools not just for calculation but for "data processing" and the analysis of "large amounts of detailed experimental results" as well. In 1957, when surveying their possible uses in biology and medicine, he gave equal weight to calculation ("numerical solutions to partial differential equations" and "simulation of biological systems") and to data processing ("bio-medical processing and reduction" and "bio-medical information retrieval").[100] In this, his vision of the field and its future was atypical. Other surveys on the uses of computers in biology and medicine mainly emphasized calculation (equation solving and numerical simulations).[101] Ledley, in contrast, highlighted the promise of computers for data processing. He went so far as to outline a vision in which scientific data would be published electronically. Instead of "publishing articles in journals, research results might be transmitted to a central information centre," he suggested. The creation of the *Atlas* represented a first step toward accomplishing this vision. Given Ledley, Dayhoff, and Eck's backgrounds in using computers to analyze sequences and to organize information, it is not surprising that the *Atlas* first emerged as a computerized system.

Independently, Eck and Dayhoff began thinking of how computers might be applied to questions of evolution. In 1964, at a conference on engineering in biology and medicine, Eck presented a "cryptogrammic" method to trace the "evolution of proteins." As he had done with his earlier speculations about the genetic code, he now suggested that "the publication of the amino acid sequences of many proteins" made it possible to "treat the whole of evolution . . . as a cryptogram." He used data from several hundred amino acid substitutions in homologous proteins and a digital computer to calculate the probability that one amino acid had been replaced by another. Based on the results, he suggested that one could calculate the "degree of relatedness of each protein" with reference to its ancestors, and from there draw "a family tree of proteins . . . to scale," the distances between the branches of the tree representing a "numerical measure of relatedness." Even though he did not actually present a phylogenetic tree, he outlined the possibility of constructing "a detailed phylogenetic tree of the vertebrates," provided that enough protein sequence data became available.[102]

Eck had originally become interested in evolutionary biology through cryptanalysis; Dayhoff also entered the field from a side door: the search for extraterrestrial life, which was a research agenda of the Cold War.[103] In the post-Sputnik era, NASA was actively supporting investigations into the physicochemical conditions that could have led to the creation of organic compounds on earth, and eventually to life. By capturing taxpayers' imaginations, this pursuit helped to legitimize NASA's use of their money. Dayhoff climbed on the bandwagon with the chemist Ellis R. Lippincott and the astronomer and science popularizer Carl Sagan.[104] Collaborating with Eck, they used the IBM 7090 available at Georgetown University to simulate the evolution of the prebiological atmosphere and examine under which conditions "biologically interesting compounds, such as amino acids, were generated."[105] This work followed up on the 1959 discovery by Stanley L. Miller and Harold C. Urey that amino acids could form spontaneously from the inorganic chemical compounds believed to have been present on earth before life appeared.[106]

Dayhoff's interests in chemical evolution and in amino acid sequences converged and became mutually reinforcing.[107] The early chemical conditions on earth suggested an important role for proteins such as ferredoxin in the origins of life. Certain amino acids were likely to have been comparatively stable under these conditions and might have been present in the ancestral sequences of protein. Eck and Dayhoff, using a computer program that searched if the ferredoxin sequence was in fact composed of smaller repeated sequences, found that the protein had evolved through the duplication of a very short primitive protein.[108] This compelling demonstration of how computers could reveal evolutionary information was published in *Science* in 1966. In a letter to her NASA sponsor, Dayhoff pointed out that "the biochemists who published the sequence missed the evolutionary implications entirely."[109]

The *Atlas of Protein Sequence and Structure*

The publication of the *Atlas of Protein Sequence and Structure* in 1965 resulted from the growing interest in collecting, comparing, and computing sequences as outlined above. It was meant as a tool that would be used to produce new knowledge about protein structure, function, and evolution. As Dayhoff put it later, in a letter to a colleague, "there is a tremendous amount of information regarding evolutionary history and biochemical function implicit in each sequence and the number of known sequences is growing explosively. We feel it is important to collect this significant information, correlate it into a unified whole and interpret it."[110] In a private letter, Dayhoff put her vision of the *Atlas* in a broader perspective:

> I realized that the answers people were giving to social problems were very shallow and naïve—often only palliative in nature. Our knowledge of ourselves is quite medieval. . . . I like to think that the Atlas and related research are going to help in the gigantic endeavor to solve these vexing problems. Species differences, race differences, sex differences, and individual differences, are largely controlled by protein differences. Motivation and mental capacity, goals and satisfactions, as well as diseases may be linked to proteins. We shift over our fingers the first grains of this great outpouring of information and say to ourselves that the world [will] be helped by it. The Atlas is one small link in the chain from biochemistry and mathematics to sociology and medicine.[111]

The first edition, authored (or edited) by Dayhoff, Eck, and two collaborators at the NBRF, was just under one hundred pages long and contained the sequences of around seventy proteins, mainly cytochromes c, hemoglobins, and fibrinopeptides from various species. Each page gave the name of a protein and an organism ("Hemoglobin beta—gorilla," for example), followed by the protein's amino acid sequence symbolized both in the three-letter abbreviation and in a custom one-letter abbreviation system. Each page also listed the relative amount of each amino acid, remarks on how the data had been obtained (when available), and a reference (usually bibliographic) to the source of the data. Other data collections published in book format, such as the medical geneticist Victor A. McKusick's *Mendel Inheritance in Man* (1966), a catalogue of inherited traits and the supporting references, or more generally the National Library of Medicine's *Index Medicus*, a catalogue of the biomedical literature, were similar to the *Atlas* in their format and function.[112] They served as book-format indexes to the scientific literature. But the *Atlas* went beyond that simple indexing function (performed by any phone book), by including alignments of sequences of a single protein such as hemoglobin, taken from several organisms presented on multiple-page foldouts to facilitate reading with a single gaze. As Nick Hopwood has argued in the case of comparative embryology, Haeckel progressively adopted a layout of embryo images on the printed page along a grid to facilitate the comparisons of developmental stages in time (horizontally) and between organisms (vertically).[113] Similarly, the presentation of sequence alignments on large foldouts in the *Atlas* allowed the users to grasp the locations of conserved regions at a glance, giving an essential clue to the presence of a part of the molecule that played some essential function. Users such as the crystallographer Richard E. Dickerson complained that foldouts "may be tolerable in more popular magazines, but are as difficult to handle in [the] Atlas as a road map on a freeway."[114] His solution was to organize the sequence alignments like an "orchestral score," with each homologous sequence representing a different instrument, over several pages. However, he did not reflect on the fact that the patterns researchers might be interested in could fall on different sides of a page, making their recognition difficult (just as travel destinations seem to fall systematically on the fold of the map). The *Atlas* was more than a (phone) book, it was a research tool stretching the material constraints associated with the traditional book format.

The editors of the *Atlas* thanked the sequence contributors by dedicating it to them, "to all the investigators who have developed the techniques necessary for the grand accomplishments represented by this tabulation, and to all those who have spent so much tedious effort in their application."[115] They firmly positioned their work in the experimental tradition by claiming that the *Atlas* "voluminously illustrates the triumph of experimental technique over the secretiveness of nature," and stated that their goal was to make apparent the information that was "hidden in the amino acid sequence." That information was important to studies of the conformation of proteins, the sequence of the underlying genes, and "the record of the many thousand mutational steps by which we can quantify a phylogenetic tree." The editors asserted that as a basis for phylogenies, sequences would be far superior to traditional taxonomic criteria, which they deemed to be "extremely vague and uncertain,"[116] thus siding—unsurprisingly—with the new molecular evolutionists and taxonomists rather than with their organismic counterparts.[117] The authors of the *Atlas* also invited their readers to cooperate with the project by submitting additional sequences and corrections. They hoped to base their collecting efforts on a "gift economy," through which researchers would contribute unpublished sequences and other data and receive a copy

Fig. 3.5 Margaret O. Dayhoff (second from left), behind Robert S. Ledley, at the inauguration of their new computer, an IBM 360 model 44, with IBM representatives, April 1968. Its magnetic tapes, which stored the data from the *Atlas*, could be copied and shipped more easily than stacks of punch cards. NBRF Archives. Printed with permission of the NBRF.

of the *Atlas* in return. Writing to the crystallographer David C. Phillips, Dayhoff asked if he could make the unpublished coordinates available for the *Atlas* and how she could "collect this rich gift."[118]

The *Atlas*, like most books, was copyrighted, including the sequence data that it contained and that could thus not be freely redistributed. Individual sequences were not copyrighted, but the collection as a whole was. When asked for copies of the punch cards containing the *Atlas* data, Dayhoff often ignored requests, turned them down, or agreed to share only a small fraction of her collection.[119] Beginning in 1969, however, she agreed to a very limited distribution of magnetic tapes containing the data in addition to printed copies (figure 3.5). The recipients had to agree not to redistribute the data set, protected by a copyright.[120] Dayhoff reminded buyers that "this information is proprietary."[121] Reluctant NIH support and the rising costs of data collection led Dayhoff to seek revenues from sales of the collection in print and magnetic formats. In 1977, she sold the tape containing her database for $400.[122] The same year, the Protein Data Bank (chapter 4) sold the tape containing theirs for less than $35.[123] Even though the price charged by Dayhoff remained modest by all standards, it put her project on the side of commercial ventures rather than publicly available resources. The fact that she copyrighted the data included in the *Atlas* and also resisted distributing it in a computer-readable format reinforced that impression and irritated some users. One wrote to Dayhoff and asked, rhetorically: "So much of your work is a collection and collation of other people's work, that I do not know when I have to give credit for a sequence or when I have to obtain a copyright release, nor from whom I have to obtain it. In this respect, you are in somewhat the position of a folksong collector who copyrights his published material; do I have to pay him if I sing John Henry?"[124]

A Work of Compilation?

Dayhoff sent out the *Atlas* to more than seventy scientists in the United States, Canada, Japan, and Europe. The list of recipients included all the researchers who had determined sequences included in the *Atlas*, those who had analyzed sequences, editors of major scientific journals, and Nobel Prize-winning scientists.[125] Reactions were generally enthusiastic. Nobel Prize-winning chemist Melvin Calvin believed the *Atlas* would "ultimately prove to be a veritable dictionary of biological activity."[126] Another Nobelist, geneticist Joshua Lederberg, stressed that the *Atlas* would become "an important contribution to the next stage of molecular biological architecture" and would be a crucial tool in the "computer search for active site configurations" in proteins.[127] Ernst Mayr was delighted by this crucial new asset, even though it would be of most use to molecular evolutionists—with whom he still battled on a regular basis. He stated that the *Atlas* would be "of immense value to an evolutionist . . . who is interested in the origin and the meaning of diversity."[128] Understandably, most pleased by the *Atlas* were the biochemists, who had been lacking a way to keep up with all the known sequence data. Emanuel Margoliash stressed the role of the *Atlas* as a repository of sequences: "It will clearly become a most valuable compilation, particularly as this sort of information accumulates and one's memory begins to be overburdened with all the details."[129] Another biochemist confessed, "I can hardly tear myself away from reading it."[130] Other researchers underlined how indispensable the *Atlas* had become for their research: "It is the most heavily used book in our lab," reported a researcher, while another confessed to Dayhoff: "We use your book like a bible!"[131]

After a trial volume printed in 1965, Dayhoff published four more editions up to 1969. The fifth volume, the last complete edition, came in 1972 (supplements containing new sequences were published until 1979). One can track the growing use of the various editions through the fact that they were widely cited in the scientific literature. In 1970, the *Atlas* was cited 138 times; by 1974 the number had surpassed 300. Between 1965 and 1978 the combined editions were cited over 2,500 times, making it among the fifty most-cited scientific items of all time.[132] Sales also attest to its growing popularity: The 1972 edition sold close to three thousand copies, two-thirds of which were purchased in the United States.

The favorable reactions to the *Atlas* were tainted by a prevailing view that, while useful, it represented little more than a collection of existing knowledge. The "compilers," as Nobel Prize-winning chemist Richard Synge addressed Dayhoff and her team,[133] had simply gathered data that was available in the published literature and reprinted it, something that could hardly qualify as a scientific contribution. When Dayhoff applied to become a member of the American Society of Biological Chemists, the biochemist John T. Edsall answered, a bit embarrassed: "Personally I believe that you are the kind of person who should become a member of the American Society of Biological Chemists. [However, the candidate must] demonstrate that he or she has done research which is clearly his own. The compilation of the *Atlas of Protein Sequence and Structure* scarcely fits into this pattern."[134] Another potential supporter of Dayhoff's application, the biochemist William H. Stein, who would win the Nobel Prize in Chemistry in 1972, discouraged her because she did "not do experimental work."[135] Emanuel Margoliash similarly declined to support Dayhoff because her work did not belong to the "common experimental variety."[136]

Compiling the *Atlas* required far more expertise (and effort) than most users realized—or were willing to admit. First, the scientific literature had to be systematically surveyed, either manually or using bibliographic indexing systems such as MEDLARS and the American Chemical Society Abstracts search service.[137] One of the challenges after having located an article was a careful proofreading of the sequence it reported. This task was necessary because, as Dayhoff complained, "there is scarcely a paper published that doesn't have at least one typographical error in the data."[138] Because the errors were not immediately obvious and there was no dictionary of sequences available for comparison, the proofreading process actually required a careful evaluation of the experimental data from which the proposed sequence had been deduced. And as biochemist Christian Anfinsen had admitted, a "certain amount of personal judgment is frequently involved" in interpreting sequencing data.[139] Dayhoff and her team also compared the new sequences with similar sequences in the *Atlas* in order to spot possible errors. They resolved discrepancies with the help of the authors and published what was believed to be the correct sequence. In addition to collecting and evaluating sequences, Dayhoff and her team made important decisions about the kind of data to be included and how to represent it in a way that would be most useful to researchers. These decisions reflected a serious engagement with the scientific research based on protein sequences, and not just a passive collection of existing data.[140]

The *Atlas* was an encyclopedic collection of experimental knowledge, primarily of protein sequences that had been determined biochemically and a few protein structures determined by crystallography. But this information had to be linked to taxonomic information regarding the various organisms from which the proteins had been collected. Given her interest in using protein sequences to understand evolution, and her desire to make the *Atlas* as useful as possible for molecular evolutionists, Dayhoff took great care to assure that species names were accurate. Unfortunately, the biochemists who supplied the data generally did not show the same concern. One of them submitted a sequence from an organism simply described as "monkey" (there are over 250 species of "monkeys"). A zoologist from the Yale Peabody Museum of Natural History drew Dayhoff's attention to the fact that it was unclear whether the fibrinopeptide sequence of "rabbit" provided by Birger Blombäck originated from the "common laboratory Oryctolagus cuniculus" or from some other species; "badger" and "fox" could similarly refer to several species.[141] The insulin sequences from "elephant," provided by Leslie Smith, could originate from the Asian or the African elephant, and the sequences of cytochrome c from "grey kangaroo," provided by Emanuel Margoliash, were most likely "determined from material representing [two different] species."[142] The hybridization of field and laboratory practices was not always a smooth process.

The professional legitimacy of Dayhoff as a curator of experimental data was constantly challenged by experimentalists who considered her work as simply clerical. When they did have to acknowledge its scientific value, they labeled it "theoretical." Biochemists sometimes had a disparaging attitude toward the intellectual, as opposed to the practical, value of the *Atlas*; this reflected their more general attitude toward theoretical approaches in science. As Thomas Jukes would put it in 1963, a theoretical approach "is not acceptable to many biochemists, being inductive, rather than deductive."[143] The German biochemist Gerhard Braunitzer, whose many hemoglobin sequences were included in the *Atlas*, told Dayhoff bluntly, "I am not a theorizer," but he nevertheless valued the data compiled in the *Atlas*.[144] The American biochemist John Edsall made the same point when he wrote to Dayhoff that he had used the *Atlas* "primarily as a source of data and [had] not read very much of [the] interpretative material."[145] Another biochemist drew the same line between theory and experimentation when he wrote: "although your theoretical predictions are very elegant and indeed very helpful . . . as an experimentalist I feel that a word of caution should be added to the kind of experiments you recommend."[146] He added that "in real life," some of Dayhoff's assumptions were not warranted. The biochemist Frederick Sanger would later express his preference for the empirical even more clearly: " 'Doing' for a scientist implies doing experiments"[147]—not, he might have added, collecting facts determined by others or engaging in theoretical speculations. Dayhoff, however, was carving out a new role as neither an experimentalist nor a theoretician but a curator of experimental knowledge. Yet the norm was set by the experimental sciences, and it was in relation to that benchmark that Dayhoff defined the nature of her work. She positioned her team as an arbiter of experimental knowledge: "Since we have no experimental research effort of our own, we are in a particularly good position to evaluate impartially the work of competing laboratories."[148] The idea that the compilation of sequences, unlike their experimental determination, did not count as a scientific contribution would plague the development of sequence databases for the decades to come, and explains a great deal about why science funding agencies resisted funding them. Almost twenty years later, after the NIH had turned down one of her grant requests for the *Atlas*, Dayhoff complained once again, "databases do not inspire excitement."[149]

Excited or not, funding agencies were embarrassed by the requests to fund the *Atlas*. They simply did not fit within the standard categories of science funding. Dayhoff's work qualified neither as experimental nor as theoretical, yet it did produce scientific knowledge. For NASA and the NIH, Dayhoff emphasized the research aspects of her work with the data included in the *Atlas*. For NASA's exobiology program, she emphasized how her work on the evolution of protein sequences allowed her to determine the properties of ancestral forms of life that had inhabited earth and thus make a contribution to the search for life on other planets. For the National Institutes of General Medical Sciences (NIGMS), Dayhoff highlighted how sequence comparisons would contribute to the understanding of human health. Yet these research projects represented only one-third of the total effort contributed by Dayhoff and her team in connection with the *Atlas*. Collecting the data and preparing the *Atlas* manuscript required twice as much effort (and funding).

In 1970, Dayhoff thus turned to the National Library of Medicine, which had a mechanism to support the production of scientific publications. After rejecting Dayhoff's request the first time, the National Library of Medicine eventually gave her modest support for a period of four years.[150] A request to renew the grant in 1974 was turned down. Arguing that this "lack of funds creates an emergency,"[151] Dayhoff asked for a modest total of $9,300 to continue the production of the *Atlas* until other funds became available. Her request was once again turned down. Finally, the NIGMS bent and reluctantly agreed to provide the necessary funding.

To be sure, federal science funding was unusually tight during this period. Dayhoff's requests came just at the moment that, as a result of the escalation in the Vietnam war, federal budgets for scientific research were tightened for the first time in the United States in the postwar period. From 1967 to 1971, the appropriations of the NIH declined in constant dollars.[152] They briefly resumed their rise afterward, only to drop sharply again in 1974, just when Dayhoff submitted her proposal for renewal. The comments of the reviewers of Dayhoff's grants make it clear that the real issue was their unease at funding an activity that did not easily enter into the categories of scientific research. As medical geneticist Victor McKusick told her: "It seems to me that what you are doing represents a library function" (one wonders if he would have described his activity as editor of *Mendelian Inheritance in Man* in the same way).[153] Dayhoff's adoption of a proprietary model for the *Atlas* and her attempts to cover the costs of the publication through the sales of volumes and other services were an unfortunate result of her funding status.

Dayhoff's tight financial situation was not the only menace for the *Atlas*. Another threat resulted from the "explosion" of sequence data itself that Dayhoff promised to collect. From the late 1960s, the pace of protein sequencing had become "fast and furious."[154] In 1968, an editorial in *Science* made the point that the determination of protein sequences was "one of the most important activities today."[155] The "explosion" in sequence data resulted from several factors, including the development, in 1967, of Pehr Edman's

"Sequenator," a rapid and efficient automatic protein sequencer. The availability of this machine emboldened researchers to challenge larger and more difficult proteins. As Emanuel Margoliash put it, the Sequenator had a "psychological effect": "you do not hesitate to start a formerly difficult sequence because you feel that now more than 75% of the work will be done by the sequenator."[156] The rising interest in molecular evolution also led a number of researchers to sequence proteins from ever more diverse species. The *Atlas* itself facilitated these sequencing efforts by offering researchers a number of homologous sequences with which they could compare their partial experimental results, thus stimulating an even greater growth of the sequence data it was supposed to tame.

Dayhoff hoped to alleviate the growing burden of collecting sequences by relying on the voluntary submission of the data by researchers. Each *Atlas* began with a plea, printed in large characters, for new data—including unpublished data. Dayhoff's team regularly sent out hundreds of letters to researchers asking for contributions to the *Atlas*. A few, such as Russell Doolittle, obliged and submitted unpublished data, but overall this collecting system failed. By 1972, only 14 percent of the sequence data had been directly submitted to the *Atlas*, and that was counting generously, including corrections of published sequences whose errors had been resolved between Dayhoff's team and the authors.[157] The remaining 86 percent of the sequences had to be found in laborious searches through the scientific literature. Although the majority of published sequences appeared in just a few journals, the goal of the *Atlas* was to be comprehensive. So Dayhoff and her team regularly searched for sequences in over ninety journals.

It is important to understand the causes for this failure because they reveal deep tensions that accompanied the kind of science Dayhoff was proposing. First of all, some of the reasons that researchers did not bother to submit their data were mundane; it was considered a clerical task that was certainly not a prime concern in most laboratories. More fundamentally, however, the system proposed by Dayhoff clashed with established practices regarding the way credit and authorship were assigned in the experimental sciences. Indeed, in a move that would have dire consequences for the future of their project, Dayhoff and her team warned the authors of sequences published in the *Atlas* that they did not want to become involved in questions of "history or priority."[158] The fact that publication in the *Atlas* did not establish priority made experimentalists reluctant to submit their data because they would get no credit for it. Worse, the disclosure of sequence data could provide important hints to competing groups that were working on the same sequence. In the experimental sciences, publication—and thus the attribution of priority and authorship—was the main motivator for scientists to make their data public. Authorship, in turn, brought recognition and scientific credit, the key social rewards for producing knowledge in science.[159] As a molecular biologist and member of the Committee on Data for Science and Technology later lamented, "scientists are fierce individualists who consider themselves lone seekers of new knowledge. . . . The idea that they are part of an unorganized community of minds involved in a collective effort to seek knowledge may be foreign to most of them."[160]

When James Watson in his 1968 tell-all account of the discovery of the DNA double helix revealed that he and Francis Crick had used some of Rosalind Franklin's unpublished crystallographic data to build their model, reviewers almost unanimously condemned their behavior, as Franklin's data was considered to belong to her.[161] In his review of the book, the geneticist Richard C. Lewontin drew broader conclusions about the norms and values of the experimental sciences: "What every scientist knows, but few will admit, is that the requirement for great success is great ambition. Moreover, the ambition is for personal triumph over other men, not merely nature. Science is a form of competitive and aggressive activity, a contest of man against man that provides knowledge as a side product."[162] If Lewontin was right about the science of his day, it is understandable that some experimentalists were unwilling to give away the product of their work. In short, Dayhoff's system of data collecting ran against one of the essential norms of the experimental sciences' moral economy, namely, that the production of knowledge deserves individual, not collective, credit.

The fact that Dayhoff conducted research on sequence data provided to the *Atlas* also provoked "resentment within the scientific community,"[163] especially among those who had determined the sequences in their laboratories. As one researcher put it, Dayhoff seemed to consider that the sequences in her collection constituted her own intellectual "private hunting grounds."[164] But the molecular biologists and biochemists who had identified sequences, a painstaking effort of many months or even years, had a strong sense of ownership of their work and were not ready to give it up to a sequence collector to analyze.

It was Dayhoff's proprietary model that received the most consistent challenge by experimentalists. The fact that she had copyrighted her database, limited its redistribution, sought revenues from it, and used data submitted to the *Atlas* for her own research was considered by some of them as a violation of the moral economy of the experimental sciences. Dayhoff's standards of knowledge ownership were unacceptable to many experimentalists who considered the data they produced as their own and therefore to be published, distributed, and used only with their agreement. This tension would continue to plague Dayhoff's collecting enterprise in the years to come.

The Gender of Collecting

Gender also played a part in the valuation of Dayhoff's *Atlas*. The work for the *Atlas* was almost exclusively carried out by women. Indeed, with the exception of Richard Eck, all of Dayhoff's collaborators were women. Microbiologist Minnie R. Sochard and applied mathematician Marie A. Chang, both co-authors of the first edition of the *Atlas*, biologist Lois D. Hunt, and four other women assisted in the project. The *Atlas* project thus squarely fit the model of earlier scientific endeavors where groups of women were employed to perform repetitive tasks, as "computers" or as surveyors.[165] Dayhoff, advising a young female colleague about her career, warned about participating in "the 'masculine' scientific world," but argued that those who eventually decided for a scientific career brought "to this desert a range of feminine concerns that have been completely overlooked."[166]

Dayhoff saw some advantages for women in the emerging field of bioinformatics. For those like her who had interrupted a professional career for several years to raise children, the field provided an especially good reentry point, "since there is no literature to catch up with."[167] Yet as she wrote to Russell Doolittle, the world of science remained difficult for women: "The 'system' does not meet the needs of women very well. I always felt that if I had not been able to get my doctorate at 23, I might not have gotten it at all."[168] A few years earlier, the NIH had given Dayhoff a clear reminder that science did not tolerate profession interruptions very well. In turning down a grant application where she was to be a principal investigator, they noted that she "has apparently been out of really intimate touch for some time . . . with this complicated and rapidly advancing area."[169] On another grant proposal, which became the major source of funding for the *Atlas*, Robert Ledley put himself as the principal investigator until the grant was awarded and then asked permission to be replaced by Dayhoff.[170] Ledley later claimed that the grant would never have been awarded had a woman been the PI on the initial proposal.[171]

By the early 1970s, Dayhoff felt her gender had led to discrimination in getting her work published and in getting hired. She also came to resent more subtle mechanisms that were putting women at a disadvantage in science. In a survey on women in science, she mentioned the fact that "men tend to notice the work of other men and to reference them preferentially."[172] She was perhaps thinking about the fact that Fitch and Margoliash were widely acknowledged as having pioneered the reconstruction of molecular phylogenies in a paper published in 1967, although their results were almost identical to those published by Dayhoff in her *Atlas* a year earlier (publishing in a book rather than a journal might have also played a role). Dayhoff also mentioned that "men tend to talk to other men about current results and to exclude women" and that "men tend to assume that the man is head of a project involving a man and a woman." Dayhoff's closest collaborator, Winona C. Barker, voiced a similar complaint in her resignation letter to the president of the American Society of Physiology: "the Society has every appearance of being an 'Old Boy' group. I have seen no efforts to bring women physiologists into the hierarchy. The 'pure democracy' method of nominating and electing officers and council members almost assures that women, minorities, and young scientists will never be represented."[173]

Overall, these comments provide evidence of the difficulties Dayhoff and her female collaborators faced in receiving support and recognition for the *Atlas*. Although the extent to which the *Atlas* endeavor should be linked to Dayhoff's gender is debatable, it is nonetheless evident that she operated in a framework in which "passive" collecting was perceived, by her and by others, to be gendered in way that made it less authoritative than "active" experimenting.[174]

Research with the *Atlas*

In addition to producing the *Atlas*, Dayhoff made a number of scientific contributions using the data it included. Beginning with the second edition, published in 1966, the *Atlas* included introductions to current knowledge about the structure of proteins, new methods to analyze them, and the inferences that could be drawn using these methods.[175] Most of these contributions concerned the evolution of proteins and the evolutionary relationships between species. These methods became part of a series of computer programs developed by Eck and Dayhoff to analyze the data contained in the *Atlas*.

One of these programs addressed the key problem that faced those who wished to measure evolutionary relations through the differences in protein sequences: namely, to determine the real number of amino acid changes that had occurred in a protein over time, as opposed to the apparent number of changes, which might be lower owing to multiple changes in a single position. Using a program called ALLELE to analyze all the data contained in the *Atlas*, Dayhoff and Eck constructed a matrix of probabilities of amino acid exchanges that they could then use to estimate, from the observed changes, the actual number of replacements that had occurred over time in a given protein. In the case of cytochrome c, two sequences exhibiting 52 differences were estimated to have actually undergone as many as 92 changes. This information made it possible to draw a phylogenetic tree to scale with the length of the branches representing the actual number of changes that had occurred between two proteins since their divergence.[176] Dayhoff refined this approach over the years; it became know as the "Dayhoff matrix" or PAM (point accepted mutation) matrix, and was widely used by molecular evolutionists.[177]

Dayhoff and Eck also attempted to find a computational method to determine the topology of a phylogenetic tree, not just the length of its branches, by inferring ancestral sequences.[178] In the 1966 edition of the *Atlas*, the authors noted that a mathematical procedure had not yet been presented "in the detail necessary for a computer program," but that "arguments based on this approach [had] been used by Pauling, Zuckerkandl, and others,"[179] probably referring to an approach that minimized the number of mutations in a topology (a "parsimony method"), or that presupposed a constant rate of mutations. The program developed by Dayhoff and Eck compared multiple topologies compatible with the data and picked the most likely. However, as the authors noted, "since there are usually several millions of possible topological configurations, it is impracticable to try them all in the search for a minimum, even on a high-speed computer."[180] The program thus proceeded stepwise, beginning with three sequences and suggesting a possible ancestral sequence from which they had derived. Then it considered one additional sequence and repeated the operation.[181] At each step, researchers could decide whether to continue the path suggested by the computer or to make manual adjustments based on physicochemical or other considerations. As with other programs developed at the NBRF, computers did not replace humans, they assisted them. Once a topology was chosen and the number of mutations minimized, another computer program determined the length of the branches in geological time, assuming a constant rate of mutation for each set of homologous proteins, and estimating the elapsed time in proportion to the number of inferred mutational changes. In multipage foldouts included in the *Atlas*, Dayhoff and Eck presented the results of their calculations as phylogenetic trees and provided the alignments of the sequences they had used to produce them.

The 1966 edition included a phylogeny based on cytochrome c, comprising organisms ranging from yeast to human and including tuna fish and kangaroo. It was topologically identical to that published a year later by Walter Fitch and Emanuel Margoliash which is often mistaken as the first phylogeny based on protein sequences produced using a computer.[182] However, the methodology of the two groups differed in a significant way. Whereas Fitch and Margoliash weighted the amino acid differences with a value of one to three, depending on the number of nucleotide changes that were required according to the genetic code, Dayhoff and Eck, using all the data contained in their *Atlas*, weighted the amino acid differences based on the empirically observed frequency of amino acid change, thus obtaining a more realistic estimate of the actual numbers of mutations.[183] These differences may have reflected different theoretical assumptions about the effects of natural selection at the molecular level, but they also reflected the possibilities offered to Dayhoff and Eck—and not to others—to derive statistical regularities by using a computerized collection of protein sequences, an essential step for reconstructing the course of evolution from molecular data. Indeed, the punch cards that were used to store the data and print the *Atlas* could be directly fed into the computer and analyzed with the programs developed by Dayhoff and her collaborators.

As more sequences became available, Dayhoff and her collaborators proposed more ambitious phylogenies covering the entire evolution of eukatyotes from simple unicellular organisms to humans.[184] Comparing thirty-three cytochrome c sequences, from tomatoes to bullfrogs and pumpkins to snapping turtles, they estimated the time of divergence of protists, plants, animals, and fungi. They also attempted to reconstruct the ancestral sequence of the cytochromes of the first eukatyotes. This last step, which many molecular evolutionists did not take, was most likely a result of earlier work by Dayhoff. Before starting on the *Atlas*, under NASA sponsorship, she had hoped to discover the most primitive forms of life that might have existed on earth and on other planets. These results were eventually published in 1978 in a widely cited article in *Science*. This earlier work on evolution had already caught the attention of the German television station ZDF, which conducted an interview with Dayhoff, focusing on "how

Fig. 3.6 Staged interview with Margaret O. Dayhoff (the filming of Dayhoff had taken place prior to the staged interview), aired on German ZDF television, 1970. ZDF, *Querschnitt—Reise in die Vergangenheit*, Jan 18 1971. Printed with permission of the ZDF Archives.

one would go about bringing a dinosaur to life" (interestingly, this was twenty years before the publication of *Jurassic Park*) (figure 3.6).[185] Even the Russian embassy in Washington, DC, seemed to pay attention; it sent "an unexpected visitor" to Dayhoff to discuss her research on evolution. Under obligations stemming from the Cold War, Dayhoff reported the conversation to NASA.[186]

Dayhoff also developed the concept of the "protein superfamily" as a method of organizing the *Atlas*. Proteins with a sequence similarity of 50 percent or more were assumed to be homologous, i.e., they descended from a common ancestor, and were grouped together in the *Atlas*. Their similarity in sequence was accompanied by similar structures and functions. But many proteins with only moderate sequence similarity (sometimes as little as 10 percent) might also share a common origin. The plant protein leghemoglobin is distantly related to human hemoglobin, although its function differs markedly. Dayhoff coined the term "superfamily" in 1974 as a higher classification level to express this relationship. She estimated that there were probably over fifty thousand different proteins in the human body alone. Dayhoff claimed that her new classification would contribute to making "the totality of biological knowledge much simpler to comprehend."[187] From then on, each protein in the database was sorted into a family and superfamily, similar to the classification of organisms into hierarchical taxonomic groups. Thus the *Atlas* was not a phone book, as some seemed to believe—a mere alphabetical list of protein sequences—but rather a structured collection of data whose organization reflected theoretical assumptions about evolution.

In order to assess their relationships, Dayhoff compared each new sequence submitted to the *Atlas* to every other sequence in her collection.

Sometimes this produced surprises such as the relationship between a viral protein that caused cancer in chickens and a common enzyme in mammals. This finding supported the idea that the cancer-causing virus originated from a mammalian gene.[188] After conducting a systematic, "comprehensive examination" of all the sequences in the database, Dayhoff's collaborator, Winona Barker, found many duplications and was in a position to discuss their generality as a mechanism of evolution at the molecular level.[189] None of these contributions could have been made without the *Atlas of Protein Sequence and Structure*.

Whose Data? Whose Database?

Dayhoff's collection was of a very different kind than the collections of organisms and blood examined in the previous two chapters. Yet similar issues were at stake: how to assemble the collaborative efforts of numerous researchers into a resource that could benefit the community as a whole. By 1972, over 2,500 researchers had made contributions to the *Atlas*. Collections have almost always been collective enterprises, and the success of databases has always rested on the identification of researchers with the collective project to which they were contributing. In the case of the stock collections of model organisms (chapter 1), this identification came naturally, as the providers and the users belonged to the same self-identified community. The strong communitarian ethos that emerged in these communities, while they were still small and organized around charismatic figures, was effective in assuring the participation of its members in the stock collections. In the case of the *Atlas* the situation was considerably different: there was only a partial overlap between the communities of users and providers of data, and they did not identify strongly with a specific set of problems. Many biochemists provided the data that would go on to be analyzed by other researchers, such as evolutionary biologists who did not contribute experimental data themselves, which helps explain the reluctance of biochemists to participate.

Another difficulty pertains to the fact that Dayhoff handled the data the way museum collectors handled specimens: namely, as the property of the collector. Natural history collectors owned specimens that they could sell, lend, or offer to other collectors.[190] The collections themselves were considered museum property rather than that of the individual naturalists who had collected the specimens in the field.[191] In the experimental sciences, researchers had a strong sense of ownership over the data they produced and were not ready to give up ownership to collectors that would allow others to do the analysis. Dayhoff's proprietary arrangements regarding the sequence data she collected and the knowledge she produced might have been acceptable to naturalists but provoked resentment among the experimentalists who had provided the data, limiting their participation in the *Atlas* project.

In the 1960s and 1970s, federal funding agencies such as the NIH and the NSF in the United States did not seem prepared to fund databases to serve the entire scientific community. Again, community resources simply did not fit the policy model for science funding, which was based on individual researchers or laboratories. Furthermore, the very status of these databases was as yet undefined. Most often they were assimilated to library resources, largely underestimating the scientific expertise that went into the organization and evaluation of their data. It took until the 1980s for the NIH to recognize the essential role of databases in the experimental sciences and begin to provide significant funding.

Although the computer was essential for the maintenance of the *Atlas* and for the analysis of the data it contained, it was not yet used as a significant means of data distribution. In part, this resulted from Dayhoff's resistance to redistribution of her database. But it also reflected the limited access to computers on the part of life science researchers. Only in the 1970s did computers become more widespread, at which point an increasing number of researchers insisted on having the *Atlas* in a computer-readable format. The remote access to computerized databases in the life sciences was introduced only in the mid 1970s by the Protein Data Bank (see next chapter).

The persistence of print as a way to distribute digital data (molecular sequence data is digital) until the end of the 1970s is quite remarkable. Long after magnetic tapes were introduced as a way to store digital data (initially in 1952) and even as "personal computers" were becoming a common fixture in laboratories (and everywhere else), sequence data continued to be printed on paper and shipped around the world in book format. For computer users of that time, digital storage did not yet possess the permanence and gravitas that a reference work should possess. Only the book—a heavy book—seemed to have the material and symbolic qualities required for a definitive and authoritative source of information about nature. But a growing number of researchers were moving from the print to the electronic culture, reaping all the computational advantages of this new medium, making the printed *Atlas* a relic of a bygone age.

Despite the numerous obstacles she faced, Dayhoff succeeded in turning the *Atlas* into a respected database of experimental knowledge whose quality was rarely challenged. It contributed to the "informational turn" in the life sciences by focusing researchers' attention on protein sequences and the knowledge that could be derived by comparing them. More than a decade before DNA sequencing, the comparison of proteins sequences offered insights into the evolutionary history of organisms and the relationships between the structure and function of molecules. Sequences were particularly amenable to computational treatment because of their linear nature and the discrete (or "digital") nature of variations. The *Atlas* presented carefully verified empirical data in a standardized format that could easily be processed by computers. Additionally, it provided examples of computer programs, developed by Dayhoff and her collaborators, to analyze the data; the results obtained by these approaches demonstrated the value of the computerized treatment of biological information. The *Atlas* thus contributed in no small measure to the rise of "bioinformatics."

The introduction of computers into their research reinforced molecular biologists' focus on biological sequences, now considered as "information." Although this transformation represented a historically significant rupture in the history of the life sciences, there are obvious, deep continuities with earlier research practices. In particular, the production of knowledge by the *Atlas*, whether to gain insights into protein function or evolution, was broadly comparative, embracing numerous species. It inherited this approach from comparative biochemistry, which, unlike the rest of biochemistry and most of the experimental life sciences, did not focus on some ideal model system. The production of knowledge through the collection, comparison, and computation of data rested on the comparative and the experimental ways of knowing; the *Atlas* played the role of the natural history museum. As with so many custodians of natural history collections, Dayhoff was driven by an ideal of completeness, hoping not only to obtain all known sequences but also to cover the entire range of living forms. When the first cytochrome sequence of a primitive photosynthetic bacterium was determined, Dayhoff wrote enthusiastically to the author thanking him for the "feeling of completeness" he had given her.[192] The addition of this sequence made it possible to explore the phylogeny of all living organisms from bacteria to humans. Unlike morphological comparisons, which were limited to organisms with similar anatomical structures, molecular comparisons could encompass the entirety of life. For Dayhoff, this comprehensive knowledge of protein sequences was essential in understanding not only evolution, but all biological processes. As Dayhoff put it: "the key to a great advance in understanding in the life sciences is knowledge about all of the protein superfamilies through which living things are controlled."[193]

4

VIRTUAL COLLECTIONS

In the postwar decades, bacteriophage geneticists used "fun" as their favorite description of their everyday research practices.[1] Crystallographers much preferred the word "tedious."[2] This distinction reflects both the self-fashioning of these different disciplines and the public images that researchers wished to project.[3] More important, it reflects the fundamentally different nature of experimental practices in the two fields. Phage researchers typically carried out genetics experiments in the afternoon, left their petri dishes containing phages and bacteria in an incubator overnight, and returned in the morning to the excitement of analyzing the results.[4] Crystallographers, on the other hand—at least those working on large macromolecules such as proteins—often required many years to obtain a rough structure of a protein and many more to reach a more precise one. Chemist Max Perutz needed over thirty years to determine the structure of hemoglobin using X-ray crystallography.[5] Together with John C. Kendrew, he earned the 1962 Nobel Prize in chemistry, before he had even finished determining the structure at high resolution. Crystallographers also had fun, as their memoirs make clear, at least between the tedious parts of their work, or when they finally worked out a structure; perhaps they were more gifted than other scientists at "delayed gratification."

The elucidation of the double helix structure of DNA through X-ray crystallography, by James Watson and Francis Crick, shaped the public image of this scientific practice but in a highly misleading way. Watson, who "grew up" among phage researchers, widely publicized the story of how he and Crick discovered the structure of DNA in Cambridge, in the same unit where John Kendrew and Max Perutz were carrying out their research on proteins.[6] In his tell-all autobiography, *The Double Helix*, Watson made X-ray crystallography sound "fun," just like phage experiments.[7] Yet because he and Crick were working on a perfectly regular and relatively simple structure, their task was far easier than that of Perutz and Kendrew, who were exploring the highly irregular, complex, and huge molecules hemoglobin and myoglobin. Furthermore, Watson and Crick knew the chemical composition of DNA, whereas protein crystallographers did not even know the exact amino acid sequence of the proteins they were studying.[8] And, of course, the pair appropriated the hard experimental work of others, Maurice Wilkins and Rosalind Franklin (without her knowing). In the case of proteins, no guessing, no shortcuts, nor any bravado would lead directly to the right structure, the trick that Watson and Crick had so brilliantly managed with DNA. Protein crystallographers had to measure, tinker, calculate, calculate some more, and all over again, for years on end. Patience was a key virtue.[9]

The "tediousness" of the protein crystallographer's job is essential in understanding how and why they developed the Protein Data Bank in 1971. This computerized collection of protein structures was intended to exploit more fully, through the systematic comparison of protein structures, the potential for interpreting structural knowledge that had been so difficult to acquire. It was also intended to provide shortcuts on the long road to the determination of new structures. The tediousness of the crystallographer's work also explains why this community was much less willing than phage researchers, for example, to share the results of their work and make it public through the Protein Data Bank.

The Protein Data Bank also reflected the growing importance of computers in protein crystallography and especially the development of three-dimensional computer graphics and computer networks. These two technologies had a strong impact on crystallography as they alleviated some of the most tiresome aspects of structure determination, particularly the need for heavy calculations and physical models. The Protein Data Bank relied crucially on these two technologies and became the first major biological collection to be conceived entirely as a shared electronic resource. Again, the tediousness of protein crystallography, resting on the performance of repetitive procedures, which could be carried out automatically, explains why these researchers embraced computers long before other experimentalists in the life sciences.

The determination of a protein structure by crystallographers usually began with the attempt to obtain a crystal of the molecule they wanted to study, often with the help of a chemist.[10] If they were successful, they projected X-rays at different angles at the crystal ("clearly a tedious procedure").[11] They were diffracted into a complex patterns of spots that, like "shadows" of the atoms present in the crystal, were recorded on a photographic plate. The intensity of each spot—there could be thousands—was measured carefully (a "slow and tedious [task]").[12] The measurements constituted the raw data from which calculations were carried out. However, reconstructing a structure from the data alone was like trying to reconstruct the three-dimensional shape of an object from its shadow alone. The so-called "phase problem" was solved by a technique called "isomorphous replacement," i.e., the measurement of two or more molecules in which some atoms had been replaced by heavier ones. Crystallographers used all of this data to calculate a three-dimensional Fourier map, involving the summation of many trigonometric functions (a practice unsurprisingly described as "tedious"),[13] to obtain contour maps representing the density of electrons in the molecules. They looked like topographic maps whose peaks represented the highest electron densities and thus the presence of an atom. A series of maps of various sections through the protein were then drawn on transparent plexiglass sheets. When they were stacked on top of each other, it was possible to obtain a rough representation of the three-dimensional shape of the molecule. To obtain the precise position of the atoms in this structure, crystallographers built three-dimensional "Kendrew models" consistent with the electron density maps, using kits of metal rods and balls to represent each atom. The model could then be "refined" using optimization calculations to adjust the position of each atom precisely, taking into account its interactions with its neighbors. The structure was considered solved when the position of each atom composing the molecule was known with relative certainty (for example, at a resolution of 2 angstroms). Finally, the positions of the metal balls in the model, representing individual atoms, were measured (a "highly tedious [process]")[14] and listed as long sets of atomic coordinates (positions of atoms along *x*, *y*, and *z* axes), which constituted the final result of the crystallographer's work.

Fig. 4.1 John Kendrew and Max Perutz with the "forest of rods" model of myoglobin, Laboratory of Molecular Biology, Cambridge, ca. 1960. These models were static, fragile, and very time-consuming to build. Printed with permission of the Medical Research Council Laboratory of Molecular Biology.

The numerous calculations were often performed by hand, or using rudimentary technologies such as Beevers-Lipson strips, paper strips containing trigonometric calculations that could then be added up using a simple mechanical calculator.[15] The heavy need for calculations in many phases of structure determination explains why crystallographers were among the first researchers to adopt computers. In the 1950s, John Kendrew made extensive use of the EDSAC, an early stored-program electronic computer, to calculate electron density maps of the myoglobin protein.[16] By the 1960s, computing was still mainly performed on very large "mainframe computers" such as the IBM 7090s or the 360s, installed in central facilities of research institutions.[17] Individual researchers "bought time" at hourly rates on the central computer to perform their calculations. This limited the use of computers to researchers who knew how to program them, in institutions that possessed them.

Alongside the immense number of calculations, structure determination required crystallographers to use their chemical and visual intuition to project a proposed structure onto a contour map. The models built by Kendrew and others were entirely static (when they didn't fall apart), and the forest of rods supporting the model atoms made it difficult to visualize the structure, let alone tinker with it (figure 4.1). In 1968, crystallographer Frederic M. Richards devised a tool that would give researchers more flexibility.[18] The "Richards Box" was a simple optical device containing an oblique mirror that reflected the contour maps into a space where the researchers could build their physical model until it corresponded perfectly to the map. For large molecules, this did not make it any easier to interact with the physical model, which was still limited by mechanical constants and the persistent attempts of gravity to take it apart. Computers, however, could help alleviate the problem.

From Physical to Virtual Models

In the 1950s, digital electronic computers were starting to be widely used to perform calculations in crystallography.[19] A decade later, thanks to advances in computer hardware, they began to be used to display and manipulate molecular models. Building physical models from metal rods and balls was so time-consuming that trying to build them virtually looked like an attractive alternative. More important, computer models could solve the key problems posed by physical models. Physical models were not just representations of a structure but a crucial tool to arrive at a structure and to interpret it: model building "was an integral part of protein crystallographer investigative practice, and an essential step in their attempts to interpret the diffraction pictures of their molecules."[20] These physical models were manipulated, twisted and tweaked, assembled and reassembled continuously until they formed what researchers believed was a correct molecular structure. Being able to manipulate virtual rather than physical models would ease that process greatly.[21]

But computer models could also help with a second problem: how to identify the position of each atom (there could be over one thousand) in the structure. With physical models, researchers had to tediously measure the relative position of each atom with a ruler, unless they did not bother to do so, like John Kendrew, who received the Nobel Prize for determining the structure of myoglobin *before* he eventually published the atomic coordinates.[22] Virtual models, on the other had, could give the coordinates of each atom in a molecular structure automatically.

Finally, the third problem was perhaps the most serious: physical models could not easily make their way onto the flat pages of scientific journals and textbooks. Therefore, researchers published photographs of their models, stereoscopic drawings (to be viewed with a pair of red-green glasses), or artistic renderings, but these were always unsatisfying compromises. A reader could see the model only from the perspective chosen by the authors and could never look at the "other side" of the molecule. Virtual models, based on atomic coordinates stored in a data bank, could constitute a convenient way to make a structure public and allow other researchers to manipulate and explore the structure. For all these reasons, a number of researchers attempted, starting in the early 1960s, to develop the computer tools necessary for the representation and manipulation of virtual molecular models.

The development of interactive computer tools for the construction and analysis of molecular models began at MIT as a part of Project MAC and, starting in 1963, funded by the Department of Defense.[23] The project made a computer with graphic capabilities available for time-sharing by academics who wished to produce computer applications. Physicist Cyrus Levinthal took this opportunity to develop "programs that would make use of a man-computer combination to do a kind of model-building that neither a man nor a computer could accomplish."[24] In the system he developed, a user could rotate a virtual, three-dimensional molecule on a screen by manipulating the orientation of a half-globe with one hand. In 1966, Levinthal described the system in *Scientific American*, claiming that "with a little practice, the coupling between hand and brain becomes so familiar that any ambiguity in the picture can easily be resolved."[25] The interactive possibilities of this system allowed researchers to "see" and "manipulate" molecules as if they were real objects, allowing "investigators to use their 'chemical insight' in an effective way."[26] This insight would help them find the correct structure and interpret it. Another and even more ambitious goal of the system was to predict how a chain of amino acids would fold into a three-dimensional structure, a task that neither calculation performed by computers nor humans using "chemical insight" had been able to achieve alone. Levinthal hoped that the combination might succeed (a goal that, to the present day, remains elusive). However, Levinthal's system had little immediate impact on crystallographers, most likely because none of them had access to the unique kind of computer required to run his system (it cost about $2 million).[27] Even so, by the end of the decade some of Levinthal's students were pursuing a technologically more modest version of his vision.[28] And within just a few years, cheaper, smaller minicomputers and computer networks with graphic capabilities were beginning to make molecular graphics much more accessible to research institutes and easier to use.

One of those students, the chemist Edgar F. Meyer Jr. (1939–2015), would become a founder of the Protein Data Bank. He spent two years in Levinthal's group writing computer programs to display small molecules, then joined Texas A&M University in 1967. There he continued to develop computer graphics systems that could be used in crystallography. Starting in 1968, he spent several summers as a visitor at Brookhaven National Laboratory on Long Island, New York, working with a new computer system that had been developed for high-energy physicists. This allowed a computer to display tens of thousands of points at thirty frames per second on a commercial color TV monitor (figure 4.2).[29] Meyer developed a number of programs to exploit these possibilities for crystallographers working on the structure of small molecules.[30] But these tools required that the crystallographic data be available in computer-readable format. As Meyer put it in 1969, a "vast amount of structural information exists in the scientific literature, with more being added each year. If this information is to be readily accessible, it must be stored and retrieved automatically, with the aid of digital computer techniques."[31] Luckily, the British crystallographer Olga Kennard had begun in 1965 to set up such a database, the Cambridge Crystallographic Data Centre, containing the atomic coordinates of small molecules and making the data available to researchers on computer tapes.[32] Meyer used his DISPLAY program to produce a "ball-and-stick" stereo image of a small molecule on a television screen from the data contained in Kennard's collection. Colored glasses gave the image a three-dimensional appearance. The program also allowed the user to rotate the model in real time, as had Levinthal's system a few years earlier, permitting investigators to view the molecule from any angle, unobstructed by the supporting elements that had cluttered physical models. Meyer's system thus made it possible to display "three-dimensional structural and

Fig. 4.2 Edgar Meyer adjusts red and green controls of the 3-D color television monitor of the BRAD system (1968) to produce a 3-D image of a complex molecule at Brookhaven National Laboratory. Printed with permission of Brookhaven National Laboratory.

chemical information in a visual, meaningful manner."[33] Instead of having to enter the coordinates of the molecule in the computer, the user could directly retrieve them from a storage format such as a magnetic tape containing the entire collection of small-molecule coordinates contained in the Cambridge Crystallographic Data Centre. It was a promising development for the average crystallographer. Unlike Levinthal's system at MIT, Meyer's relied on small computers and common television displays.[34] The system still required an initial investment of $50,000, but each additional terminal would cost only $3,000, and the price of graphics systems was declining rapidly.[35] As Meyer noted in 1970, "it may be expected that as graphics terminals become more commonly available their use in the various stages of structure analysis and the retrieval of structural data from the scientific literature will become much more frequent and facile."[36]

Meyer's remark indicates that he clearly distinguished two roles for computers in crystallography. On the one hand, computers were useful as a "method of interacting with the graphical model . . . in the study and analysis of molecular structures." On the other, they could become tools for the "rapid retrieval of structural information from the scientific literature in the form of graphical models."[37] The two roles went hand in hand. Computer tools that virtually manipulated molecular models were the key to making the computerized storage and retrieval of information from the scientific literature useful. But attaining these two goals required bridging the gap between the paper world of scientific publications and the digital world of molecular graphics. That was precisely the intent behind the creation of the Protein Data Bank.

The Systematic Study of Protein Structures

The idea of creating a computerized collection of proteins was first publicly articulated by two young researchers, Edgar F. Meyer Jr., 35, and Helen M. Berman, 29, at a meeting of the American Crystallographic Association in Ottawa, Canada, in August 1970.[38] Berman's interest in crystallography dated from her days as an undergraduate student in chemistry in the laboratory of Barbara Rogers-Low, a student of Nobel Prize winner Dorothy Crowfoot Hodgkin, at Columbia University in New York. Berman then embarked on a PhD in this field at the University of Pittsburgh under the crystallographer George A. Jeffrey, an early promoter of digital computers in crystallography. As a requirement for her degree, Berman prepared a mock grant proposal for a research project addressing one of the most vexing problems in crystallography. For more than a decade, it had been clear that the three-dimensional structure of a protein, which is responsible for its function, was determined by its amino acid sequence.[39] In principle it should thus be possible to predict the structure of a protein from its sequence, but nobody had yet succeeded in this task. What made the idea so attractive was that the determination of protein sequences was becoming simpler, quicker, and cheaper, especially after automatic sequencers became available in 1967,[40] while the determination of protein structures through X-ray crystallography remained complex, time-consuming, and expensive (and, yes, "tedious"). A solution to the "protein folding problem" could thus give a formidable boost to efforts to describe the structures and understand the functions of proteins.

The "protein folding problem" attracted a number of researchers, including Cyrus Levinthal, who hoped that the number-crunching power of digital computers would bring a solution. Others approached the problem empirically, trying to find regularities between sequences and structures by comparing many proteins. That was Berman's approach. She was particularly concerned with the sharp turns (or "corners") of the polypeptide chain, which were most important in determining the final structure of a protein. She proposed "a method of elucidating some common features of the corners" and of confirming that these features were partially "responsible for the large scale folding of proteins."[41] Her strategy was to adopt "a systematic procedure employing a computer" to compare sequences and structures of different proteins.[42] But before such a procedure could be carried out, it was "necessary to obtain the fractional atomic coordinates of all the proteins whose structures have been solved."[43] In other words, one had to have a computerized collection of protein structures.

Berman did not elaborate in 1967 on how such a collection should be constructed. The few proteins whose structure had been determined were published as verbal descriptions, drawings, or photographs of physical models. The actual coordinates and other parameters of the structure were generally not publicly available for calculations; crystallographers kept the numerical data to themselves.[44] Berman's dissertation committee gave her mock research proposal a rather chilly reception. "What, are you crazy?" a member of the committee apparently asked her, pointing to the fact that only three proteins had been determined so far, undermining any hope to learn something from a systematic approach until many more structures had been solved. (Berman was nevertheless authorized to defend her thesis in 1967.)[45]

The experimental component of the thesis consisted of the determination of the structure of several sugar molecules by X-ray diffraction.[46] Her work represented an example of experimental virtuosity. However, Berman also approached the determination of structures using a comparative method that went back to the earliest days of crystallography, a time when the study of crystals was an integral part of natural history alongside work with animals and plants.[47] Berman systematically compared all the known structures of simple sugars. By tabulating the lengths of the bonds between the atoms of fifteen different molecules, she discovered "a similar pattern of bond shortening" between two atoms in almost every case. This significant result resulted in Berman's first publication, in *Science* in 1967.

Berman's comparative approach had been adopted by her supervisor, George A. Jeffrey, in determining other properties of molecules. Recognized as an uncompromising leader—a sign on his desk read "Be reasonable, do it my way"[48]—Jeffrey promoted this comparative approach within his laboratory and throughout the crystallographic community. He believed that "comparing the results [of] several structures, and including other chemical and physical information" would reveal "overall patterns" constituting "much more meaningful contributions to chemistry than just an isolated structure."[49] In 1965, a year after Berman joined his laboratory, he compiled all the known structural data from pyrimidine rings (found in DNA and many other biomolecules) and entered it on punch cards in a format "suitable for a uniform representation" and for "information retrieval" by a computer.[50] The cards (or a magnetic tape) containing the data were made available to crystallographers for a nominal fee through the University of Pittsburgh's Knowledge Availability Systems Center.

The creation of this institution was a result of the Sputnik shock and reflects the growing interest in the sharing of scientific information. After its founding in 1958, NASA established the center and several other institutions to facilitate the dissemination of knowledge.[51] Another factor in its establishment in 1963 reflected the enthusiasm of the chancellor of the University of Pittsburgh for the emerging field of "information science."[52] In 1962 he noted that new knowledge was "transmitted to users too slowly" and existing knowledge was "insufficiently mobile."[53] The Knowledge Availability Systems Center was established to overcome this problem by distributing crystallographic and other kinds of scientific information in a computer-readable format.

Berman left Jeffrey's laboratory in 1969 to join the group of Jenny P. Glusker, another of Dorothy Crowfoot Hodgkin's students, at the Fox Chase Cancer Center in Philadelphia (figure 4.3). There Berman conducted experimental studies on the structure of nucleic acids such as DNA and RNA. Even though she did not work on proteins or pursue a comparative approach at the time, her experience at Pittsburgh in Jeffrey's laboratory had made her realize the potential value of a comprehensive collection of protein structures for the production of knowledge through the search for regularities and patterns.

In 1970 Berman attended a meeting of the American Crystallographic Association (ACA), where she discussed the idea of setting up a protein data bank with Edgar Meyer Jr. He was interested in the project for different reasons. Like Berman, he had worked only with small organic molecules, not proteins. His interest in collecting coordinates resulted from his unique awareness of the potential of computer graphics in crystallography, including for proteins. Provided one could obtain a set of atomic coordinates, one could

Fig. 4.3 Helen M. Berman posing in front of a structural model of DNA, a standard portrait format for crystallographers, at the Fox Chase Cancer Center, Philadelphia, ca. 1970. Helen Berman personal collection. Printed with permission of Helen Berman.

use the computer to visualize the molecule in three dimensions, creating a unique tool for crystallographers. He had already been relying on the Cambridge Crystallographic Data Centre, an electronic collection of the coordinates of small molecules. Now he thought that protein molecules might be tackled with his molecular graphics programs, but he had no data—again, coordinates were generally not published. Early in 1971, he began "assembling a library of protein structures" from colleague crystallographers, "for the three dimensional display."[54] He eventually collected the atomic coordinates of eleven (partial) structures, which became the initial datasets of the Protein Data Bank.

Obtaining data, however, was no simple matter. As several crystallographers recalled, it was necessary "to have a friend, who would be willing to share" to obtain the data.[55] Meyer thanked a crystallographer who had shared his data for his "present."[56] Given this reality, the project of assembling a collection of all known structures as envisaged by Meyer and Berman was rather unlikely to succeed; it was doubtful that crystallographers would be willing to share their precious data with two relatively unknown researchers in their thirties. To obtain backing, Meyer and Berman sought support from professional organizations of crystallographers.

Meyer initially thought about approaching the International Union of Crystallography (IUCr), but became discouraged when told he would run into the opposition of "certain blocking groups."[57] He then suggested that the American Crystallographic Association (ACA) sponsor the project. The first collective discussion about protein data collection involved two dozen crystallographers, including Berman and Meyer, and took place at the ACA meeting held in February 1971 in Columbia, South Carolina.[58] The issues that were discussed ranged from the eminently practical—such as how to fund a data bank—to quite sensitive topics such as finding ways to get crystallographers to share their data. A pragmatic suggestion included the idea of getting funding from the National Science Foundation or the National Institutes of Health, while also raising the possibility of charging users for the retrieval of data.

But the bulk of the discussion centered on data collection. Crystallographers were notoriously secretive with their data, for utterly understandable reasons: The time and resources invested in solving a structure were so substantial that researchers' careers seemed to depend on exploiting it to the maximum—the description and interpretation of a single structure ought to be worth several papers—before it was turned loose for someone else to use. Small-molecule crystallographers had reached a sort of common understanding whereby researchers were "prohibited" from tackling a molecule that another investigator was working on. One participant noted, "90% [of] small mol[ecule] crystallographers adhere to the ethics—someone else is working on it, hands off,"[59] but would protein crystallographers behave in the same way? For the participants, the most obvious way to convince researchers to make their data public was to make the deposition of a structure in a data bank a condition for the publication of an article describing it. That idea might backfire too, however, if as a result authors delayed publication to avoid having to disclose all their data. The participants could not solve this problem and decided to stick with the prevailing trend: "no one will be forced" to share data; instead, it would continue to be obtained through "personal contact."[60] The participants reiterated their beliefs in the traditional moral economy of the experimental sciences, based on the voluntary disclosure of data among colleagues, rather than acknowledge that increased competition in their field had made adherence to this norm rather inconsistent.

Fig. 4.4 Rebellious crystallographers Sung-Hou Kim, Helen M. Berman, Joel L. Sussman, and Nadrian C. Seeman (left to right), in front of MIT, on their way (uninvited and mostly unregistered) to the Cold Spring Harbor Symposium on Quantitative Biology, "Structure and Function of Proteins at the Three-Dimensional Level," 1971. Helen Berman personal archives. Printed with permission of Helen Berman.

Any solution to this problem would require a broad international consensus. Fortunately, a unique opportunity soon arose to discuss the data bank project with the international crystallographic community. In June 1971, the Cold Spring Harbor Symposium on Quantitative Biology was devoted to the "Structure and Function of Proteins at the Three-Dimensional Level." Organized by James Watson, the list of attendees of this select meeting read like a "who's who" in protein crystallography, including (future) Nobel Prize winners Dorothy Crowfoot Hodgkin, Max Perutz, Aaron Klug, and William N. Lipscomb. Although the meeting was by invitation only, a few scientists who were too junior to be on the list decided to participate anyway and "kind of crashed the meeting."[61] Helen Berman and three friends, self-described "hippies" who valued communitarian ideals (figure 4.4), drove from Philadelphia to Long Island to attend the meeting and present the idea for a crystallographic data bank.[62] One participant at the meeting, Walter C. Hamilton, became particularly enthusiastic about the proposal and decided to help Meyer and Berman make it happen. The meeting saw the revelation of the structures of seven new proteins "in atomic detail" for the first time.[63] The Cold Spring Harbor Symposia were unique in that they brought together elite scientists in a beach resort atmosphere conducive to informal discussions.[64] The photographs of the meeting show small children playing quietly on the beach and young crystallographers playing more turbulently in the Cold Spring Harbor bay.

The Creation of the Protein Data Bank

Walter Hamilton was an imposing figure in crystallography. Tall, charismatic, and loud-spoken, he was elected president of the ACA in 1969, at the age of only thirty-eight.[65] At the time he was co-chairman of the chemistry department at Brookhaven National Laboratory, where he led a group engaged in the determination of molecular structures. In 1965, Brookhaven had started to operate a state-of-the-art nuclear reactor, the High-Flux Beam Reactor, which had been designed to produce large amounts of neutrons. Neutrons, like X-rays, could be used to study the structure of molecules through diffraction techniques. One advantage of neutrons over X-rays was that they could help locate hydrogen atoms, which play an essential role in the binding of biological molecules such as the two strands of the DNA double helix.

Hamilton was an expert in the field of neutron diffraction and had coauthored a book on hydrogen bonding in 1968. He had been a graduate student in Linus Pauling's Division of Chemistry at the California Institute of Technology, making it no coincidence that he decided to replicated Pauling's systematic studies of all the amino acids (carried out using X-ray diffraction in the 1930s), now using neutron diffraction at Brookhaven. In the first of a series of fifteen papers, Hamilton announced how his comprehensive study would be useful for all the researchers interested in the "morphology of biomacromolecules."[66] Hamilton also produced a "Stereoscopic Atlas of Amino-Acid Structures" that presented drawings of all the known amino acids. This compilation of data was intended to pave the way for a "detailed comparison of bond lengths and angles" and a discussion of "the general features of the hydrogen bonds."[67] Hamilton was thus seriously committed to systematic and comparative studies of macromolecules and was ready to understand the potential of a protein structure data bank such as the one Berman and Meyer discussed with him at Cold Spring Harbor in 1971.

Hamilton was also acutely aware that the number of known structures would soon explode. Just a year before the Cold Spring Harbor Symposium, he published an article in *Science* outlining what he believed amounted to a "revolution in crystallography." He noted that for small molecules, it was now possible to carry out "a complete crystal structure determination to obtain results of chemical significance in less than 2 weeks," although a timeframe of a few months was more typical. Even in the case of proteins, the determination of a structure now formed only "part of a Ph.D. thesis," whereas just a decade earlier, such an accomplishment had been worthy of a Nobel Prize.[68] This dramatic increase in speed was made possible by the fact that crystallographers now used "automatic diffractometers" to collect data and a "high-speed, large-memory digital computer" to determine and refine the structure of their models. Yet while protein crystallography was becoming slightly more efficient, it still remained much slower than many other experimental approaches and thus continued to be perceived as relatively tedious. The new possibilities offered by computers had also had some unintended consequences. Hamilton deplored that "it has unfortunately become traditional among many crystallographers to squeeze the last possible item of information out of their data [before making it public] even when this item of information may be basically uninteresting—or at least not worth the cost."[69] So this tendency still hindered the prompt release of crystallographic data to the community, which was one of the main goals of the proposed protein data bank.

Hamilton's proposal to host the data bank at Brookhaven was well received by crystallographers attending the Cold Spring Harbor meeting. The fact that Hamilton was not a protein crystallographer might have reassured those concerned that the managers of a data bank might use the data for their own research, a recurring concern with Margaret Dayhoff and her *Atlas of Protein Sequence and Structure* (chapter 3). After the Cold Spring Harbor Symposium, Hamilton noted that there seemed to be a "general agreement" in favor of having a "depository" in the United States, as well as another one at Olga Kennard's Cambridge Crystallographic Data Centre in the United Kingdom.[70] This support also resulted from the fact that for the first time, enough protein structures had been solved to make it worthwhile collecting them. David C. Phillips, who had determined the first structure of an enzyme (lysozyme) in 1965, concluded the meeting by noting that "so much structural information" was now available that the time was "almost ripe for the jump in our understanding that will bring all the evidence together."[71] This was precisely the goal of the envisioned protein structure data bank.

A few days after the meeting, Hamilton contacted Olga Kennard to explain his plans, hoping they could reach an "agreement on format and modes of cooperation."[72] A month later, while visiting England, he made the trip to Cambridge to meet with her and discuss the data bank project. Kennard was also an authority in crystallography, and obtaining her cooperation seemed to be necessary for any data bank project in the field.[73] In the fall of 1971, *Nature New Biology* published an announcement outlining the agreement between Hamilton and Kennard. The Protein Data Bank, operated jointly by the Cambridge Crystallographic Data Centre and Brookhaven National Laboratory, would store "atomic coordinates, structure factors, and electron density maps" and make them available on magnetic tapes to the community at no charge beyond handling costs. The announcement made clear that the success of the proposed data bank would "depend on the response of the protein crystallographers supplying the data." However, the data bank, like Dayhoff's *Atlas of Protein Sequence and Structure* established six years earlier, should not be considered a "substitute for the publication of the results of structural investigations in a scientific journal."[74] Thus the submission of data to the bank would not result in the attribution of priority, authorship, or other forms of credit for the submitter; the data bank would have to rely on communal ethos or generosity alone.

But to understand why the Protein Data Bank came to be viewed as essential to crystallographic research, it is necessary to emphasize that computers were becoming essential not only in determining new structures, but also increasingly in analyzing existing ones, as demonstrated so clearly by Meyer's molecular graphics programs, in part developed at Brookhaven. Hamilton was also an "expert programmer" and saw how the fast computers at Brookhaven and the expertise he had developed in computer networks could be crucial for the success of a protein data bank project.[75] In order to make Brookhaven's computing power (two CDC 6600 mainframes) and the visualization tools developed by Meyer more widely available to crystallographers, Hamilton and Meyer created, in the fall of 1971, CRYSNET, a computer network for crystallographers. The network spanned "from New York to Texas" but in fact contained only three nodes initially: New York (Hamilton at Brookhaven), Philadelphia (Berman at the Institute for Cancer Research), and Texas (Meyer at Texas A&M University).[76] Remote users could retrieve atomic coordinates stored at Brookhaven, visualize the corresponding molecular structures on the screen of their microcomputer (DEC PDP 11/40), and analyze the data using some of the tools developed by Meyer. The real life attempt with this system took place in September 1971, between Texas and Brookhaven, and a month later between Philadelphia and Brookhaven.[77]

For the first time, crystallographers had access to remote computing. The more protein structure data was available in the Brookhaven library, the more valuable this new system of remote computing would be. For this reason, Meyer pursued his efforts to collect all known structure data from protein crystallographers.

By February 1972, Hamilton, who had decided to put a "major emphasis" on the protein data bank, had outlined a project more ambitious than a simple repository that, at the time, included just nine structures, collected by Meyer.[78] Hamilton intended the data bank to become an essential tool for a broad community of researchers. He planned to develop different kinds of computer programs to display the structures, including through "computer generated motion pictures," in a way that would be helpful to the "biochemical community."[79] Although the data would come exclusively from protein crystallographers, Hamilton hoped that the data bank would help make knowledge about protein structure available to a broader audience, including biochemists, who believed that structural information would cast new light on protein functions. Meyer and Berman, who both held appointments at Brookhaven, were respectively in charge of developing programs for the display of structures and formatting the data uniformly. When they returned to their home institutions in College Station, Texas, and in Philadelphia, Hamilton made sure Meyer and Berman were connected to the Brookhaven computers through the CRYSNET network. They could thus experience firsthand how remote users would interact with the data bank.

The following months were spent trying to put the data in a standardized format and obtaining funding for the data bank, after a first proposal was rejected by the NSF. The project developed swiftly until December 1972, when Hamilton was diagnosed with cancer. Hamilton suggested that Thomas Koetzle, a postdoc who had been working with him on the determination of amino acid structures, could take over the data bank project. Hamilton died just one month later, aged only forty-one.[80] Koetzle, who had been a student of William Lipscomb at Harvard, began managing the Protein Data Bank alongside his own research. Hamilton's death had been "a great personal blow and a tremendous shock" to all those working on the project.[81] After this initial setback, however, Koetzle was able to bring the project back on track.

One of his main tasks was to obtain formal permission for distribution of structures from the authors who had shared them with Hamilton. Most authors, who had been friends of Hamilton, agreed to release the data. However, not all must have been entirely comfortable with their decision. One crystallographer authorized the release of his data but asked that he be informed of who would access it.[82] Others, including Max Perutz, wished to hold back data until it was further refined.[83] Koetzle tried to persuade him to release it based on the argument that "coordinates at any stage of refinement will be extremely interesting and very useful to many people,"[84] and Perutz eventually agreed, but only after a year and after having published several more papers on the structure of hemoglobin.

By May 1973, the data bank was "about ready to begin distribution," and a formal announcement was published in *Acta Crystallographica*; another followed three months later in the *Journal of Molecular Biology*.[85] The Protein Data Bank contained the coordinates of just nine proteins, and anyone could obtain the entire data bank on a magnetic tape for a very modest sum to cover the cost of the tape and shipping. The announcement repeated the call made two years earlier that the "usefulness of the system" would depend on "the response of the protein crystallographers supplying the data." In the following months, researchers who determined structures began to acknowledge in their publications that they had deposited the atomic coordinates in the Protein Data Bank. The number of available structures grew slowly, however. In January 1974, there were just 12 structures available, a year later 15, and the following year 23.[86] The slow pace at which data was deposited was not due to lack of interest among the crystallographic community, as shown by the distribution of the data sets: it grew rapidly from 119 to 375 in the same period. The problem was that while crystallographers were enthusiastic users of the Protein Data Bank, they were much more reluctant to deposit the structures they had determined. They published papers describing new protein structures but voiced a range of motives for not depositing the data in the Protein Data Bank.[87]

The most obvious reason was simply a lack of incentive. Preparing the data for deposition in the Protein Data Bank could be time-consuming, even though the Protein Data Bank accepted a variety of formats. More fundamentally, the researchers who used the Protein Data Bank were not necessarily the crystallographers who had produced its data. In a limited survey carried out in 1980, the organizers of the Protein Data Bank found that only half of the users were crystallographers; the other half comprised computer scientists, physicists, chemists, and biologists who did not perform experimental research on protein structures. In other words, crystallographers were asked to provide data that would largely benefit users outside of their own professional community. But the most important reason for withholding experimental data was crystallographers' fear that others would use it before they could fully exploit it. A single set of experimental data usually spawned a series of papers describing the structure at increasing resolutions obtained with different refinement procedures. More papers explored the structural, functional, and sometimes evolutionary significance of protein structure. The creators of the Protein Data Bank had been aware of this problem all along, and had wondered in 1971 if the ethics of "someone else is working on it, hands off" would hold for protein crystallography. Yet with the increasing size of the crystallographic community and the growing number of noncrystallographers who worked on protein structures, it was unlikely that these unspoken community-based rules would persist.

The Natural History of Macromolecules

The scientific community did not wait until all known structures were deposited in the Protein Data Bank to begin tapping into this unique collection; they quickly began using it in a variety of ways. Some were interested in the atomic coordinates of a single protein to help determine the structure of a similar protein. Others took advantage of the systematic comparisons made possible by the fact that the Protein Data Bank (PDB) was a uniformly formatted collection of data. Most important, this comparative approach was used to establish taxonomies of protein structures and to understand how proteins acquired their three-dimensional structures.

Although the determination of protein structures by X-ray crystallography had become much easier in the 1970s, thanks to improvements in instrumentation (diffractometers and computers), it remained far more difficult than sequencing. Since 1967, automatic sequencers had been revealing the amino acid sequence of a large variety of proteins in numerous species. In 1976, Dayhoff's *Atlas of Protein Sequence and Structure* contained 767 sequences, but the PDB contained only 84 structures.[88] What came to be known as the "sequence-structure gap" was widening fast as the automatic sequencers became a common fixture in laboratories.[89] The failure of all the attempts, beginning with Levinthal's, to predict protein structures from sequences alone led researchers to explore other strategies. One approach, known as "comparative modeling" or "homology modeling," consisted of fitting the amino acid chain of a protein of unknown structure onto the three-dimensional model of a homologous protein whose structure had been determined by X-ray crystallography. This approach was first carried out successfully in 1969, when a group of researchers determined the structure of a whey protein present in cow's milk (α-lactalbumin) based on the structure of an enzyme present in chicken egg white (lysozyme). In these two proteins almost half of the amino acids were identical, leading researchers to suspect that they would fold into a similar structure. Using a wire model of lysozyme several feet tall, they replaced one after another the amino acids that differed in α-lactalbumin. The authors were optimistic that their approach could be applied to many more proteins. In the paper describing their model, they concluded: "It may be possible to derive the structure of all the members of a family relatively easily once one or two have been analyzed in detail." The authors believed their approach to be superior to that of Levinthal and others searching for the rules of protein folding: "Certainly [comparative modeling] seems more promising at the moment than does the prediction of unique protein structures from chemical information alone."[90]

The structure of α-lactalbumin was determined crystallographically twenty years later and was found to be very similar to the model proposed through comparative modeling.[91] Yet before this long awaited confirmation, researchers began applying comparative modeling to a broad range of proteins. The availability of structures in electronic format from the PDB made this possible using computers graphics programs such as those developed by Edgar Meyer Jr. instead of physical models built from wires.[92] The first structure determined through comparative modeling was deposited in the PDB in 1978, alongside other structures determined directly. The PDB organizers were delighted as this demonstrated that the data bank was not just a repository but was becoming a crucial tool in the production of new knowledge. The editors of the *Protein Data Bank Newsletter*, after announcing the inclusion of the "first coordinate set for a protein which has not been studied crystallographically," highlighted how the "increasing availability of good graphics devices coupled to powerful computers allows intelligent, interpretative 'synthetic' use of the rapidly increasing body of knowledge of macromolecular structure."[93] Unlike the computers available to crystallographers in the 1950s, modern machines made it possible to perform real-time manipulations of graphic images instead of "batch processing," where it sometimes took hours to obtain the results of calculations. The inclusion of this structure in the data bank was also significant in that it constituted an important step in the convergence between knowledge derived from experiments and computer modeling. It also represented an example of how evolutionary considerations, specifically regarding homology, were playing an increasingly important role in the determination of protein structures.

One of the main difficulties with the comparative modeling approach was that although sequences were believed to determine structures, the sequences of proteins were far more variable that their structures. As Perutz had already pointed out in 1961, hemoglobin and myoglobin folded roughly in the same way, even though their sequences were rather different. The biochemist Cyrus Chothia and the physical chemist Arthur M. Lesk relied on the PDB to examine this finding more broadly and discovered that there was nevertheless a direct relationship between variations in amino acid sequence and protein structure among homologous proteins. This result was expected, but the PDB made it possible to demonstrate it empirically. It also made clear that "the degree of success to be expected in predicting the structure of a protein from its sequence using the known structure of a homologous protein [i.e., comparative modeling] depends upon the extent of the sequence identity."[94]

Besides serving as a tool to determine new structures, the PDB was widely used for bringing order into the great diversity of structures it contained. Proteins were classified into hierarchical categories, families, and superfamilies, following a standard taxonomic practice. Through systematic comparison of protein structures, researchers built taxonomies of proteins based on their common structural features, much as naturalists had done for centuries with organisms. Chothia compared sets of proteins to find "structural invariants." He hoped that this study would help overcome the "aesthetic shock of the first protein structure," which was, in Kendrew's words, "almost nothing but a complicated set of rods sometimes going straight for a distance then turning a corner and going off in a new direction," a structure that was "much more complicated and irregular than most of the early theories of the structure of proteins had suggested."[95] But by comparing fifteen protein structures, Chothia found that there were many regularities, such as the observation that all proteins were "closely packed." In a follow-up paper he compared thirty-one proteins, focusing on the positions of basic structures, α-helices and β-sheets, revealing that it was possible to classify all proteins in "only four clearly defined classes."[96] Chothia adopted a similar approach in comparing all the structures of antibodies available in the PDB. Once again he found that there were a small number of "canonical structures" within the region responsible for their specificity, and just a few amino acids that determined these structures.[97]

Most classifications of protein structures were based on visual comparisons. One of the most comprehensive attempts to establish a protein taxonomy using this method was carried out by Jane Shelby (later Richardson).

Despite the lack of a graduate degree in science—she had a masters in philosophy, supplemented by courses in plant taxonomy and evolution at Harvard—she obtained a position as a technician at a chemistry laboratory at MIT. There she met David C. Richardson, her future husband, who was determining the structure of a bacterial protein, and became interested in protein structure as well. After elucidating several protein structures on her own, she concentrated on the problem of protein structure classification. In the mid-1970s, she began a systematic survey of all known protein structures and visually identified distinct patterns of β-sheets. She compared these patterns to geometric motifs common in Greek and American Indian weaving and pottery and used them as a basis for her classification system, featured on the cover of *Nature* in 1977.[98] Her work culminated a few years later in a nearly two-hundred-page review of the "anatomy and taxonomy of protein structure," which made extensive use of the data contained in the PDB. She grouped all known proteins into classes according to their structures and provided simplified representations of each to make the common features more visible. She sketched a three-dimensional arrow to represent β-sheets, a representation that soon became a standard in protein science (figure 4.5).

Richardson explicitly acknowledged the debt her approach owed to natural history: "The vast accumulation of information about protein structures provides a fresh opportunity to do descriptive natural history, as though we had been presented with the tropical jungles of a totally new planet. It is in the spirit of this new natural history that we will attempt to investigate the anatomy and taxonomy of protein structures." Her taxonomic approach relied not only on the visual inspection of structure, but also on an intimate, personal—and perhaps intuitive—grasp of similarities, like that of George Gaylord Simpson and other systematists (chapter 2). She later explained that she believed in the importance of "exhaustively looking, in detail, at each beautifully quirky and illuminating piece of data with a receptive mind and eye, as opposed to the more masculine strategy of framing an initial hypothesis, writing a computer program to scan the reams of data, and obtaining an objective and quantitative answer to that one question while missing the more significant answers which are suggested only by entirely unexpected patterns in those endless details." Contrasting her own approach with that of her male colleagues, who relied heavily on computers to process and analyze data, she confessed her "love of complex primary data and what is essentially a new kind of natural history."[99] The objects Richardson classified might have been the product of experimental virtuosity rather than field collection, but

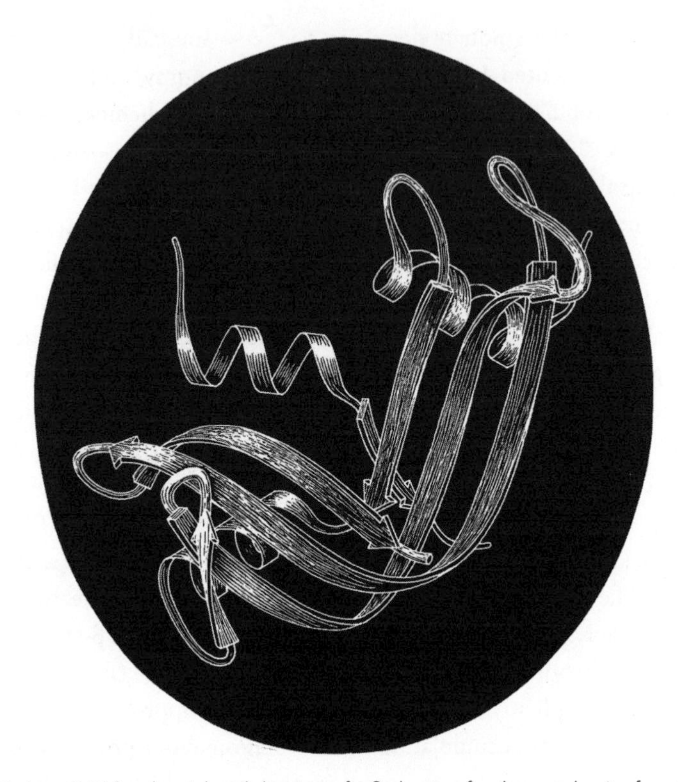

Fig. 4.5 Jane S. Richardson's hand drawing of a β-sheet, a fundamental unit of protein structure. Better than an automatic computer rendering, a hand drawing could help the viewer understand the important structural aspects of the protein. Richardson, "Anatomy and Taxonomy of Protein Structure," 168. Printed with permission of Elsevier.

the ways in which she approached them were clearly in line with the comparative way of knowing, so prevalent in the natural history tradition, even though she worked increasingly on a computer (figure 4.6).

Protein taxonomists became engaged in the same debates as those who classified organisms, among others, over the value of intuition and objectivity, of visual and numerical methods (chapter 2). They faced skepticism about the validity of using visual methods to classify proteins. As Chothia and Richardson built their taxonomies, other researchers, such as the biochemist Michael Levitt and his colleagues, developed a method that aimed to "analyze automatically and objectively" the coordinates of proteins to identify protein domains.[100] They criticized those who relied on visual inspection of three-dimensional models. This approach was "generally very satisfactory," they conceded, but "somewhat subjective, as different groups of workers can use different criteria to identify secondary structure."[101] Automated approaches were far superior, they argued, because they were "objective" and provided "the ability to process objectively and reproducibly the secondary structure of a large number of proteins."[102] Levitt's definition of "objective" as a form of "mechanical objectivity" might have been the result of contingent historical circumstances; they were widely shared and he would receive the Nobel Prize in Chemistry in 2013.[103] These concerns regarding objectivity were widespread, as numerous authors emphasized their "objective method" and criticized the "subjective definitions" of their colleagues.[104] In subsequent years, numerous algorithms were developed with the same ambition, namely, to reach the "goal of a fully objective definition of domains" that could be used for "the automatic classification" of proteins.[105]

These automatic methods were used to develop new classifications of proteins, organize the protein structures present in the PDB, and establish new databases. In 1992, using automatic methods, researchers at the European Molecular Biology Laboratory in Heidelberg, Germany, established a

Fig. 4.6 Jane S. Richardson and David Richardson virtually manipulating and discussing the structure of a protein, stored in the Protein Data Bank, on a PS300 Evans & Sutherland display system in their laboratory at Duke University, early 1980s. Jane and David Richardson personal papers. Printed with permission of Jane and David Richardson.

classification called Families of Structurally Similar Proteins (FSSP).[106] Three years later, Chothia and collaborators at the MRC Laboratory of Molecular Biology in Cambridge, UK, established a competing Structural Classification of Proteins Database (SCOP), which was based on the "visual inspection and comparison of structures" instead of algorithms.[107] Shortly afterward, Christine A. Orengo and Janet M. Thornton, at University College, London, established yet another classification, based on class (C), architecture (A), topology (T), and homologous superfamily (H), CATH, a semi-automatic hierarchical classification of proteins based on their structural domains.[108] It grouped all proteins into categories such as "clam," "sandwich" (and "distorted sandwich"), "propellor," and "barrel," according to their overall structure. The authors resorted to an "initial manual approach based on visual recognition of protein architecture, akin to the early strategies for biological classification of organisms," but only because "the ideal of a completely automatic approach" was not yet feasible.[109] While the FSSP, SCOP, and CATH systems relied on different principles (automatic, visual, and semi-automatic), the classifications of proteins they produced were, overall, very similar.[110]

These were attempts to bypass the problem Levinthal had been unsuccessful at solving: how to predict the structure of a protein from its amino acid sequence alone. None of the purely physicochemical approaches had succeeded. A number of researchers continued trying to determine "rules of thumb" by finding correlations between amino acid sequences and structural patterns. Biochemist Gerald D. Fasman and his postdoctoral colleague Peter Y. Chou surveyed the coordinates of all proteins present in the PDB in 1977. They identified hundreds of structural patterns called β-turns. Then, using information from these proteins' sequences, available from Dayhoff's *Atlas of Protein Sequence and Structure*, they developed an algorithm to predict whether β-turns would form from an amino acid sequence. The "Chou-Fasman method" became widely used and the paper describing the method a citation classic. Unknowingly, they had used the PDB in precisely the way that Berman had envisioned a decade earlier in her dissertation proposal.

Similarly, in an attempt to determine the role of certain amino acids in protein folding, Janet Thornton and Malcolm W. MacArthur surveyed all the proteins in the PDB to see if a specific amino acid (proline) played any special role. They discovered that it did; it was found overwhelmingly in β-sheets.[111] The increasing number of entries in the PDB made it conceivable that all basic structural patterns found in nature might eventually be represented. Chothia estimated that there were no more than one thousand different protein families, most of which were already represented in the PDB.[112] As in taxonomies of organisms, some families comprised numerous examples, while others had only a single member. And like naturalists who used a type specimen to define a species, protein taxonomists defined a class of structures on the basis of a single protein. Once a class contained enough proteins with slightly different sequences, it might become possible to predict which structure a protein would adopt from any given amino acid sequence, thus solving protein structures without the help of crystallographers.

Privacy, Priority, and Property

The different applications of the PDB outlined above all required a comprehensive collection of protein structures and thus rested on one precept: that crystallographers would continue to deposit new ones in the data bank. Thomas Koetzle and his team were aware that a number of newly published structures were not deposited. They made numerous attempts to encourage crystallographers to do so, including "actively soliciting depositions from laboratories thought to have coordinates available" and "asking journal editors to suggest to authors of papers describing crystallographic studies" to deposit them in the PDB.[113] In 1978, the *Journal of Biological Chemistry*, edited by Frederic Richards (a member of the first advisory board of the PDB), told authors they were "strongly encouraged to deposit" the coordinates of their structure in the PDB, as a part of its submission instructions. Some of the means of encouragement were more creative. That same year, Koetzle and his team awarded prizes to the authors of the hundredth structure deposited in the data bank and the hundredth structure distributed through it. The first received a copy of the artist M. C. Escher's *Fantasy and Symmetry* and the second a "gold plated bent-wire model" of their structure.[114]

By 1980, the PDB had grown to 145 entries. However, it was not comprehensive; at least 60 other structures described in the literature were still missing.[115] To make matters worse, only about a quarter of the 145 structures in the PDB included the raw data, the so-called structure factors, along with the atomic coordinates.[116] Evidently, not all crystallographers felt compelled to share the data they had produced, even though it was crucial for many of the projects relying on the PDB. Researchers were most reluctant to share their raw data (structure factors) because it could be used to challenge the structures they proposed in print and because they hoped to be able to exploit this data further and publish other papers. More generally, a structure was never really "finished," as researcher continuously "refined" their structure for years, increasing the precision of the spatial localization of each atom. Researchers were thus wary of publishing atomic coordinates that made it sound definitive, when it was just an approximation.[117] Nevertheless, the size of the data bank continued to grow. In 1983 a prize was distributed for the two hundredth data entry (due to the "tight budget," the prize was more modest—a free copy of the database).[118] The staff of the PDB was becoming increasingly concerned with the growing gap between the production and the public availability of knowledge. Obviously the communitarian ethos was not a sufficient incentive for crystallographers to share their data.

In 1981, a commission of the International Union of Crystallography (IUCr), which edited some of the main journals publishing protein structures, formulated a new policy stating that "all structural papers, including low resolution protein structures, should be equally subject to the requirement of deposition of atomic coordinates and lists of structure factors."[119] The commission also made clear that the "final acceptance" of the manuscript required the data to "have indeed been satisfactorily processed" by the PDB. The policy did not, however, specify how this requirement would be enforced. Less than a year later, it stepped back and relaxed the policy significantly in order not to "deprive the investigator of a hard-earned advantage." It allowed "an author who expects to be disadvantaged by having his list of structure factors made generally available to request that this list be granted a privileged status for a period no longer than four years from the date of publication."[120] In other words, the data would be submitted to the PDB but kept confidential for a period of up to four years.

This policy affected the journals edited by the IUCr, but other journals publishing crystallographic results, such as *Science* and *Nature*, lacked a similar policy. They neither required the deposition of data in the PDB nor included data in the published papers. Jenny Glusker, who shared a laboratory with Berman in Philadelphia and headed the US National Committee for Crystallography, urged the editor of the *Journal of the American Chemical Society* to at least publish the crystallographic data along with the article describing the structure.[121] The editor agreed to request coordinates, but to make them available only on the microfilm edition of the journal to which specific libraries subscribed.[122] Glusker's response was to make the matter public. In an editorial dramatically titled "Lost Data," published in *Accounts of Chemical Research*, she posed a rhetorical question: "Did you ever get excited about an article only to find that the data you needed to study the system further were 'available from the author on request' or not offered at all?" She warned against the "loss of essential data" and exhorted authors to include all critical data in their article. Failure to do so would "diminish the value" of their work and could even "force someone else to duplicate it," she warned.[123]

Berman, who had stepped back from daily activities with the PDB for almost a decade to run her own laboratory, returned to help the PDB persuade scientific journals to require the deposition of crystallographic data. By 1985, Berman was president-elect of the ACA and a respected crystallographer, having published more than sixty papers, mostly describing new molecular structures. She contacted key scientists to raise awareness of the problem of data deposition. In a letter to the editor of *Science*, Daniel E. Koshland Jr., she suggested that the journal adopt new guidelines requiring that coordinates of large molecules be deposited with the PDB.[124] But *Science* was reluctant to move ahead before other journals did the same. At the time, Koshland Jr. was rebuilding the organization of the journal in order to compete with better-standing journals, such as *Nature* and *Cell*.[125] A decision to enforce mandatory data submissions could have hampered his efforts.

Berman also shared her concerns with the NIH, where she found a more receptive audience. The program administrator of the National Institute of General Medical Sciences readily acknowledged the growing number of structures that had failed to be deposited in data banks, a worry for the NIH. The standing policy was that all "results of NIH-sponsored research should be published and made available to other scientists." Yet the policy had not been "rigidly enforced" owing to the "difficulty in deciding at what point a structure is sufficiently complete" to be submitted, a somewhat "subjective judgment."[126]

Another leading figure of American protein crystallography, Richard E. Dickerson, was thinking along similar lines. As a postdoctoral researcher in Kendrew's laboratory in Cambridge in 1958–59, he had developed the computational methods used to elucidate the structure of myoglobin, the first protein structure ever solved. Dickerson described the situation in no uncertain terms to the president of the ACA as an "intellectual scandal." Dickerson explained that a good paper should consist of data, results, and interpretations. Interpretations were based on results, which in turn derived from data. Thus, results were necessary for the reader to evaluate the soundness of the interpretations and data were necessary to evaluate the results on which the structure was built, possibly even to estimate whether another structure could be built using the same data. The fact that researchers "almost never" replicated structures determined by others was "irrelevant," the key point being that such replication should be possible for the "skeptical reader." For Dickerson, "results without data are unproven, and interpretations without results are hearsay." In "almost every" branch of science these standards were "faithfully adhered to," he stated, but not in macromolecular crystallography, where laboratories published only interpretations with few or no data. He wondered how "standards managed to fall so low in macromolecular crystallography."[127]

In support of his claims, he showed that among the 35 structures of synthetic oligonucleotides in the public domain, only 11 included coordinates and only 7 original data. Thus by "the standards normally applied in other branches of science," the remaining 28 structures were "not really published at all, in the literal sense of making the information public."[128] Dickerson thus articulated the crucial point that what made scientific knowledge published had less to do with which medium supported the publication, paper or electronic, than with whether the data, results, and interpretations were made public. It was a difficult argument to hear for journal editors, who had been accustomed to being the gatekeepers of scientific publications, and thus of almost the entire reward structure in science.

Dickerson attributed the secrecy of crystallographers "at best to inertia or laziness," or "at worst to a desire to hold back from the scientific community" their results and data "out of fear either of being found incorrect or of being scooped in some aspect of interpretation," neither of which being "a worthy motive, or one that should be allowed to prevail in science." Dickerson's argument rested on the premise that science was defined by universal norms and values (the sociologist Robert K. Merton would have agreed),[129] in particular the possibility of replicating research and the critical evaluation of interpretations. Yet, as the example of crystallographers made clear, these norms and values were not intrinsic to science but only characteristic of particular scientific communities at particular times. In macromolecular crystallography, "a custom of non-publication" had been "allowed to grow from an idiosyncrasy, to an inconvenience, to an outright scandal." For Dickerson it was now time to "put our house in order."[130]

This could be accomplished with the help of professional organizations such as the National Academy of Sciences (NAS), funding agencies such as the NIH, and scientific journals. Dickerson did not have much faith in the NAS, since it counted among its members "both conscientious adherents of data publication and blatant offenders"; membership seemed "to be no guarantee of integrity." Agencies such as the NSF and the NIH seemed more promising, because they could withhold funding from researchers who did not disclose their results upon publication. Dickerson argued not that all scientific data constituted a common property and should be disclosed, but that disclosing the scientific data on which results rested was a price to pay for the credit obtained through publication. In the end, the disclosure of scientific data and results was a matter of accountability: "if you want the credit for your research, you must be ready to put your results on the line." In this respect, scientific research was no different from politics, Dickerson argued. Evoking President Truman's famous desk sign, "The buck stops here!," Dickerson quipped that federal agencies might "want to find out just where the buck stops in any proposed research, as a prerequisite for deciding where the bucks stop."[131]

Alongside funding agencies, scientific journals had the greatest leverage to bring researchers to disclose their data. Of particular importance were journals edited by professional associations such as the IUCr and the American Society of Biological Chemists. These associations could decide to enforce a disclosure policy for authors publishing in their journals. But unless all major scientific journals adhered, the situation was unlikely to change. Indeed, as Dickerson argued, "if adherence is spotty, the less-principled protein crystallographers will simply publish in nonconforming journals."[132]

There seemed to be a consensus in the crystallographic community that making the acceptance of a manuscript for publication dependent on data sharing through the PDB was the most promising incentive to bring scientists to the table. However, the devil being in the details, enforcing this link would become a matter of intense debate within several committees. In the summer of 1987, a new committee of the IUCr restated the consensus that all journals publishing crystallographic structures "should require" that the data be deposited with the PDB "as a condition of publication by the journals," without, however, specifying how this requirement would be enforced.[133] The proposal that the authors should provide "documentary evidence" that they had deposited the data was met by vigorous opposition. A crystallographer claimed there was "virtually no support" for such a proposal, calling it unnecessarily "draconian," and maintaining that proof of deposition should be based on trust alone.[134] The committee also argued that authors should be able to request that the data be withheld "for a given period of time before public release." How long this period should be, anywhere between two to four years, provoked "hours of debate," resulting in a consensus for the longest period of four years.[135]

Another committee, led by the crystallographer Frederic Richards, played a parallel role in defining the mechanisms that would tie the publication of an article in a journal to the deposition of the data and results in a database. Richards was best known for being the third scientist, after John Kendrew (myoglobin) and David Phillips (lysozyme) to solve the structure of a protein, that of ribonuclease S in 1967, which he immediately deposited in the PDB. Bovine ribonuclease had become one of the most-studied proteins in science, not only for its biological activity, but because the meat-packing firm Armour and Company had purified a large amount of it and made it available to researchers.[136] Richards had been on the first advisory board of the PDB and was also the inventor of the "Richards box" used by all protein crystallographers. His committee drafted guidelines for authors, which he hoped journal editors would adopt. These guidelines were similar to those of the IUCr committee, except that they required authors to provide documentary evidence from the PDB that data had actually been deposited before an article could be published. In 1988, the committee's guidelines were sent to the editors of about forty journals, accompanied by the signatures of 173 supportive researchers.[137] A year later, eight journals had changed their policies and a few others were considering doing so.[138] The *Journal of Biological Chemistry,* which was still edited by Frederic Richards, immediately adopted the policy requiring proof that the data had been deposited before a paper was published. Other journals, such as *Science,* accepted authors' statements that they had deposited their data. At the other end of the spectrum, *Nature* did not have any requirements at all. John Maddox, its editor in chief, attempted to justify the journal's policy in an editorial. "It is splendid and entirely consonant with the doctrine that the scientific enterprise is a communal enterprise," Maddox wrote, noting "that data arising in the course of discovery should be generally available." Yet he claimed that "journals have no right to adjudicate upon a contributor's subsequent conduct," i.e., whether or not data was effectively deposited and made available after an article was published. Maddox concluded that if "there must be policemen, grant-making agencies are better placed."[139] Maddox's editorial drew heavy fire. In a letter to *Nature,* Nobel Prize-winning molecular biologist Richard Roberts wrote he was "appalled" by the editorial, which he believed contained a "pot-pourri of excuses for inaction."[140] In milder terms, the head of the PDB, Thomas F. Koetzle, urged *Nature* to "reconsider its policy of not requiring deposition of data."[141]

The longer history of *Nature* illuminates Maddox's position. When John Maddox became editor (for the first time) of *Nature* in 1966, he did much to speed up the reviewing process and build the reputation of *Nature* as a journal that could publish quickly. Adding hurdles to a prompt publication, such as mandatory data deposition, would have conflicted with that goal. When Maddox came back to head *Nature* in 1980, he did not shy away from using his editorial privileges, for example to publish papers against the opinion of reviewers, but apparently did not want to exercise the same authority to require data deposition from authors.[142]

Maddox's attempt to explain the position of his journal was most likely prompted by a news piece published in *Science* two weeks earlier that explored at length the problem of "the missing crystallographic data." Dickerson described the situation as "pretty sickening," while Richards renewed his call for making the release of data a condition for publication. The problem of "missing crystallographic data" was caused not only by secretive academic scientists and journals wanting to "put as few impediments" as possible "between authors and publication," as Maddox told the *Science* reporter.[143] It also resulted from the institutionally ambiguous position of the growing biotechnology industry. Biotechnology companies, such as Genentech, attracted scientists by offering working environments resembling academia, including the possibilities of gaining scientific credit through publications, yet the protection of corporate interests required certain limitations as to the type of information that could be published.[144] The vice-president of Genentech claimed that his company "believed in publication" but would not publish articles describing structures if it were forced to release its data at the same time. Richards and Dickerson were unmoved by this objection, restating that "publication means making public," and if private companies did not want to release their data, they should not publish.[145] The issue at stake with the database deposition was the very meaning of a scientific publication. The question of the exact standards for publication, and especially of the kind of data that had to be made public, defined what constitutes a scientific contribution. Given the fact that the granting of authorship through publication constituted the cornerstone of the reward system in science, the standards were no insignificant matter.

Because of the reluctance of some journals to adopt a policy of mandatory submission of data, the NIH grew increasingly concerned that it would be funding research whose results would not end up in the public domain. The NIH's National Institute of General Medical Sciences (NIGMS) had even begun to take steps to push researchers toward more openness. When a review committee of the NIGHM visited Dickerson's laboratory, the chairman asked him whether he could guarantee that he would release his "coordinates and other findings rapidly to all parties" in case he solved the structure of an

AIDS-related protein. The urgency of the AIDS epidemic and its high public visibility had made federal funding agencies particularly wary of not appearing to fund scientists whose self-interests could hinder the progress toward lifesaving drugs.[146]

Jim Cassatt, a program administrator at NIGMS, took a particularly active role in ensuring that the results of NIH-funded research would become available to the scientific community in a timely fashion. In doing so, he was implementing the NIH-wide policy requiring that the results of NIH-sponsored research be "published and made available to other scientists."[147] The NIH was particularly sensitive to the availability of crystallographic data concerning proteins related to diseases of great public concern, such as cancer or AIDS. In 1988, a group of researchers published a paper describing the structure of a protein from a cancer-causing gene. By January 1990, the coordinates were still unavailable in the PDB. When questioned about this omission, the author argued that he still needed to resolve problems with the structure before depositing the data. Other researchers resorted to the same argument, leading Cassatt to conclude that an attitude seemed to prevail where "data are good enough so that conclusions that are drawn from them can be published but not good enough to see the light of the day."[148] Shortly afterward, the NIGMS passed a resolution recommending that all grantees make their crystallographic data available within one year of publication, as recommended by the IUCr, and that funding be withheld from those who did not comply.[149]

As a result of the growing pressure from funding agencies and professional societies, by 1990 a number of journals had adopted policies mandating the sharing of crystallographic data via the PDB. However, the policies included a provision allowing authors to keep their data secret for a period of up to one year after publication for coordinates and four years for primary data. More than 75 percent of the authors depositing data took advantage of the withholding policy.[150] By the end of the decade, technical improvements had made the determination of protein structures so much faster that an increasing number of crystallographers felt that the holding period was excessive and constituted an impediment to the growth of the field. Together with several other prominent crystallographers, the new head of the PDB, Joel Sussman, proposed abolishing the hold period altogether.[151] Six months later, *Science* and *Nature* simultaneously adopted this new policy. To publish a paper based on crystallographic data, the journals required proof, provided by the PDB, that the atomic coordinates had been deposited and would be made available to the public no later than at the time of publication.[152] By this time Maddox had been replaced as *Nature*'s editor by Philip Campbell, who would become a champion of open access in scientific publishing, and the NIH followed suit.

There persisted some debate in the scientific community about whether these new policies would make industry more reluctant to share the structural data it had produced, but overall the deposition of coordinates was becoming an accepted procedure. Concerns remained about the spotty deposition of primary data (the structure factors), necessary to verify coordinates. But an increasing number of users of the PDB were not interested in them. As a crystallographer put it in 2007, "many of the current depositors might be proficient in running crystallographic programs, but have little or no understanding of crystallography as such."[153] The broader environment had dramatically changed since the 1970s, from a situation where the sharing of protein structure data was the exception to a situation where it was the norm.

A New Tool for Research

In 1998, the PDB was moved from Brookhaven to Rutgers University, after a team headed by Helen Berman, the computational biologist Philip Bourne, and the medical researcher Gary Gilliland won the NSF grant supporting the PDB, which had been open for renewal.[154] The transition occurred soon after the major journal editors had finally agreed to make deposition mandatory, solving the major hurdle in the constitution of a comprehensive data collection, the initial dream of the creators of the PDB twenty-five years earlier.

The story of the PDB is historically significant for at least two reasons. First, it transformed the practices of protein structure research, making possible a convergence of the experimental and the comparative way of knowing. Second, it reaffirmed the meaning of a scientific publication as an act of making data accessible to the public.

The determination of a protein structure has long been considered an illustration of the exceptional power of the experimental method in unlocking the secrets of nature. The elucidation of the first protein structures was rewarded by the attribution of Nobel Prizes, and subsequent structures were celebrated as major scientific achievements. These successes also played an essential role in the formation of the discipline and institutionalization of molecular biology, particularly in Cambridge, UK.[155] But there is another story too that does not focus primarily on the knowledge produced through experimentation, but highlights the deep historical transition that occurred within crystallography. Until the early twentieth century, crystallography was considered part of natural history, as it dealt primarily with the collection, comparison, and classification of crystals (Reichert was a late example, chapter 2). With the development of X-ray diffraction methods beginning in 1913 through the work of William Henry Bragg and William Laurence Bragg (father and son), crystallographers disposed of a powerful experimental tool for the determination of the structures of the molecules that composed the crystal. Successfully applying this technique to the study of macromolecules such as DNA and proteins took several decades and required numerous mathematical (Patterson functions) and experimental (isomorphous replacement) developments. But once a few protein structures became known, researchers resorted again to collecting, comparing, and classifying them. In this crucial task, the Protein Data Bank became as important for crystallographers as the natural history museums were for naturalists working with organisms. The Protein Data Bank, wrote Edgar Meyer Jr., "is our museum, with models of molecules reflecting the wonders of nature and complex shapes that may be as old as life."[156]

The Protein Data Bank was used in many of the same ways as museums. It helped establish classifications of the items it contained and to identify new ones. The different taxonomies were all based on the systematic comparison of the structures deposited in the Protein Data Bank. Individual structures were also used to model homologous proteins that most likely had conserved a similar structure owing to the pressure of natural selection. By the beginning of the twenty-first century, some protein scientists were ready to acknowledge the similarity between their work and that of naturalists. In "The Natural History of Protein Domains," a review published in 2002, protein researchers drew these parallels explicitly: "For over a century zoologists have classified organisms using the Linnean system in order to provide insights into their natural history. Biologists are beginning to appreciate the benefits of hierarchical domain classification systems based on sequence, structure, and evolution. The numerous parallels between these systems suggest that domain classifications will prove to be key to our further understanding of the natural history of domain families."[157] This is not to say that structural biology, as the field came to be known, had simply returned to its natural historical origins, but rather that the comparative and the experimental ways of knowing have become intimately connected.

The development of this natural history of molecular structures was made possible by the development of new technologies, especially computers and interactive graphics. Natural history has never been simply about the collection of objects but has consistently involved the invention of specific technologies to identify, organize, and classify specimens. Techniques of note-taking, of mounting plants on loose paper sheets, of cumulatively labeling them, and of sorting them in herbarium cabinets were all essential to the rise of the natural history of plants.[158] A number of these techniques were meant to allow investigators to manipulate specimens easily, to look at them under any angle, to gather several specimens in the same viewing field, and to let the naturalist identify similarities, differences, and patterns. This is precisely what computer graphics made possible for protein molecules. From Levinthal's attempt to couple "hand and brain" on huge mainframe computers to contemporary visualization software running on smartphones, the virtual display and manipulation of molecules became an essential practice in the study of molecular structures.[159] It offered researchers ways to look at structures as if they were objects of the size of plants and animals and use their "intuition," chemical and otherwise, to understand their structures and functions. Coupled with the Protein Data Bank's extensive collection of structures, new software made it possible to systematically compare structures for taxonomic and identification purposes.

The Protein Data Bank developed as an institution for the preservation and distribution of knowledge. But above all, it became an instrument for the production of new knowledge. Alongside laboratory instruments, the Protein Data Bank and its associated software became indispensable tools in exploring the structure of nature. The possibility of manipulating protein structures on small personal computers and of accessing the Protein Data Bank through computer networks (and today the internet) allowed these research practices to be carried out in laboratories and many other places. So spaces traditionally specialized in the production of experimental knowledge came to host comparative practices as well.

Protein researchers who adopted the comparative way of knowing were also confronted with tensions similar to those faced by naturalists, the result of divergent notions about credit and authorship. Because natural historical research rested on collections that were generally the product of collective work, it was essential to delimit precisely who could claim credit and authorship over knowledge that had been produced. It was crucial that knowledge about a new species be attached to the deposition of a specimen (the "type specimen") in a publicly accessible collection. The impressive collections in museums of natural history were largely achieved due to the requirement of depositing a specimen in order to get credit for the naming of a new species. Similar rules existed in the experimental sciences, in that authors were expected to describe methods, data, results, and interpretations if they wanted to gain authorship and credit through publications. Yet what exactly constituted "data" and "results" was open for negotiation. Almost any point along the way in the long chain of transformations stretching from nature to knowledge could be considered data, results, or interpretations.[160] Protein crystallographers came to adopt a very conservative understanding of this notion and were often reluctant to share what others considered essential data (structural factors) and results (atomic coordinates). In general, only a drawing of the overall shape of the molecule and a verbal description of its essential features were required in a publication to support the claim that the structure had been solved.

The secrecy of some crystallographers might also have resulted from their background in chemistry. Academic chemists, far more than biologists or physicists, had a long tradition of collaboration with industry and the development of propriety knowledge. Unlike biologists, chemists were not attached to the idea that knowledge should be freely accessible. Since 1907 the American Chemical Society has sought significant revenues from its Chemical Abstracts Service and its scientific journals. It also led the opposition to the open access initiatives that emerged a century later (chapter 6).[161] In this context it is no small achievement that the Protein Data Bank succeeded in making protein structure data publicly available and free, unlike the Cambridge Crystallographic Data Centre, whose audience, unsurprisingly, consisted mainly of chemists.

It required considerable efforts on the part of idealistic individuals such as Helen Berman, Frederic Richards, and Richard Dickerson to convince journals editors, professional organizations, and individual crystallographers that sharing data was in their best interest. The appeal to a communitarian ideal, personal reputation, or individual trust did not suffice, so they had to persuade journal editors to tie the sharing of data to the rewards brought by authorship. Many crystallographers complied only when the disclosure of data became mandatory and enforced through documentary proof that data had been deposited in the Protein Data Bank. Science might be a collective endeavor, but it is carried out by individuals. It is understandable that scientists, being neither morally special nor uniquely virtuous, defended their individual interests.[162] Sharing data was often clearly against those interests, so the mechanisms for the mandatory submission of data can be understood as an attempt to balance an opposition between individual and collective interests.

The matter of submission policies concerned not only science and scientists but society at large. The fact that protein structure determination constituted an essential step in understanding diseases and developing new drugs made the broader public (and taxpayers) key stakeholders, and the Protein Data Bank was created at a time when science was under pressure to be of greater relevance to the public. Part of the countercultural criticism of science had dwelt on the notion that science, despite a heavy investment of resources, had insufficiently benefited society.[163] The AIDS crisis made the demands even more pressing. The pressure to make data-sharing mandatory and to keep the price of access to the Protein Data Bank at a minimum should be understood against that background. The Protein Data Bank came to exemplify disinterested science, community service, and open access. But, as this chapter has tried to show, there was nothing inevitable about this outcome; it was the result of many struggles that redefined the very meaning of scientific knowledge and the moral economy of experimental research.

5

PUBLIC DATABASES

"Almost the number of stars in the Milky Way." Through this stellar comparison, the National Institutes of Health proudly announced in 2005 that the content of their computerized collection of nucleic acid sequences called GenBank had reached one hundred billion nucleotides, the building blocks of DNA. Only five years later, it contained twice as many, and it continued to double every five years. The creation of GenBank, like that of the heavens, was no small achievement, and similarly represented a significant historical turning point.[1] In astronomy, researchers had long been familiar with the coordination of large collective research enterprises to accomplish grand challenges such as the mapping of the skies.[2] Molecular biology, on the other hand, was a "'little science' par excellence," as a researcher described it in 1980.[3] There were few precedents for a collective infrastructure based on the sharing of data through a centralized institution (chapter 3), most likely owing to the experimental sciences' strong individualist ethos. Yet within just a few years, sequence databases became common fixtures in molecular biology laboratories and essential tools for the production of knowledge. Their content was provided by tens of thousands of researchers and represented the most frequently accessed collection of experimental knowledge in the world. The debates leading to their creation—concerning the collection and distribution of data, the attribution of credit and authorship, and the proprietary nature of knowledge—illuminate the different moral economies at work in the life sciences in the late twentieth and early twenty-first centuries. They lend perspective to the recent rise of open-access publishing and data sharing in science.

It is paradoxical that sequence databases were created as public, open, and free resources precisely at a time when scientific knowledge was increasingly considered proprietary and its circulation ever more restricted by intellectual property rights. The fact that sequence databases came to embody the opposite values was historically all the more significant because they went on to serve as models for numerous other scientific databases. Furthermore, they became a weapon in the rhetorical arsenal of those spearheading a more general movement toward open access to scientific knowledge. Although sequence databases came to embody the values of open access—and this is the second paradox of this story—their creation revealed that researchers were actually reluctant to share their data. It was only when a proper reward system adapted to the individualist ethos of the experimental sciences was adopted that researchers fully participated in data collection. Finally, the development of databases contributed to the computerization of biological knowledge. It made possible—and legitimate—a new set of theoretical research practices based on the comparison and computation of sequence data. These diverse practices soon coalesced as the field of "bioinformatics," "computational biology," or "*in silico* biology," with databases at their core.

The last paradox of this story resides in the fact that computerized databases had an impact on the "wet laboratory" at least as important as on the theoretical approaches carried out in bioinformatics. At first sight, databases can be thought of as a "new, 'theoretical' way of doing biology," but their impact goes far beyond the field of bioinformatics or computational biology.[4] Sequence databases were one of the causes (rather than a consequence) of the introduction of computers in molecular biology and have become an integral part of the practice of the experimental life sciences ever since. Like previous collections, they were not mere repositories; they were tools for producing knowledge. Researchers routinely compared the sequences they had determined in their laboratories with those present in databases using sophisticated software to infer, by analogy, the function of genes or the evolutionary relationships between species. By 2019, sequences represented more than 430,000 different species, in striking contrast with the handful of model organisms on which molecular biologists had traditionally focused.[5]

It is easy to misunderstand the characterization of databases as "collections," because many would associate collections with passive, static repositories.[6] It is precisely because databases are used as *tools* for research, as we will see again in this chapter, that they can usefully be compared to other collections, such as natural history museum collections that served a similar productive function. Indeed, as the abundant literature summarized in the introduction makes clear, and as Lorraine Daston forcefully argues more generally in *Science in the Archives: Pasts, Presents, Futures*, scientific collections were never "just" repositories for the preservation or display of specimens but tools for the production of taxonomic, evolutionary, or anatomical knowledge. Precisely for this reason it is heuristic to look at the creation and use of databases, such as GenBank, in the light of the history of so many other biological collections.

GenBank and its sister databases, the European Molecular Biology Laboratory Nucleotide Sequence Data Library and the DNA Data Bank of Japan, became the central repositories for sequence data. This chapter focuses on the creation of GenBank in 1982, in part because it was the result of a competitive process that illustrates the moral tensions at work between collection and experimentation particularly well. Its especially rich archival record offers a unique window into how GenBank took shape as a public, open, and free resource. Even though the emergence of GenBank was closely interwoven with the computer revolution, some of the most significant challenges for its establishment were not technological but intellectual, social, and cultural. Clashes between different moral economies—issues of credit attribution, data access, and knowledge ownership—go a long way in explaining how GenBank acquired its particular characteristics. Again, comparing its creation with that of other collections, especially in the natural history tradition where similar issues were at stake, will help bring out the salient features of the late-twentieth-century databases.

Information Overload on the Horizon

In the sixteenth century, the expansion of European travel led to accumulations of previously unknown specimens and to the rise of natural history collections throughout the continent.[7] Collections were a practical means to bring order to the "information overload" resulting from a burgeoning diversity of new natural forms. They made possible the direct comparison of widely different organisms for the purpose of identifying individual specimens, producing general knowledge, or even ultimately making sense of the Creator's plan.[8] Finally, collections were often created by patrons or nation states as displays of power and wealth. Early modern wonder cabinets and nineteenth-century natural history museums clearly attest to this.[9]

The impetus for the creation of GenBank in 1982 was parallel to that for the founding of so many natural history collections. It was a reaction to a perceived "information overload," augmented by a new recognition of the scientific promise of the knowledge it would contain, and the potential for individual and institutional prestige that would accompany its development. In the preceding decade, key scientific and technological developments had radically transformed the intellectual landscape in the field of DNA sequences. From the determination of the first sequence of a protein, insulin, in 1953 to the 1970s, protein (not DNA) sequencing held center stage (chapter 3). The development of the automatic sequencer in 1967 resulted in an "explosive" growth in the number of known protein sequences.[10] By the end of the decade, it had reached into the hundreds.[11]

Sequencing long stretches of DNA, on the other hand, remained technically impossible until 1975.[12] That year, Frederick Sanger devised a new method that made DNA sequencing relatively easy; two years later, the American molecular biologists Allan M. Maxam and Walter Gilbert at Harvard devised a second such method. (Sanger and Gilbert received the Nobel Prize for their sequencing methods in 1980.)[13] As a result, the number of known DNA sequences also began to climb exponentially, leading to the feeling among molecular biologists that they would soon be overwhelmed by new data.[14] In 1976, fewer than ten papers reporting nucleic acid sequences were published; in 1979, more than one hundred.[15] The bulk of known sequences began to shift from proteins to DNA, and it seemed clear that the number of DNA sequences would continue to grow at an increasing rate.

One contemporary observer was particularly struck by the exponential rise in sequence data: science historian Derek J. de Solla Price. His 1963 *Little Science, Big Science* was based on the observation that scientific knowledge, as measured by the number of published papers, was growing exponentially. So when he read in *Science* that DNA sequences were accumulating at a rate of 15 percent per month, far higher than earlier estimates, he explored the matter further with Margaret Dayhoff, who acknowledged that this rise was indeed "extraordinary in the history of science."[16]

The significance of molecular sequences was also undergoing a radical transformation. Originally, they were themselves considered objects of scientific interest, and their determination represented a considerable achievement, demonstrating great experimental virtuosity. With automation, however, sequences came to be considered highly prized pieces of data that could be used to draw new biological conclusions or develop new hypotheses for experimental exploration. The greatest excitement about DNA sequences focused on the structure and function of genes. Whereas the function of a protein was almost always known before its sequence was determined, the new DNA methods produced vast amounts of data that at first seemed meaningless. However, if the sequence of a DNA fragment could be matched against another sequence from another organism, scientists could infer that they probably had similar functions, provided that they were of common evolutionary origin (homologous). The first publication of this type appeared in 1978 and indicated that the DNA sequences of two virus proteins were similar.[17] Furthermore, the comparison between many sequences could reveal a common pattern, suggesting that it might have a functional role. The discovery in 1977 that genes were often composed of segments ("introns" and "exons") and surrounded by several functional elements (such as "TATA boxes") also heightened interest in the analysis and comparison of large sets of DNA sequences.[18]

A comprehensive database seemed indispensable in making sense of the abundant new DNA sequences that were being determined. As two molecular biologists would put it soon after, "the rate limiting step in the process of nucleic acid sequencing is now shifting from data acquisition towards the organization and analysis of that data."[19] These concerns converged in March 1979 at a crucial meeting at the Rockefeller University, New York, which resulted in the first call from the scientific community for the creation of a centralized sequence database.

This meeting was convened by molecular biologists Carl W. Anderson, Robert Pollack, and Norton Zinder to "discuss ways to collect, verify and make available to the world wide scientific community nucleic acid sequence information."[20] The organizers explained the necessity of such a gathering on the basis of the "rapidly increasing rate" of DNA sequences and the "wide range of biological questions that can be asked using a sequence data base."[21] Representatives were present from the European Molecular Biology Laboratory (EMBL), the National Institutes of Health Division of Research Resources (DRR), and the National Science Foundation (NSF), which sponsored the meeting. The participants included more than thirty scientists with special expertise in the fields of computation in the life sciences, the management of biomedical databases, or molecular biology.[22]

A look at the list of participants gives a good view of not only the fields with an interest in the project but also the tools and resources available at the time. Joshua Lederberg, Nobel Prize-winning molecular biologist and president of the Rockefeller University, who opened the meeting, was best known for his discovery of bacterial sex. However, he had also vigorously promoted the use of computers and artificial intelligence in the biomedical sciences since the 1960s at Stanford, where he had founded the shared computer resource SUMEX-AIM. Another participant, chemist and computer scientist Howard S. Bilofsky from Bolt, Beranek and Newman (BBN), the company that had developed the ARPANET for the Department of Defense in 1969, was working for the PROPHET project, another shared computer resource for pharmacologists established in 1973.[23] Mathematician Michael S. Waterman and physicist Temple F. Smith were developing algorithms to analyze sequence data at Los Alamos Scientific Laboratory.[24] In the field of database management, Margaret Dayhoff, creator of the *Atlas of Protein Sequence and Structure* (chapter 3), as well as biochemist Elvin A. Kabat, who had assembled his own specialized collection of immunoglobulin sequences at the NIH in Bethesda, had significant experience with protein collections. Crystallographer Olga Kennard was maintaining the Cambridge Crystallographic Data Centre she had founded in 1965 to collect and distribute structural data on small organic molecules. Carl Anderson from the Brookhaven National Laboratory was well aware of the progress of the Protein Data Bank hosted there (although he was not directly involved), which had been collecting and distributing the atomic coordinates of protein structures since 1973 (chapter 4). Molecular biologists included such luminaries as Walter Gilbert, Richard J. Roberts, and Sydney Brenner.

In addition to heated discussion about launching a DNA database, the participants engaged in practical demonstrations of how computers could be used in a future database. Dayhoff showed how her sequence database at Georgetown University could be accessed remotely; another participant demonstrated access to the SUMEX-AIM computer facility at Stanford University. A third participant showed how sequences could be compared using an "inexpensive 'personal' computer produced by Radio Shack."[25] These technical possibilities were new to many of the experimental biologists at the meeting, who were more familiar with wet laboratory instruments such as electrophoresis equipment and ultracentrifuges than with computers and networks. At the end of the meeting, the participants concluded that a "centralized data bank" of nucleic acid sequences was "highly desirable and essential for the organized and efficient use of nucleic acid sequence information."[26]

However, a number of concerns remained unresolved.[27] First, some worried that a single centralized facility would jeopardize the collecting efforts of individual laboratories. Whereas physicists had long been familiar with the centralized facilities intrinsic to postwar big science, many biologists were reluctant to emulate them, taking pride in the smaller scale of their laboratories. It was no accident that the Protein Data Bank was hosted in Brookhaven National Laboratory, an institution devoted to physical research, rather than in an academic biomedical laboratory. Second, the participants wondered how the privacy of preliminary data could be maintained in a publicly accessible database.[28] The issue was how to protect later priority claims of those who had determined sequences. Third, they wondered how to make the content of the database available on an equitable basis without giving an unfair advantage to the laboratory that hosted the database.[29] If it were located in an academic institution, its hosts might be tempted to exploit the contents of the database before they became publicly available.

This point was forcefully made by Olga Kennard. Her experience with the Cambridge Crystallographic Data Centre gave her, alongside Dayhoff, the most experience in data collecting, and she was thus speaking authoritatively when she pointed out that the database organizers themselves must be well-recognized figures in that community in order to gain the "interest and confidence of the scientific community," which was essential for the success of data collection.[30] But at the same time, in order to allay any doubt that the organizers might appropriate the content of the database for themselves, it would be crucial that "every assurance" be given that the content of the database would be "distributed world wide" and at "a minimum cost" for individuals.[31]

Kennard's perceptive analysis pointed to an essential contradiction in the requirements for a sequence database: The collector had to be a recognized figure in the field of DNA sequences yet not display any personal interest in the data it contained. Most great natural history collectors of the past—Joseph Hooker at Kew Gardens, Alphonse de Candolle at the Geneva Botanical Garden, or George Gaylord Simpson at the American Museum of Natural History—had been keenly interested in the items they had collected and did not practice the separation of collecting and studying that Kennard saw as necessary. So as much as the participants at the 1979 Rockefeller workshop favored collaboration, preserving individual interests remained a key issue. The moral tensions between different concepts of credit attribution, data access, and knowledge ownership had a key influence on the way the debates on the establishment of a centralized database were structured. More than the legal forms of intellectual property such as patents and copyrights, it was the "informal" modes of appropriation that were the major preoccupation of the participants.

The impact of the workshop was multifaceted, but above all it made clear that there was a strong desire in the scientific community for a single, computerized, nonproprietary database.[32] Two American institutions were particularly well positioned to take the lead in developing such a facility in the United States: the National Biomedical Research Foundation (NBRF) and the Los Alamos Scientific Laboratory.[33] The distinct natures of these institutions, the very different personalities they hosted—Margaret Dayhoff at the NBRF and Walter B. Goad at Los Alamos—and their various research trajectories at the interface of computers and biology resulted in contrasting perspectives on the collection of biological data. Even though none of the key actors had any significant connection with the natural history enterprise, it is enlightening to compare their efforts at collecting sequences with other collecting enterprises in the natural history tradition. It reveals how all of these undertakings relied on similar strategies.

Margaret O. Dayhoff vs. Walter B. Goad

Margaret Dayhoff had by far the most experience in the field of sequence databases at the Rockefeller meeting (chapter 3). At the time, she was managing the largest collection of protein sequences worldwide, containing more than one hundred thousand amino acids.[34] Her collection also comprised a small number of nucleic acid sequences, essentially transfer-RNA that had been included in the *Atlas* since 1966, and she was "deeply involved" in increasing the size of her DNA collection. In 1978 she had released her first computer tape exclusively devoted to nucleic acid sequences (i.e., DNA and RNA); it contained 24,000 nucleotide residues.[35]

Even though Dayhoff had pioneered some of the early methods for sequence comparison and the construction of phylogenetic trees, increasingly complex computational methods were being developed in various places, including the Los Alamos Scientific Laboratory in New Mexico.[36] Two frequent visitors to Los Alamos, mathematician Michael Waterman and physicist Temple Smith, were present at the Rockefeller meeting and returned to New Mexico bearing the news about a projected national database. It struck a chord in the Los Alamos Theoretical Biology and Biophysics (T-10) group established in 1974 by George I. Bell, a physicist who had "converted" to the field of theoretical immunology.[37] Los Alamos had hosted a small research group devoted to medical aspects of radiation since the Manhattan Project.[38] The controversy over the effects of fallout from atmospheric nuclear testing had made radiation genetics the main focus of biological research carried out at Los Alamos during the Cold War. A number of physicists and mathematicians who had been involved in the Manhattan Project and subsequent weapon projects, including Stanislaw M. Ulam and George Bell, had decided (out of guilt, boredom, or curiosity) to turn their minds to more peaceful ends. Radiation genetics and more general themes in biophysics, theoretical biology, and computational biology represented ideal outlets for their skills.[39] The efforts to build a DNA sequence database at Los Alamos were thus a direct result of the changing research agendas of the Cold War.

One of the members of the Los Alamos Theoretical Biology and Biophysics group, theoretical physicist Walter Goad (1925–2000), became particularly interested in the prospects for the computerized sequence database outlined at the Rockefeller meeting (figure 5.1).[40] He had received his PhD in physics from Duke University in 1954 but had been a member of the national laboratory since 1950; there he would spend his entire career, eventually becoming associated with the team who developed the first thermonuclear bombs.[41] In the 1960s, he started to become interested in problems of theoretical biology, leading to yearlong stays at the University of Colorado Medical Center and the MRC Laboratory of Molecular Biology in Cambridge, UK. Since the creation of the Theoretical Biology and Biophysics group, Goad had devoted his entire time to biological problems.[42] His biological research seemed to follow no clear direction; he picked up new problems as they came along, sometimes applying the expertise in digital computers that he had gained while working on thermonuclear weapons to biological problems. Unlike Dayhoff, Goad had no experience in collecting sequences, but when he heard about the prospect of developing a national nucleic acid database, he thought that Los Alamos was the right place to host it.[43]

The projects to develop computerized collections of DNA sequences mounted by Dayhoff, Goad, and their respective teams should be examined against the backdrop of the rising use of computers to share biomedical information. If experience with data collection was unequivocally located at the National Biomedical Research Foundation, and mathematical virtuosity in sequence analysis at Los Alamos, then expertise in the interactive use of computers resided at Stanford University. Joshua Lederberg had founded the Department of Genetics there in 1958, the year he received the Nobel Prize

Fig. 5.1 Walter Goad with student at Los Alamos National Laboratory reviewing a paper printout of the GenBank database, undated, ca. 1983. Even though most data curation was performed onscreen, this group portrait emphasizes the continuing importance of printed paper. Walter Goad Papers. Printed with permission of the American Philosophical Society.

for his discovery of bacterial sex. In the immediate post-Sputnik years, his interest in the exploration of life in space and particularly on Mars led him to think about ways in which computers could replace humans to perform experiments in unmanned spacecrafts.[44] Together with the computer scientist Edward A. Feigenbaum and the chemist Carl Djerassi, Lederberg applied concepts of artificial intelligence to make computers emulate chemists' reasoning in the analysis of data.[45] The program they produced, known as DENDRAL, was created in 1965 and made available locally to researchers.[46] In order to reach a wider audience, Lederberg conceived a shared computer resource that could be accessed remotely, for example through ARPANET, the computer network sponsored by the Department of Defense, or through TYMNET, a commercial network.[47] With NIH funding from the Department of Research

Resources, his vision of a shared computer facility was realized in 1973 under the name SUMEX-AIM (Stanford University Medical Computer–Artificial Intelligence in Medicine).[48]

By the mid-1970s, Lederberg was hoping to do for molecular biology what he had done for analytic chemistry with DENDRAL. He brought in Peter Friedland, a graduate student who would later participate in the Rockefeller meeting, to develop a computer program providing "intelligent assistance to the scientist designing an experiment in a biological laboratory,"[49] a project known as MOLGEN.[50] Working in collaboration with biochemist Douglas L. Brutlag and molecular biologist Laurence H. Kedes, who was also present at the Rockefeller meeting, Friedland also developed a program for sequence analysis as a side project.[51] Along with other tools for sequence analysis and a small collection of nucleic acid sequences, the program was made available in February 1980 on the SUMEX-AIM system, under an account called GENET.[52] More than 250 users had registered by the end of the year. Participants could use the software through a user-friendly interface, exchange messages, and post on electronic bulletin boards. Although molecular biologists were considered a "computer-naive community" by the organizers of SUMEX-AIM,[53] GENET made them more familiar with electronic communication and created a sense of community among those using computers for biological research.

Another shared computer resource for the biomedical community, named PROPHET, had been developed on the opposite side of the country, in Cambridge, Massachusetts, by the private company Bolt, Beranek and Newman (BBN).[54] With PROPHET, users could remotely run specialized software on the BBN central computer much as they could on the SUMEX-AIM system at Stanford University. The BBN system was also supported by the NIH's Division of Research Resources and aimed primarily to assist pharmacologists studying the biological effects of small molecules. Some scientists used the PROPHET system for other purposes. Immunologist Elvin A. Kabat organized, analyzed, and distributed his collection of immunoglobulin sequences on this system.[55] PROPHET became accessible in 1972; five years later, approximately thirty laboratories were using it.

Europe Takes the Lead

The participants at the Rockefeller meeting recognized that the National Biomedical Research Foundation and Los Alamos Scientific Laboratory could each potentially host a nucleic acid database.[56] They also identified the European Molecular Biology Laboratory (EMBL) in Heidelberg, Germany, as a possible candidate but didn't foresee that EMBL would soon move on in creating a centralized nucleic acid database.

European researchers had also begun to collect nucleic acid sequences. The molecular biologists Kurt Stüber at the University of Cologne and Richard Grantham at the University of Lyon had assembled small collections for their personal use.[57] EMBL, however, had greater plans. In January 1980, the European laboratory announced that it was hoping to collaborate with "whatever group in the USA" would become "responsible for computer storage and analysis of nucleic acid sequences."[58] Because the Rockefeller meeting ten months earlier hadn't been followed by any indication of which institution would take the initiative in setting up a national facility, EMBL decided to take the lead and convened its own meeting on "Computing and DNA sequences" near Heidelberg in April 1980.[59] The goal of the meeting was to discuss the "use of the computer as an aid to sequence determination, . . . the utilization of data banks . . . and possible role for EMBL in these matters."[60] The agenda was thus very similar to that of the Rockefeller meeting, and was similarly aimed at positioning its hosting institution with a view to the future development of a centralized facility. Attendees included a large number of European researchers and several American scientists who had been present at the Rockefeller meeting. Like the Rockefeller gathering, the EMBL meeting ended with an agreement that a sequence database should be centralized, that it should be computerized and available free of charge, and that it was urgently needed.[61]

Crucially this time, the results of the discussions of a small group of scientists were made public. The following week, *Nature* dedicated its main editorial to "Banking DNA sequences."[62] The author reflected on the recent increase in the number of sequences that had been published and contemplated future "grandiose sequencing" projects that would include the human genome. The editorial stressed that the need for a computerized DNA sequence data bank that would make sequences "freely available" was "becoming urgent."[63] "Although number, or rather letter, crunching is no substitute for thought," the author argued, computers would be an essential aid for a sequence collection. A consensus seemed to be emerging on both sides of the Atlantic around the necessity of such an undertaking.

Only two months later, in June 1980, EMBL announced its decision to make its nucleic acid database publicly available,[64] a striking contrast to the slow pace at which the foundation for a national database was developing in the United States. The idea for a DNA library had been promoted by the British molecular biologist Ken Murray, who had just cloned the hepatitis B virus and was temporarily working at EMBL. The project was warmly received by John Kendrew, its director general.[65] EMBL had been created in 1974 under the assumption that molecular biology, like high-energy physics, would need an expensive piece of equipment that could be provided only by an international laboratory similar to CERN.[66] The prospect of hosting a nucleic acid sequence database on a centralized computer seemed thus perfectly in tune with this idea.

Yet EMBL's announcement might have been a bit premature, as the institute was not ready to distribute its collection of sequences and had nobody in charge of running a database. In October 1980, it hired Gregory H. Hamm, an American computer programmer with a bachelor's degree in biology, who began working with researchers such as Kurt Stüber and Richard Grantham to integrate their personal sequence collections. It took a considerable amount of time to resolve issues related to the format, collection, and distribution of the data, especially as EMBL sought input from the scientific community.[67] The database didn't become available until two years later, in April 1982. It was "freely available" and "open to everyone." Furthermore, in contrast to the Dayhoff protein and nucleic acid collection, it was not "subject to any restrictions on use or redistribution."[68] Since it was first announced in 1980, more than one thousand requests for information had been received, testifying to the eagerness of the scientific community to access a collection of DNA sequences.[69] Molecular biologist Allan Maxam from Harvard, who was partially responsible for the outpouring of sequences after inventing one of the two sequencing techniques, congratulated Hamm for the new library and assured him it was considered "a feather in EMBL hat." He also asked for a copy of the database, noting it would be used "strictly for academic (noncommercial) research."[70]

In the United States, Dayhoff, Goad, and their teams were preparing to compete for an eventual national contract for a DNA database. Only weeks after the Rockefeller meeting, Dayhoff had outlined a large-scale project to develop a nucleic acid sequence database and applied to the NIH for support.[71] She put great emphasis on verifying the accuracy of the data and on having sequences "certified" by several experts, including the original authors. She argued that a carefully verified collection was "more economical in the long run than a 'quick and dirty' collection," a clear allusion to other sequence collectors who didn't put the same effort into verification.[72]

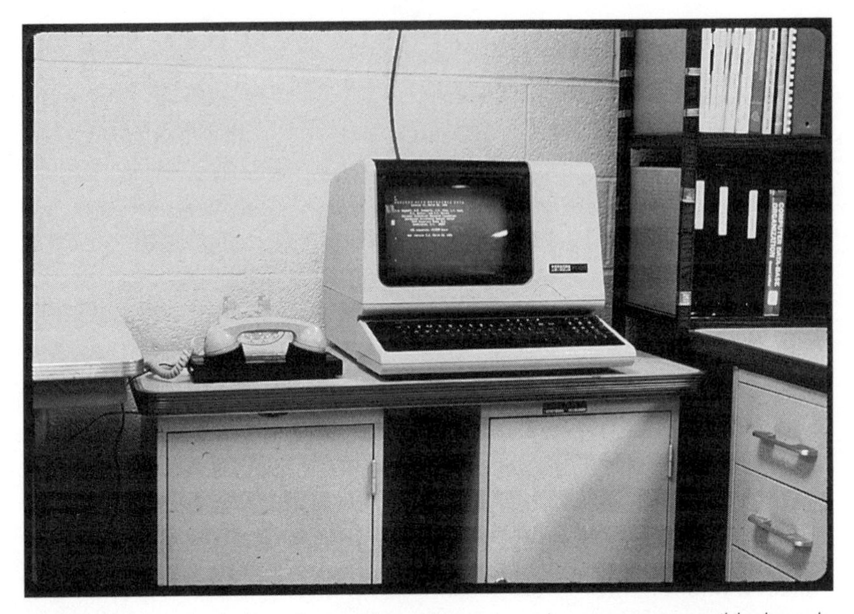

Fig. 5.2 Margaret Dayhoff's online database of nucleic acid sequences, accessible through a regular telephone connected to a modem, NBRF, 1979. Unlike previous images from the laboratory, emphasizing sophisticated computers and electronic equipment, this one highlighted the simplicity of the mundane technology necessary to access the database. NBRF Archives. Printed with permission of the NBRF.

Dayhoff simultaneously turned to NASA, a longtime sponsor of her activities, to seek funding for a "demonstration project" that would convince the NIH study committee to provide further support. This computerized database would be of crucial importance "to the NASA projects on the origins of life," Dayhoff argued.[73] Indeed, her work on the evolution of proteins such as ferredoxins or cytochromes led her to infer ancestral sequences that might have been present in the first forms of life on earth. She also approached major biotech and pharmaceutical companies for support. Access to the database would give these companies "a competitive advantage," she claimed, and she was able to receive contributions of between $5,000 and $10,000 from Genex, Merck, Eli Lilly, DuPont, Hoffman–La Roche, and Upjohn, while Pfizer Medical Systems provided "computer time."[74]

On September 15, 1980, after publishing an announcement in *Science*, Dayhoff made her nucleic acid sequence database available free over the telephone network, using a modem connected to a personal computer (figure 5.2).[75] It comprised over two hundred thousand residues and was the largest sequence database worldwide, containing more than twice the amount of data in the second-largest DNA sequence database hosted at Los Alamos Scientific Laboratory.[76] Dayhoff's DNA sequence database was modeled after her protein sequence collection and included one sequence per entry, with annotations regarding the molecule's structure and the corresponding protein's function. The database was an immediate success; in the first month of operation, over one hundred scientists requested access.[77] What Dayhoff did not mention in the published announcements was that users had to request a password from the NBRF and sign an agreement not to redistribute the data in order to access it.[78] Whenever they did so remotely, they received the following notice on their screens: "Welcome to the NAS [Nucleic Acid Sequence] Reference Data System. You are licensed to use this data for your own research. As a licensee, you are legally obliged not to redistribute the data or otherwise make it available to any other party."[79]

In letters to the NIH reporting on her progress, Dayhoff stressed that the data was being made freely available, but that benefit came at a price, namely, substantial funding through research grants.[80] Yet even NIH and NASA funding was not sufficient to make the database self-supporting. Thus only two days after making the database available at no charge, Dayhoff was engaged in negotiations with Laurence Kedes at IntelliGenetics, a small private company that had just been founded in Palo Alto by several Stanford University faculty members to sell computer software to the emerging biotechnology market.[81] Dayhoff, confessing her "immediate cash problems,"[82] asked IntelliGenetics to distribute her database commercially, thus abandoning her pledge of distribution free of charge.[83]

In June 1981, given the NIH's uncertain support and the fact that the negotiations with IntelliGenetics did not pan out (she may have withdrawn out of concern that the company might become a competitor), Dayhoff decided to sell access to the database through a subscription. Commercial users were asked to contribute anywhere between $3,000 and $10,000 and noncommercial users between $750 and $1,000 per year.[84] Despite the modest amounts, the charges made a crucial symbolic difference between a free public good and a commercial product—as had happened with her protein database. Dayhoff put this unambiguously: "[W]e have tried to get the database on a businesslike basis."[85] Indeed, when she made her database available on the SUMEX computer, she hoped that the increased visibility would help her find "new customers."[86]

The size of Dayhoff's collection increased at a rapid rate. Three months after its launch, it had grown from over 227,000 to over 340,000 residues, and eight months later to more than 500,000 residues.[87] Funding, on the other hand, remained extremely tight. In July 1981, the NIH informed Dayhoff that it would stop supporting the project.[88] In a letter to the NIH, she pointed to a direct connection between the lack of public funding and her decision to market the database on a commercial basis.[89] In the fall of 1980, Gregory Hamm, the scientist in charge of EMBL nucleic acid collection, went on a tour of the United States, visiting the main actors involved in nucleic acid data collections. He was most impressed by Dayhoff's collection: "I don't think anyone else can match [Dayhoff's collection] without several years startup time," he reported back to EMBL. But he also understood her difficulties: "The problem with Margaret Dayhoff at present is that she is extremely short of funds. . . . This has had the effect that she is extremely reluctant to allow people unrestricted access to her data since she feels she must use the data as a bargaining chip to get funds."[90] Even though Dayhoff's expertise and the quality of her *Atlas* were undisputed, the proprietary model on which she based her collecting enterprise was consistently challenged by experimentalists (chapter 3).

Over the summer following the Rockefeller meeting, Goad and Bell at Los Alamos moved ahead on the possibility of contracting for the national database, trying to convince other scientists that Los Alamos was "the natural place to locate a center for sequence analysis of DNA," primarily because of the national lab's unique "computer facility."[91] The argument that computer power was essential for the success of a sequence database would be one of the cornerstones of the Los Alamos campaign to host the central facility.[92]

Starting from almost nothing at the time of the Rockefeller meeting, Goad and a small team including computer scientist Minoru I. Kanehisa, mathematician James W. Fickett, and molecular biologist Christian Burks put a great deal of effort into creating a comprehensive database of DNA sequences.[93] In May 1980, the Los Alamos group invited its collaborators for "cake and coffee to celebrate 100,000 bases now in the DNA sequence library."[94] At the time the collection was about the size of Dayhoff's[95] but fell far behind following her intense efforts in the fall of 1980. With little experience in data collecting and no staff trained to scan the literature for published sequences, it was unclear how Goad and his team could possibly catch up with Dayhoff's rapidly growing database. Thus two days after Dayhoff's collection became available, Temple Smith, soon to be a consultant for Los Alamos, asked her for a copy of her entire collection. Smith did not hide the fact that he was to become one of her "future competitors,"[96] and Dayhoff turned him down. In January

1981, Goad boldly wrote Dayhoff again, hoping to obtain her most recent collection. After stating that access to her data had been useful in correcting errors in his own, he asked somewhat hesitantly, "I wonder if at some point you would consider the possibility of allowing your database to be resident in our files?"[97] Since his collection was just a subset of Dayhoff's, Goad did not have much to offer in exchange. Dayhoff could hardly refuse to share data that she herself had not produced, and to which she thus had little proprietary claim—yet she did.[98] Goad had more success acquiring smaller collections from other researchers by proposing an exchange for his own.[99] EMBL also asked Dayhoff to contribute her collection to be redistributed with the other data from the European laboratory's growing DNA sequence database. Once again Dayhoff refused to cooperate, unless her data would be protected by a nonredistribution clause. Gregory Hamm, who was in charge of the project at EMBL, confessed that he was "somewhat puzzled" by this response, explaining that he could not accept data into the EMBL database that was "subject to restriction, since this defeats the whole purpose of our effort."[100]

The contrasting attitudes of Dayhoff and Goad toward the ownership of data collections were already apparent in their early collecting efforts. Whereas Goad treated the Los Alamos sequence collections as free for exchange, Dayhoff considered her database as proprietary. This difference reflected alternative standards of knowledge ownership, but also resulted from the unequal status of their collections. Goad's collection was much smaller than Dayhoff's, and he was more than willing to share if she would reciprocate. In one way, however, Dayhoff and Goad were very similar: both resembled what critics of museum naturalists, who worked from their office rather than in the field, called "armchair naturalists."[101] Unlike more adventurous naturalists, who actually traveled to remote places to collect specimens, armchair naturalists remained near home and focused on the organization and display of their specimens. Their collections had been built through the acquisition of others, exchanges of specimens, or the maintenance of a network of correspondent-naturalists, rather than a personal engagement in acquiring new specimens. Neither Dayhoff nor Goad had ever sequenced a protein or a piece of DNA; they relied on the efforts of others. Goad tried to acquire sequences in bulk from other collectors, whereas Dayhoff obtained them from the literature and through daily interactions with those who had determined them in their laboratory. Because of the amount of work that went into acquiring and verifying each individual sequence, Dayhoff perhaps felt more entitled than Goad to assert proprietary claims on the sequences she had assembled in her database, just as earlier naturalists had felt about the specimens in their collections.

Mobilizing the National Institutes of Health

While Dayhoff, Goad, and EMBL were each trying to lay the groundwork for a comprehensive collection of DNA sequences, the NIH began to discuss how to address the scientific community's call for such a centralized facility. When the EMBL announced in June 1980 that it would make its database available in the near future, the NIH was still preparing for its first workshop on "the need for a nucleic acid sequence data bank" to be held the following month.[102] EMBL's declaration played no small part in pushing the NIH to take the initiative in supporting the establishment of a database in the United States.[103]

Even though there was considerable interest within the scientific community, there seemed to be "little agreement as to what kind of arrangement would best serve the field."[104] Indeed, there were many possible ways to organize a database and as many possible schemes for collecting the data, verifying its accuracy, and distributing it to the broader scientific community. The speakers at the NIH meeting presented their views on these issues and debated "often sharply" over the best format for the database and its mode of distribution.[105] By the end of the first day, however, a "consensus became evident."[106] On the second day, the participants drafted recommendations defining the needs of the scientific community. Those who were expected to compete for a possible contract were asked not to participate in these discussions. Dayhoff (NBRF), Goad (Los Alamos), and Kedes (IntelliGenetics) left the room.

"Clearly, we must act now," declared the authors of the recommendation report, after noting the exponential increase of sequence data and the decision of EMBL to establish its own database.[107] The authors insisted that the long-term goal should be not simply to constitute a collection of sequences, but to construct a "sophisticated" and structured library. Its managers should "aggressively" collect and solicit sequence data, but only data that had been accepted for publication ("i.e. refereed data").[108] The "private communications" that had been included in previous databases, such as Dayhoff's, should thus be excluded. The data was to be made available through telephone and computer networks in order to provide "interactive access to the stored data."[109] Finally, the sequences should be in "the public domain."[110] Participants thus reaffirmed the principles first outlined at the Rockefeller meeting of having a computerized and nonproprietary database. In an electronic message posted on the SUMEX system, Laurence Kedes summarized the results: "A strong endorsement for the establishment of a national nucleic acid sequence data bank was hammered out today [and] the meeting adjourned with the optimistic expectation that there would never have to be another one."[111]

This last claim was overly optimistic, and numerous other meetings soon followed to work out the details. At the NIH, Elke Jordan, a phage geneticist and deputy director of the Genetics Program at the National Institute of General Medical Sciences (NIGMS), which funded most of the basic sciences at the NIH, took the lead in organizing them and tried to convince other institutes to support a sequence database, since it would serve "scientists NIH-wide."[112] Jordan and Ruth L. Kirschstein, the director of NIGMS, eventually succeeded in convincing different institutes within the NIH (NIGMS, NIAID, NCI, and DRR) to fund the project, together with the National Science Foundation, the Department of Energy, and the Department of Defense. The participation of the latter two departments in a biomedical project is less surprising given that they were trying to diversify the research priorities of the national laboratories to include topics more directly relevant to society.[113] In December 1981, the NIH finally issued a "Request for Proposals" for the development and maintenance of a national nucleic acid sequence database containing all published sequences over fifty base pairs long.[114] Most important, the database was to be up and running within a year of the awarded contract.[115]

It took almost three years after the Rockefeller meeting for the NIH to come up with a funding scheme, and by that time EMBL had already made its own sequence database publicly available. This somewhat embarrassing delay on the part of the NIH might have resulted from bureaucratic inertia, as some critics later charged.[116] More to the point, the cautious attitude of the NIH reflected the fact that it was unclear whether the NIGMS mission should include the funding of databases at all. Its stated mission was to support experimental research of medical importance, and the maintenance of a database didn't clearly fit into that category.[117] More fundamentally, many doubted the scientific potential of a sequence collection, especially at a time of triumph for the experimental approach. As an anonymous participant complained on the electronic billboard at SUMEX-AIM, there was resistance from "within the NIH among staff who feel that molecular geneticists really do not need such a facility."[118] Funding a database was the kind of project that did not "inspire excitement," as Dayhoff had complained, reflecting the priority given to experimental work. Frederick Sanger, quoted in chapter 3, expressed this hierarchy clearly: "'Doing' for a scientist, implies doing experiments."[119] Collecting and comparing were common ways of producing knowledge in natural history but were often regarded as archaic by experimental biologists, even when these practices involved sophisticated computers.

Once the funding issue was worked out, at least three proposals were submitted to the NIH: one by the National Biomedical Research Foundation (Dayhoff), one by Los Alamos Scientific Laboratory (Goad) as a team project with Bolt, Beranek, and Newman (Bilofsky), and one by Los Alamos with IntelliGenetics (Kedes). The first two were selected by the NIH for further evaluation. These proposals and the responses by the NIH reviewers offer a window into different solutions to the challenge of data collection and the problem of data ownership in the experimental sciences. The proposals were similar in many ways, reflecting the convergence of views that had resulted from more than two years of meetings among those invested in the development of a database. However, they also reflected fundamental differences over credit attribution, data access, and knowledge ownership.

Collecting Data, Negotiating Credit and Access

As discussed in the introduction, in the comparative sciences a number of different strategies have been adopted to build collections. In their twentieth-century proposals for the DNA sequence database, Dayhoff and Goad also proposed various strategies to address this problem. Dayhoff's approach, once again, reflected her idea that published knowledge belonged to the collector, whereas Goad's was more in tune with the idea that published knowledge belonged to the community as a whole.

Even though Dayhoff already had the largest existing sequence database, she believed her collection was "a mere shadow of its ultimate grandeur."[120] To realize her vision, she planned to collect data as she had done in the past, by surveying the literature, estimating that twenty-nine journals contained more than 98 percent of all published sequences.[121] She insisted on the importance of comprehensiveness, a key value in the natural history collecting tradition, just as precision was a key value for the experimental tradition.[122] "Comprehensiveness" was an ambiguous term, however. From Dayhoff's proposal it was clear that she would give priority to long sequences (over 500 base pairs) over short sequences (50 base pairs, which represented the bottom limit required by the NIH). This decision reflected the fact that she envisaged the database mainly as a tool for research in evolutionary biology, rather than as an aid for researchers in molecular genetics.

Goad and his partner at BBN, Howard Bilofsky, envisioned collecting data in a similar way, but with one crucial difference. Apparently more sensitive than Dayhoff to the fact that experimentalists had a strong sense of ownership over their sequences, they proposed relying on collaborations with journal editors, rather than on voluntary contributions from authors or scanning the published literature alone. They stressed that coordination with journal editors "on topics ranging from electronic uploading of published sequences to standards for annotation" was essential to the success of the database.[123] They suggested a mechanism that had first been proposed at EMBL to bring authors to collaborate in the collecting effort: "We believe—and a number of journal editors have already agreed in principle—that once a national centre is established, most journals will be willing to furnish or <u>require</u> authors to furnish, a copy of the original figures, or, preferably, a computer-readable copy of each sequence, to the national data bank."[124]

They expected the electronic transmission of data between journals and the databases to become increasingly common "as computer to computer transmission grows more facile."[125] The NIH reviewers judged that the reliance on journals would be an excellent mechanism for collecting data and were confident that the Los Alamos team "should have no difficulty bringing the database up to date within the first year."[126] Conversely, the NIH reviewers criticized Dayhoff's traditional approach to data collecting, which rested essentially on (wo)manpower to scan published papers and on individual relationships with the authors of sequences. They estimated that Dayhoff had given "little thought . . . to increasing the efficiency of data collection and dissemination," which raised concerns about her capacity to meet deadlines in view of the exploding number of sequences becoming available.[127] In fact she had thought about securing the collaboration of editors but was skeptical they would cooperate.[128]

Apart from the matter of efficiency, Goad and Bilofsky appealed to a different system of values than Dayhoff in relying on the authority of scientific journals and their role in the scientific reward system. In the experimental sciences, the attribution of priority and authorship by scientific journals was a key motivation for scientists to make their data public.[129] As the developmental biologist Lewis Wolpert reflected a couple of years later: "J. B. S. Haldane is reported to have said that his great pleasure was to see his ideas widely used even though he was not credited with their discovery. That may have been fine for someone as famous and perhaps noble as Haldane, but for most scientists recognition is the reward in science."[130] Thus Dayhoff's system of data collection, similar to the one she had adopted for the *Atlas*, ran against one of the essential values of the experimental sciences' moral economy, namely, that the production of knowledge deserves individual, not collective, credit. There was no reward or incentive for researchers to share their data with the future sequence data bank.

Neither Dayhoff, Goad, nor any of the participants at the initial meetings on the national database envisioned challenging the demarcation line between public and private knowledge set by publication in a printed journal. Even Dayhoff had explicitly shied away from following this trajectory in the preface of her early *Atlas of Protein Sequence and Structure*, stating that the editors did not want to "become involved in questions of history or priority," notwithstanding the fact that they accepted unpublished data.[131]

Other databases, such as the Protein Data Bank in Brookhaven (chapter 4), had taken an even more conservative route with unpublished data in order to protect individual authors and avoid challenging the authority of journals. Data could be deposited in the database without being made available to outside users for one to four years after publication in a journal of the general conclusions derived from that data, in order to protect the authors' ability to exploit it further.[132] Similar concerns prevailed among researchers in the field of DNA sequence databases, and in the absence of a mechanism to protect the privacy of their data, they hindered the data-collecting efforts. In *Science*, a reporter put some of the resistance to a centralized database this way: "many people were uncomfortable with the prospect that sequences might become freely available before principal investigators had had time to work with them and therefore benefit from their sequencing efforts."[133]

As Olga Kennard had made clear at the Rockefeller meeting, collectors needed to be recognized figures in the communities from which they were collecting data. In this respect Dayhoff and Goad were in a weak position, because they were each personally and institutionally quite peripheral to the community from which data would be collected. The NBRF was a small nonprofit research organization that, aside from Dayhoff's theoretical work in molecular evolution, was best known for the development of computer applications for medicine, not for contributions to basic scientific research. Los Alamos had drawbacks of its own: its specific culture as a national laboratory and its association with military projects. It was best known as the home of the Manhattan Project during the Second World War and of the thermonuclear weapons project during the Cold War, a project in which Goad was personally involved.[134] The fact that Los Alamos was considered an institution with major ties to the military isolated it from the biomedical community. In a résumé sent to his superiors at Los Alamos, Goad stated that from 1950 to 1969 he had been "active in all phases of the theoretical work involved in nuclear weapons design and development, including weapon effects," and had served at the same time as "consultant for the US Air Force Foreign Weapon Evaluation Group" and on several other weapon research committees.[135] Significantly, Goad omitted all these activities from the résumé he submitted with his application to the NIH. In the wake of public criticism voiced since the mid-1960s of the involvement of science, particularly physics, in the military-industrial complex,[136] biomedical researchers had become wary of being associated with the military. The participation of the Department of Defense (DOD) in the database project had caused "some practitioners a degree of nervousness," as a reporter in *Science* pointed out.[137] Elliott Levinthal of the DOD's Advanced Research Projects Agency (DARPA) explained that the DOD was not interested in "chemical or biological warfare."[138] It seems unlikely that such disclaimers would have reassured molecular biologists. Having just confronted the turmoil of the recombinant DNA controversy in the aftermath of the Asilomar conference of 1975, where they drafted self-regulations, molecular biologists seemed particularly unwilling to expose themselves to another potential source of public criticism.[139] To those contemplating a national sequence database, it was unclear whether molecular biologists would collaborate fully in a project hosted at Los Alamos. It was a critical point since, as Goad had explicitly recognized in his application, the success of a sequence data bank would depend crucially on "the level of cooperation and communication the contractor establishes with the scientific community."[140]

A related aspect of the Los Alamos identity that could threaten "cooperation and communication" with the biomedical community was simply the fact that it was a national laboratory. Secrecy and security were perceived to be key elements of its culture because of its close relationships with the military and the fact that its research often related to national security interests. The NIH referees investigated the question in various oblique ways, worrying that this might constitute an obstacle to the necessary relationship of trust. They asked the Los Alamos–BBN team "Exactly what access will users have to the Cray computers at Los Alamos?"[141] Goad and Bilofsky had to confess that the Cray computers that figured so prominently in their application would be out of bounds to the general user: they were "not accessible from outside Los Alamos because of security restrictions."[142]

In the NIH application for his project, Goad was very much aware that Los Alamos was both an asset and a liability. The national laboratory was undergoing a new security partitioning, and Goad expressed his concerns about the impression it might make on biomedical researchers when they visited Los Alamos to review its application to the NIH: "I have some misgivings about being within the secured area during the first six months of 1982. We expect to be evaluated during that time for the computer-based DNA sequence resource . . . It is important that we be perceived by the molecular biology community, and particularly by our reviewers, as offering completely free and open access to the information and programs we will be collecting."[143] The site's restricted access would be all the more damaging in that, as Goad pointed out, there were "people who already feel, however unfairly, that our openness is compromised by national security programs that demand security protection." Goad wanted to "avoid anything that unnecessarily tends to reinforce that view"[144] and made every possible effort to make Los Alamos appear more civilian and less military—more open and less secret—in order to accommodate the civilian ethos of the biomedical community. As Richard J. Roberts would explain later, "biologists didn't want to be associated with a weapons lab; biologists thought they were pure, and physicists were not."[145]

Goad and Bilofsky tried to take advantage of the unique resources offered by Los Alamos to compensate for its negative cultural resonance. They advertised the powerful computers available at the national laboratory, including four Cray-1 supercomputers that made the laboratory "one of the most powerful computing centres in the world." This tremendous number-crunching capacity was indispensable for the database, the authors argued, because "the Los Alamos approach to sequence data collection" relied heavily on sophisticated computer software to verify and annotate sequences that were submitted. The Cray computers would be very useful for the curators, who could make searches through the entire data to find if a new sequence was homologous to an existing one and annotate the new entry in the database accordingly. They would also be used to search each sequence for specific patterns that were indicative of functional elements. The scientists claimed that since some of the programs needed to accomplish these tasks were "computationally quite intensive," they could only "be operated cost-effectively on the Los Alamos Cray computers."[146]

Dayhoff, on the other hand, emphasized the human expertise of her team in verifying sequences. But in computational power, Dayhoff's "modern, high speed computer" certainly could not compete with Goad's four Cray-1s, the fastest computers available in the world. The NIH reviewers found the Los Alamos computing power "impressive and unique" but didn't question whether it was truly necessary for managing a sequence database. Aligning two sequences and verifying their statistical significance could typically take several minutes to several hours on a minicomputer such as the popular PDP-11.[147] On a large computer, such as Dayhoff's DEC VAX-11/780, the same operation would take just seconds. On a supercomputer such as the Los Alamos's Cray, it would be orders of magnitude even faster, but with no practical consequences for the user.

Distributing Data, Negotiating Ownership

The NIH's review of the database proposals took place against a background of a raging public debate on the effects of patenting the techniques and living products of molecular biology. In 1980, the United States legislature passed the Bayh-Dole Act, expanding universities' intellectual property rights over federally funded research. The same year, in the case *Diamond v. Chakrabarty*, the US Supreme Court ruled that living organisms could be patented, noting that a 1951 congressional report had concluded that "anything under the sun that is made by man" was patentable. Independently, in December the US Patent Office issued a patent to Stanley Cohen, Herbert Boyer, Stanford University, and the University of California, San Francisco, for the basic genetic engineering technique they had invented, a method to cut and paste DNA from various origins together.[148] Fears ran high in the scientific community that the rise of intellectual property would lead to increasingly secretive practices and hinder the production of scientific knowledge.[149] In such a context, it is unsurprising that the greatest concern for the NIH, even beyond the question of the mechanism of data collection, was the issue of copyrights on sequence data, and more generally the issue of ownership of information included in the database.

The NIH prompted both applicants to explain how they planned to obtain copyright agreements with the journals from which the sequence data would be obtained for the future database.[150] Neither applicant declared an intention to obtain copyright permissions from the journal publishers. This question was perhaps most embarrassing for Dayhoff, because she had been copyrighting her *Atlas of Protein Sequence and Structure* and her demonstration DNA database, including its electronic edition, from the outset. Reviewers implied that this practice might bring potential legal difficulties, but Dayhoff dismissed the argument by replying that in seventeen years, her copyright had never been challenged by journals.[151] Robert Ledley, director of the NBRF, had sought legal advice on the subject and was informed that the inclusion of sequences from copyrighted articles would constitute "fair use."[152]

However, the NIH reviewers pressed the matter further, questioning both applicants specifically on the subject of whether the NIH would "own all data in the database . . . regardless of whether it was collected prior to inception of the bank."[153] Goad and Bilofsky replied the most clearly, explaining that they did not intend "to assert any proprietary interest whatsoever in any data."[154] Furthermore, the Los Alamos–BBN team noted that Los Alamos had already made its database "freely available" to anyone, "without restriction on further distribution."[155] The NBRF made similar claims concerning the future database, at least for the period during which it would be supported by the NIH. The NBRF emphasized that the sequence data would be in "the public domain and available to all interested people" and that users would "be free to make whatever use they wish of the information, including redistribution."[156] The NBRF left some ambiguity, however, as to whether or not it would reclaim proprietary rights on data that had been collected before the beginning of the contract, once the contract had terminated.[157] Given that the NBRF had been running its database on a "businesslike basis," it seemed most likely that it would want to revert to this mode after the termination of an NIH contract. The Los Alamos–BBN team made sure that NIH referees would remember this point: the NBRF, they noted in their answers to the questions, had "sought revenues from sales of their database" and "prevented redistribution," including "to NIH users of the PROPHET system."[158] Goad was clearly aware that Dayhoff's "businesslike" database was handicapping her application to the NIH when he wrote to a colleague: "we seem to be developing an edge . . . as our principal competitor becomes increasingly enmeshed in proprietary arrangements."[159] Indeed, the NIH reviewers made clear their distrust of Dayhoff's standing "proprietary arrangement," which they found "not reassuring" for the future of the public database.[160]

The issue of data ownership was a legal one, involving copyright, but it was also a practical issue: namely, how would data be distributed physically? Dayhoff planned to distribute the DNA database as she had the *Atlas*, by sending out magnetic tapes and printing sequences in a book format. In addition, she proposed offering three dial-up telephone lines to the NBRF computer where the database was stored, a DEC VAX-11/780, which allowed remote computing.[161] Unsurprisingly, the Los Alamos–BBN proposal was technologically more sophisticated than Dayhoff's. Goad and Bilofsky emphasized the fact that the Los Alamos database had been "available to all scientists all over the world" through online connections to the Los Alamos computers as well as through the BBN-based PROPHET system and the Stanford-based SUMEX system, both of which were connected to national computer networks such as the ARPANET network and the commercial Telenet network (figure 5.3).[162] Goad and Bilofsky projected providing "on-line access to the Los Alamos facilities over national networks" to those contributing data and to those in charge of managing the database,[163] and they envisioned that the future would involve "extensive on-line user access" and "electronic communication and collaboration among users."[164] They also seemed well aware that microcomputers were becoming increasingly common in biomedical research since they proposed developing software to read the new 5¼-inch disks used on "the small word processing computer systems that have been installed in hundreds of laboratories around the country."[165] The NIH reviewers praised the extensive use of networks and microcomputers made in the proposal.

When prompted by the NIH reviewers to say whether she could also offer network access to the database in addition to telephone connections, Dayhoff replied that it was a costly option and that she did not see it as indispensable for the distribution of data.[166] Furthermore, unlike Los Alamos, she could not hope to use the ARPANET network, which was run by BBN for the Department of Defense, because it was restricted to institutions such as Stanford University and Los Alamos that carried out DOD-related work. Los Alamos had obtained special permission from the DOD to use ARPANET for the sequence database project.[167] Even though it was funded by the NIH, the PROPHET computing system was also controlled by BBN, open only to specific users. Without access to computer networks, Dayhoff relied on dial-up connections to the NBRF computer at Georgetown using only the telephone network. Connections could be made through small computers costing "less than $1,000."[168] Thus, if Dayhoff somewhat underrated network access, she seemed just as aware of the growing potential of microcomputing as Goad at a time when "computer anxiety was still strong in the molecular biology community."[169]

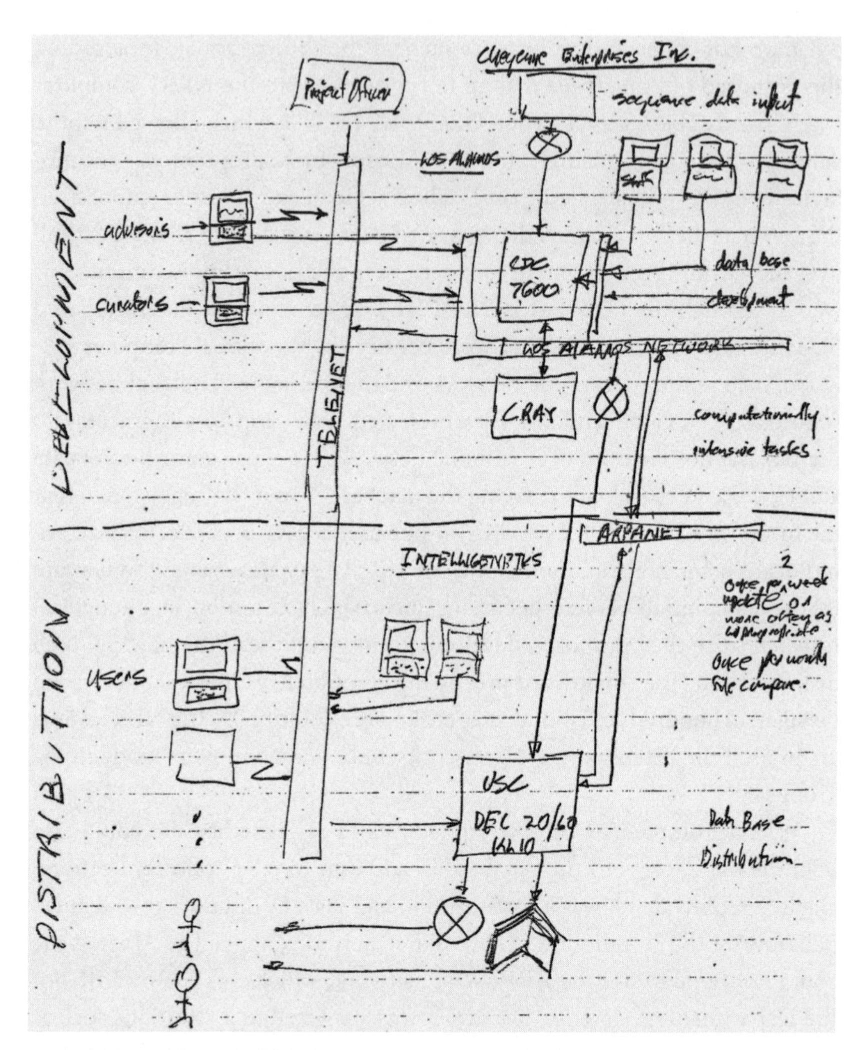

Fig. 5.3 Sketch of the data flow between users (bottom left), advisors and curators (top left), and GenBank at Los Alamos (top right), and IntelliGenetics (bottom right), undated, ca. 1981. Note the emphasis on the Cray computer at Los Alamos and the use of ARPANET to exchange data between New Mexico (Los Alamos) and California (IntelliGenetics). Walter Goad Papers. Printed with permission of the American Philosophical Society.

The problem of network access should be seen not only as a technical matter; it was also an epistemic and a cultural one. The possibility of accessing the database on growing computer networks and from microcomputers would guarantee that data could reach an increasingly broad audience and become available for wider review by the scientific community. Scientific objectivity was perceived as resting precisely on the ideal of a public review process and the open disclosure of scientific facts. As Joshua Lederberg put it in a 1978 article on "digital communication": "the claim of science to universal validity is supportable only by virtue of a strenuous commitment to global communication."[170] For Lederberg and others, computer networks were becoming central to the communication of experimental results in science, especially in view of the recent "information-explosion."[171] By the early 1980s, computer resources and networks such as SUMEX, Telenet, and ARPANET also increasingly embodied the cultural values of shared resources and free access to data, precisely the values that the NIH reviewers hoped the database would represent. Providing access to the database through computer networks became a matter of reaffirming the value of a broader participation in the production of scientific knowledge.

On June 30, 1982, at 5 p.m., three months after the proposals had been submitted, Dayhoff received a phone call from the NIH telling her that the contract had just been signed with Los Alamos and BBN, providing $3.2 million over five years to set up and maintain a nucleic acid sequence database.[172] *Science* welcomed this "long awaited" decision and, using rhetoric typical of concerns in the 1980s about American industrial decline, deplored the fact that rival facilities in Europe and Japan had beaten the NIH to the mark.[173] A few months later, in the same journal, the NIH announced that the Genetic Sequence Data Bank (GenBank) would be available to the public by October 1, 1982,[174] and started an effort to obtain a trademark on the term "GenBank," which was ultimately successful. Goad announced, as the contract required, that the database would contain all published sequences within a year. Ledley, the president of the NBRF, was staggered and thought it was "inconceivable" that his institution had lost the contract to Los Alamos and BBN,[175] while Dayhoff expressed great "surprise" at the decision[176] and privately showed "huge disappointment."[177] She had been the world's leading sequence collector for almost twenty years, at a time when data collecting was hardly considered a worthy scientific enterprise. When science funding agencies and the community finally recognized the potential of sequence databases for the production of knowledge, she lost to a physicist with little prior experience in sequence collecting. Dayhoff decided to turn her focus once again to her protein sequence collection, leaving DNA collection to the Los Alamos National Laboratory and distribution to Bolt, Beranek and Newman. She did not see GenBank develop, however, as she died of heart failure eight months after the contract was awarded.

A Conservative Revolution

The creation of GenBank reflects two major historical transformations in the experimental life sciences of the late twentieth and early twenty-first centuries: changing moral economies (the rise of open access) and changing research practices made possible by electronic databases (the rise of comparative practices). The development of GenBank, arguably the most important data collection in the experimental life sciences today, both reflected and contributed significantly to these deep historical changes, discussed further in the next chapter.

GenBank's focus on individual gene sequences was very much a product of prevalent ideas about gene action in the 1980s. The changing formats of GenBank have been interpreted as reflecting transformations in the biologists' understanding of genes and genomes.[178] The initial format of GenBank, following Dayhoff's *Atlas*, was that of a "flat-file" database, focusing on single proteins or gene sequences, at a time when these individual units served as central explanatory elements of biological processes. In the late 1980s, GenBank became a relational "database," reflecting the shift toward explanations emphasizing connections among various genetic sequences. And in the postgenomic era, GenBank became a "federated" database, including links to other kinds of biological data, reflecting an even greater integration of heterogeneous elements in biological explanations. The important point here is that database structures can be seen as reflecting particular theories about the world.[179] However, instead of seeing these structures as reflecting *successive* paradigms, it is important to recognize that databases progressively incorporate new layers (it is still possible to search for single genes in "federated" databases), offering additional ways for interacting with them and extracting data to construct novel kinds of scientific explanations. The ideal of a "federated" database was already present in Dayhoff's earliest editions of her *Atlas*, which, in addition to protein sequences, increasingly included three-dimensional protein structures and related information, such as the relation of abnormal proteins to human diseases. Recognizing that her current database was "a mere shadow of its ultimate grandeur," she would have marveled at the degree of integration and sophistication that GenBank would achieve in the internet era.

Although Dayhoff seems to have partially envisioned the course that sequence databases would take in the future, she underestimated the changing cultural sensibilities toward data ownership in her times. There are several ways to read the competition between Dayhoff and Goad and its final outcome. While gender and personality issues may have played their part, I have argued here that the key differences between the contenders related to issues of credit, access, and ownership in science. These were the major components of a shift in the moral economies of the life sciences. Dayhoff and Goad faced the complex challenge of adapting a natural history endeavor based on the collection of natural objects to the moral economy of the experimental life sciences in the late twentieth century. For Dayhoff, as for many naturalists in the past, collections and the items they contained were private property, and the collector was free to use them as commodities, gifts, or public goods. No item carried much value until it became part of a collection—an element in a system designed to preserve and produce knowledge. The relations among elements, revealed through their systematic comparison, were more valued than the elements themselves, and thus the collector could take credit for bringing them to light. Naturalists studying collections such as those in museums of natural history were entitled to appropriate the work of the numerous individuals whose contributions had filled them.[180] Dayhoff did precisely that in her *Atlas* and in numerous scientific publications that drew conclusions from the sequence data provided to her for inclusion in the database.

A very different set of norms prevailed in the experimental life sciences in the late twentieth century. There the production of knowledge rested on revealing singular facts of nature in the laboratory. The experimentalist community considered the elucidation of the structure and function of molecules a key intellectual achievement that deserved credit and authorship, and the experimental scientists who succeeded felt a sense of ownership over the knowledge they had produced. The fact that so many Nobel Prizes have been awarded for the determination of molecular structures and functions indicates that these were considered major individual scientific accomplishments. Goad built his vision of a sequence database on these premises. He laid no claim to ownership over the data it would contain and made it as widely accessible as possible, eventually taking advantage of increasingly globalized computer networks. He also stayed clear of exploiting the database's scientific content, leaving that to experimentalists who had determined the sequences and the emerging community of computational biologists who would soon rally under the banner of "bioinformatics." In doing so, he successfully adapted the requirement of the comparative way of knowing—to collect and compare objects of nature—to the existing moral economy of the experimental sciences. Goad took into account the growing resistance of some academic scientists and science administrators to the appropriation of biological knowledge and their corresponding efforts to make it publicly available. At the same time, he was keenly aware that the production of knowledge deserved individual, not collective, recognition in the reward system of the experimental sciences. Following the proposal made by EMBL, he astutely suggested that the database rely on the authority of journal editors, whose power to attribute authorship would compel researchers to share the knowledge they had produced.

In addition to these issues of credit and access, the problem of ownership also defined the debates over the creation of GenBank. In the early 1980s, under Ronald Reagan's business-friendly administration, powerful forces were at work to make scientific knowledge more relevant to the US economy.[181] It might seem surprising that in such a context the NIH and leading molecular biologists so strongly resisted proprietary models for a sequence database. This can largely be explained by the personal and professional commitments of some of the most influential figures involved in building it. When prompted to say whose advice she had taken during the establishment of GenBank, NIGMS director Ruth Kirschstein replied without hesitation: "Rich Roberts."[182] Roberts had been an unusually strong advocate of sharing data and research materials. He had established his own collection of restriction enzymes that he distributed freely to the scientific community, an uncommonly generous practice.[183] In the following years, he would become one of the most vocal advocates of open access publishing.[184] Also influential was the computer science background of some of the reviewers chosen by the NIH to examine the database proposals. Having emerged from the counterculture movement, many computer scientists resisted commercial appropriations of knowledge and valued the sharing of computer codes.[185] Most likely, they saw the genetic code as parallel to computer code, and thus as a resource to be made freely available to others for the greatest benefit of the community.[186] But implementing the ideal of open access to experimental data proved to be far more difficult than envisioned by computer scientists and theoretical physicists involved in developing GenBank, as the next chapter shows. Moral economies are so deeply embedded in professional identities and in social and cultural practices that they offer tremendous resistance to change.

6
OPEN SCIENCE

The award of the sequence database contract to Los Alamos came as a surprise to quite a few observers. Many expected Dayhoff to win the contract and were puzzled that an institution like Los Alamos, seen as a "military" institution without a particularly impressive track record in biological research, would obtain a better score.[1] These factors increased the uncertainty that experimentalists would use or contribute sequences to the database, whose success was predicated on their participation.

This point was made in a series of "Guidelines for Development of Biology Data Banks" issued in 1981 by the National Library of Medicine and the Federation of American Societies for Experimental Biology, which warned that "the success or failure of a data bank is intimately related to data acquisition."[2] Margaret Dayhoff was particularly aware of this problem and attempted to bring it to the attention of *Science*'s editor in chief: "Funding databases is difficult. People in influential places persist in thinking and acting on the presupposition that good data collections are cheap and easy to make, that perfect data comes drifting in if you just announce that you are collecting, that things organize themselves and present themselves in a useful format. All you really need is a secretary."[3]

Walter Goad, who had little experience in collecting data, other than by integrating other people's collections, almost drowned under the deluge of sequence data that was being produced in the course of the 1980s. The Los Alamos team quickly realized that it was unable to process data efficiently for its database, and the development of GenBank turned out to be far more difficult than Goad and his team had envisioned. Instead of meeting the target of a year for keeping up-to-date with all published sequences, as they had promised and as the NIH contract required, GenBank lagged increasingly behind the rising number of published sequences even though the project had teamed up with its European (EMBL) and Japanese (DDBJ) partners to collect sequences from their respective geographical areas.[4] The success of GenBank, and more generally of any community database, hinged on finding a solution to the problem of data acquisition. And if the example of crystallographers was any indication (chapter 4), that would not be easy.

Databases, Journals, and the Gatekeepers of Scientific Knowledge

From the outset, sequence databases were intended to be tools for scientific research as well as archives of the scientific literature. Their content would exactly mirror the sequences dispersed throughout the published literature printed in journals. The availability of sequences in a computer-readable medium, rather than printed pages, would open up a range of possibilities for new modes of analysis. But overall, databases did not upset the primary locus of scientific authority; publication in a journal remained the source of legitimacy of scientific data. Journals decided what counted as validated (i.e., peer-reviewed) scientific knowledge. As the gatekeepers of scientific publications, they also defined the norms of good practice in science, whether in experiments or data sharing among the community. Within a few years of their creation, the EMBL library and GenBank challenged these roles. They contributed to a blurring of the boundaries between journals and databases and to redefining some key norms of scientific behavior. These challenges came as a result of an unexpected explosion in the production of sequence data.

In 1983, Walter Goad estimated that the number of sequences would increase linearly at a rate of around 1 million bases per year.[5] This implied that after an initial effort of gathering all sequences published in the past, a constant amount of resources would be required for GenBank to keep up with newly published ones. The original sequence data were found in scientific journals, entered into a computer by a local typing firm, and annotated by student employees.[6] Based on this setup, GenBank predicted that by mid-summer

1983, all sequences would be in the database "within, at most, three months of publication."[7]

However, in the following year, it became increasingly clear to GenBank's advisory committee that the Los Alamos team was "unable to keep-up with the expanding literature."[8] Worse, there was a dangerously widening gap between the collected sequences and those available from the published literature. The number of published sequences was no longer growing linearly but exponentially. In an effort to speed up the collection process, the GenBank staff sent out forms to the authors of published papers imploring them to submit sequences and their description in a computer-readable format. However, only 20 percent of the authors fully complied.[9] Without the collaboration of the rest, the Los Alamos team proposed a doubling of the resources devoted to collecting and entering data.[10] This solution was well in tune with the culture of the national laboratory; in pursuing projects related to national security interests, resources were rarely perceived as a limitation. But even the additional resources for Goad's "crash effort" were insufficient to keep up with "data explosion."[11]

In addition to an increase in (wo)man power (a number of people entering data were women), the Los Alamos team considered technological solutions. Walter Goad hoped that emerging computer networks would offer an infrastructure to facilitate data collection. By making GenBank accessible through the Stanford-based SUMEX system, the NIH PROPHET, and the BIONET networks, he hoped that authors would contribute to the database electronically as well as access it that way. When the Department of Defense concluded that the development of GenBank was relevant to its own plans in biotechnology, it even allowed Los Alamos to give access to the database through the ARPANET, a network that was originally reserved to communications related to military research.[12] However, as GenBank's advisory committee reminded Goad and his team on several occasions, most molecular biologists were still "uncomfortable with computers" even as late as 1987.[13] Furthermore, those most interested in sequence comparisons and databases and therefore most likely to have access to computers were not necessarily those who produced sequences in "wet" laboratories.

Things did not seem to improve much in the following year, and pressure on GenBank was mounting from all sides. In 1986, an editorial in *Science* noted that DNA databases were "swamped" and lamented that only 19 percent of the sequences published in 1985 were publicly available in GenBank.[14] While EMBL faced the same problem, the article noted, it addressed it more promptly and was able to clear much of its own backlog. The situation was critical for the emerging biotechnology industry, which was developing software packages to analyze sequence data that were advertised in the same issue of *Science* (figure 6.1). These tools were addressed to an increasingly wide audience since, as the advertisements claimed, they required "no computer experience."[15] For the tools to fully benefit customers, companies needed up-to-date sequence collections. Of even greater concern to the GenBank staff, its contract with the NIH was up for renewal the following year, adding even more pressure to catch up the backlog.

The main problem was the inadequate pace of sequence collection, but another important factor was the time required to process data and annotate the sequences for interesting features. EMBL was using an incompatible format for database entries, which hampered cooperation. Finally, there were disagreements as to what exactly GenBank should be collecting, as "everyone has a different idea as to what constitutes a minimal entry."[16] The pace of data collection seemed unlikely to improve in the short run, so GenBank decided to start releasing unprocessed sequences.

The GenBank's advisory committee was growing increasingly impatient, perceiving the situation as an "emergency" and calling for drastic changes.[17] In January 1987, Richard J. Roberts, one of its members, called for a meeting at the "earliest possible time," as he was "most disturbed [by] the failure of both GenBank and EMBL to take a more vigorous role in soliciting help from the journals."[18] He advocated that journals adopt a mandatory data submission policy. Goad and his team had proposed the same approach in the initial NIH proposal five years earlier but then decided to drop it, because apparently no one wanted submission to be "made a requirement for publication."[19] Goad fell back on approaching journal editors to obtain "clean copies" of sequence data but proposed no mechanism for enforcement.[20] In 1987, the advisors were unable to reach a definitive consensus about Roberts's proposal, stating it was "unlikely that some (most) journals will ever agree" to such a mechanism because it was "contrary to most editor's policy to serve as policemen."[21] The advisors preferred "carrots" to "sticks" such as mandatory submission policies. They hoped to mount an "aggressive marketing campaign to inform authors of potential rewards for the submission of data" such as a "free analysis of his/her data including a complete search against the whole database." They quickly dropped the idea when potential costs were taken into account. Another suggestion "carrot" was a "slick computer program" that would promptly reward submitters by producing "a graphics display" of

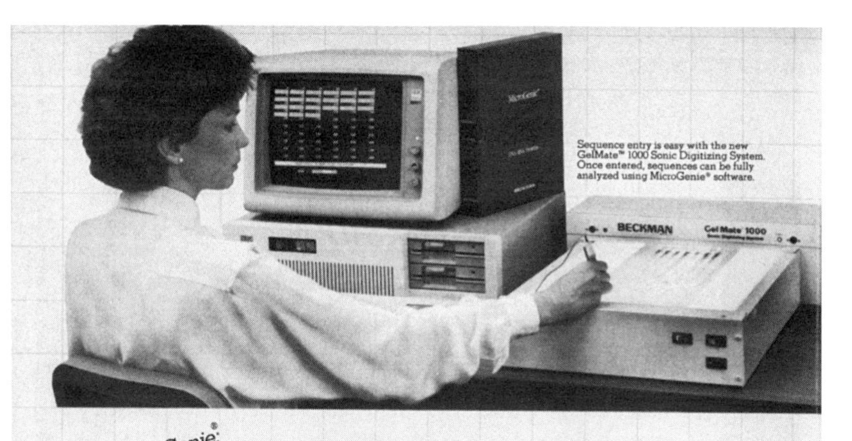

Fig. 6.1 Advertisement, running in 1984, for Beckman's MicroGenie® software analysis package: "with MicroGenie and an IBM PC/AT, you can compare a 1000-nucleotide sequence against the entire 7 million nucleotides of GenBank in just 40 minutes!" The advertisement relies on the common visual strategy of depicting a woman dressed like a secretary sitting next to a piece of scientific equipment to emphasize the simplicity of its use. *Science* 234, no. 4776 (Oct 31 1986): 548.

their data. In the end, the board concluded that the problem of data collection would solve itself spontaneously: "as the database gets better, i.e. is more up to date and accurate, more scientists will participate."[22]

EMBL, on the other hand, had been pursuing the matter differently, even before the creation of GenBank. In March 1982, it had asked journals if they would be willing to transmit sequence data automatically to EMBL when a paper was accepted and make the authors' permission a condition for publication of their manuscript.[23] Initial responses from journals were not encouraging. The editor of *Biochimica et Biophysica Acta* flatly refused to make the acceptance of an article on any other basis than its "scientific value."[24] As a scientist explained, editors "don't like being told what to do and they aren't looking for extra work."[25] Journals were also reluctant to add any hurdle that could delay the publication of articles, as speed of publication was a key criterion by which authors chose to publish in a given journal.[26]

In 1984, Graham Cameron, the new database manager at EMBL, sounded the alarm by noting that there was "no possibility of achieving the 1985 completeness deadline" because at the time the database contained fewer than 40 percent of all published sequences. Some of these were available from GenBank but could not easily be retrieved owing to format incompatibilities. Cameron thus suggested a major undertaking to make the GenBank and EMBL formats compatible and to release data before the sequences were fully annotated to speed up the process.[27] At the same time, the director general of EMBL requested the help of a number of scientists working in the European laboratory to review the entries so that they would be presented in "a biologically meaningful way."[28] Indeed, for EMBL, more than for Los Alamos, the quality of annotations was expected to reflect the laboratory's reputation in the field of molecular biology.

The major threat to the collecting efforts among all the databases resulted from the fact that journals, in order to save space, were becoming increasingly hesitant to publish complete sequences. As early as 1983, *Nucleic Acids Research* warned its authors that the journal "cannot serve as a mere repository for nucleic acid sequence information; [reported DNA and RNA sequences] must shed significant new light on basic questions of structural or functional interest."[29] Thus papers were starting to appear reporting on research that had involved important sequencing efforts, but without providing all the sequence data. This created somewhat of an emergency for database managers because if the sequences were neither published nor deposited in a database, they were likely to become irremediably lost. On the other hand, the fact that editors hoped to save space in printed journals but were still expected to make all data relevant to an article available to their readers opened up an opportunity for databases to make their case for collaborations.

At this point molecular biologist Patricia Kahn, who had been working at the bench at an EMBL laboratory in Heidelberg, was recruited to the small staff of the EMBL library and took up the mission of bringing journals to participate more fully in the collecting effort. Unlike Greg Hamm and Graham Cameron, who had been hired in 1982, she was the only "card-carrying biologist" of the team.[30] In 1986, EMBL and GenBank asked journals to send a data submission form with an accession number to authors whose papers had been accepted, and to urge them to return it to the databases with sequence data and annotations.[31] Authors could submit their data through computer networks or via magnetic tapes, floppy disks, or even printed sequences sent through the mail. This time, journals were happy to go along with the scheme since it did not involve any constraints. But participation by authors still lagged behind. Two-thirds did not cooperate, and of those who did, half sent in sequences on paper instead of computer-readable media.[32] EMBL reacted swiftly to this situation and devised yet another plan with *Nucleic Acids Research*, this time pushing for a mandatory submission policy, which would require each article based on sequence data to include an accession number provided by EMBL as proof that the sequence had been deposited.[33] As Patricia Kahn noted, this was a "bold and controversial" step.[34]

Richard J. Roberts (figure 6.2), an executive editor of *Nucleic Acids Research* and member of the advisory committee of GenBank, was a strong believer in the value of data sharing in science and was instrumental in sealing the agreement.[35] He could count on the support of another executive editor, Richard T. Walker, who understood perfectly well the contradictions of data sharing among scientists. As Walker put it: "scientists would like access to everyone else's data though they do not necessarily wish to reciprocate."[36] The agreement with *Nucleic Acids Research* was also made possible by the fact that the journal was both becoming overwhelmed by the amount of data being produced and remained unwilling to publish articles discussing the significance of sequences without making them available to its readers.[37] Authors could decide if their data was to be made available immediately or held in confidence until the paper appeared in print. The editors announced to prospective authors that they would be "guinea-pigs in an experiment to try once and for all" a system capable of coping with the "expected flood of sequence data."[38] Beginning in 1988, the journal made the inclusion of an accession

Fig. 6.2 The molecular biologist Richard J. Roberts, advisor to the NIH during the establishment of GenBank, member of the GenBank advisory board, and advocate for open access, in his laboratory (note the shaker in the foreground) at Cold Spring Harbor, ca. 1974. Cold Spring Harbor Laboratory Archive. Printed with permission of CSHL Archives.

number provided by EMBL Nucleotide Sequence Data Library a requirement for the submission of a manuscript.[39] This was the first policy of its kind in the life sciences.

GenBank eventually followed the same path as EMBL. In April 1988, its advisory board reached a consensus that the "interaction with the journals to establish author submission of data must have top priority." And it recognized that it "had not approached the editors as aggressively" as it should have.[40] Board members lobbied journals including the *Proceedings of the National Academy of Sciences* (*PNAS*) and the *Journal of Biological Chemistry* to reach agreements similar to that made with *Nucleic Acids Research*. Another favorable factor in the negotiations was the slowly changing status of sequences. The NIH and the DOE had begun funding several large-scale sequencing projects as part of the Human Genome Initiative the previous year. The increasing likelihood that the entire human genome would eventually be determined made individual and partial sequencing efforts seem less and less relevant. As Richard Walker noted in 1988, "In the future, this is going to be technician's work."[41] The amount of data that would be produced, often without precise knowledge of the biological role of a sequence, promised to change the epistemic status of the data from prized pieces of knowledge for which scientists would claim authorship to simple data, often determined by machines. Protein sequences had undergone the same transition two decades earlier.

But even the decreasing value of individual sequences did not mean that journals were ready to police submissions on behalf of databases. Instead of enforcing a mandatory submission policy, *PNAS* decided to rely on self-enforcement by the scientific community. Although it requested an accession number from authors, its editors decided not to reject papers that arrived without one. Instead, it would add a footnote to readers exposing "the author's lack of cooperation." Authors' concern for their reputations, the editors hoped, would do the rest. In addition, because "scientists who generate sequences . . . are also the users of sequences," the chairman of the editorial board argued, "self-interest should . . . dictate compliance."[42]

By the end of the decade, a number of journals had, often reluctantly, adopted some form of mandatory submission policy. But they didn't always enforce them, as the example of *PNAS* shows. These policies, however, had an immediate and dramatic effect: in 1990, 75 percent of all data submitted to GenBank came directly from authors. Journals such as *Nature* kept opposing any mandatory submission policy, and its editor in chief, John Maddox, encouraged other journals to resist "being turned into instruments of law-enforcement."[43] Maddox did not shy away from enforcing his scientific views with a strong editorial hand during the controversies over homeopathy and cold fusion and believed that journals should play an important role in shaping science.[44] But *Nature* was becoming increasingly isolated, even if authors also seemed to have some misgivings about having their data released too quickly: almost 50 percent of those who submitted data to GenBank asked that they be kept confidential until their papers appeared in print.[45] Overall, the efforts of EMBL and GenBank brought enough journals to adopt submission policies, essentially solving the problem of data collection for sequence databases.

Although it might seem that databases had finally obtained what they wanted from journal editors, it should be clear that journals were also able to preserve what they cherished most: their unique role as gatekeepers of scientific knowledge. Indeed, database managers pushed for a much more radical solution to the problem of data collection, where they would become the equivalent of journals and be able to grant authorship, although for data deposition, and thus credit. Although this scheme was never successfully implemented, it had (and still has) the potential to transform the landscape of scholarly publication.

EMBL and GenBank staffs contemplated this scheme, but it was most vigorously pursued by scientists working at Los Alamos, including molecular biologist Christian Burks and computer scientist James Fickett. Instead of having databases reflect the content of journals, and thus depend on them for the attribution of priority, authorship, and credit, they proposed that databases take up these roles themselves. After all, the curators of databases were already evaluating the scientific merits of the submitted data and could thus legitimately adjudicate the author's claim to have produced valid scientific data. Journal reviewers, on the other hand, rarely evaluated sequence data,[46] because they lacked the computer tools to do so. If database entries came to be considered equivalent to publications, they could be considered "quotable in a curriculum vitae,"[47] or cited by others, thus providing a form of recognition for the sequencing work and an incentive to submit data. The head of the NIH informatics committee predicted that "researchers will increasingly include in their curriculum vitae accession numbers given to them when they deposit data in the databanks."[48]

In the midst of the fight to convince journals to collaborate with databases, EMBL director Lennart Philipson explained that their "entire policy is to work against publication in regular journals" and that they were "gradually convincing the authors that an accession number to the databanks is corresponding to a publication." He added that "before the year 2000," "computer publication" would become common.[49] Seizing upon this idea, in 1991 GenBank collaborators proposed the notion of an "electronic publication."[50] The only difference between a journal and a database would be the medium: journals were printed and databases were electronic. Although there was no strict interdependence between the two factors, the development of computer networks such as the internet tilted the playing field in favor of electronic databases. As the authors noted, "in recent years, many people in the scientific community have become accustomed to participating in global 'conversations' as they unfold on various electronic bulletin boards around the Internet," and "computer networks have the potential to radically alter the way in which people access information."[51] If scientists increasingly came to rely on computer networks to access information, as the GenBank researchers predicted, they would likely turn to databases, instead of printed journals, as a source of data. When scientific journals also became available in electronic format in the 1990s, the playing field began tilting back in favor of journals. Even today, although journals and databases have converged, they haven't merged.[52] So far, journals have been able to maintain their role as gatekeepers of scientific knowledge.

Databases and the Production of Experimental Knowledge

Researchers did not wait for sequence databases to be comprehensive or to have solved the problem of data collection to begin using them; various communities of scientists became enthusiastic users as soon as these databases became available. First to seize the opportunity were computer scientists who developed new algorithms and software to analyze and manipulate the data. New programs were developed to compare sequences (and offer a measurement of their similarity), to discover patterns within them, or simply to discover whether the nucleotides in a sequence came in a random succession or had some sort of order. The last task was crucial in discovering the locations of genes; it required the development of a statistical benchmark, derived from many known genes and then applied to new sequences. According to the author of such a new computer method to identify genes, it was "completely objective" and did "not depend on the subjective evaluation of results by the user."[53] In 1982, *Nucleic Acids Research* began a series of special issues describing the various programs. Most ran on large computers, but others were increasingly being developed for "personal" computers, such as the IBM PC or the Apple II.[54]

A second group of researchers was less interested in the development of computer tools than in their use to discover new biological facts by conducting broad comparisons, often across many organisms, in the sequence databases.[55] For these scientists, a sequence database was "an entity onto [*sic*] itself," and they "used the database just as one would use a library."[56] Researchers looked for short patterns located just "upstream" of gene sequences that might play a role in their regulation. They also compared the frequency of the different nucleotides in various organisms and found it to vary specifically. The widening availability of databases in the 1980s fueled the growth of both approaches, and the practitioners began describing their field as "bioinformatics."[57]

Laboratory scientists involved in experimental research made up a third set of users. One of the goals of GenBank, as Christian Burks put it in 1985, was to "facilitate and extend experimental work."[58] Experimentalists could use databases to easily identify the portions of a chromosome that had already been sequenced before embarking on a lengthy project that might be redundant. And using a sequence close to an unknown region, obtained from a database, they could build a molecular probe to begin "walking" down the chromosome in the direction of their target.

The most important advantage that electronic sequence databases offered to experimentalists over a dispersed set of printed sequences was in the identification of gene functions (chapter 3). Unlike proteins, DNA sequences were determined before the exact function was known. With a computerized DNA sequence collection, researcher could compare their new DNA sequences with all other sequences available. If they found a "match," i.e., a sequence that was sufficiently similar and with a known function, they could infer that the two sequences produced proteins of similar structure, and probably similar function, which could then be further explored experimentally.[59] Databases offered a unique shortcut for experimental investigations. Although journals first accepted sequence comparisons as sufficient evidence to warrant a publication, they soon required that the proposed functions be confirmed experimentally. Even so, sequence comparison remained a crucial step in the process of producing knowledge by experimental means.

The importance of similarity searches for experimentalists triggered the very active development of better and faster algorithms in the 1980s. One of the first alignment algorithms had been developed by the biochemists Saul B. Needleman and Christian D. Wunsch at Northwestern University in 1970 and generalized by the physicist Temple Smith and the mathematician Michael Waterman at Los Alamos in 1981.[60] The problem with such algorithms was that they required a great deal of computing time. A comparison of all vertebrate sequences present in GenBank in 1982 required almost three hours on one of the Los Alamos CRAY computers, the fastest in the world.[61] As the content of databases became larger and researchers began using slower personal computers, numerous new algorithms and programs were developed to speed up the process. The medical researcher and computational biologist David J. Lipman, who would become the first director of the NIH's National Center for Biotechnology Information (NCBI), played a particularly important role in developing algorithms such as FASTA and BLAST.[62] They produced large gains in speed at the expense of some degree of accuracy. They made possible, in 1986, searches of the entire GenBank database for homologies of a 654-base sequence in just 61 minutes, using an IBM personal computer.[63]

Increasing the speed of searches was often a practical necessity, but it was also part of a "culture of speed" that was becoming prevalent in molecular biology.[64] As researchers began accessing databases remotely, rather than on personal computers, they increasingly performed searches remotely as well, on the fast computers provided by the institutions distributing the databases. By 1991, the average user with online access to GenBank was performing several searches and comparisons every day on a computer at the NIH.[65] And by 2005, two hundred thousand researchers were performing homology searches in GenBank alone every day.[66] Searching and comparing sequences in a database had become an integral part of the practice of experimental science. For most, the computer was not an alternative to the bench but an essential moment in experimental inquiry. Querying a database such as GenBank made the student of Thomas Hunt Morgan and geneticist Theodosius Dobzhansky's famous 1973 saying ("nothing in biology makes sense except in the light of evolution") more true than ever, even at the level of molecular sequences.[67]

Sequence Databases, Genomics, and Computer Networks

By 1990, the mandatory submission policies adopted by journal editors led to an unprecedented increase in the amount of data submitted to sequence databases. At the same time, large-scale efforts to sequence entire genomes were making a huge contribution to the explosion of sequence data as well. In 1995, researchers sequenced the genome of the first living organism (the bacterium *H. influenzae*), followed by the first multicellular organism (the worm *C. elegans*) in 1998, the first plant (*A. thaliana*) in 2000, and the human genome in 2003. Although a specific database, the Genome Database, was set up by the NCBI to deal with the "genome data deluge,"[68] all sequences were shared with GenBank, EMBL, and the DDBJ, leading to a tremendous increase in their size. When GenBank began operations, it contained less than one million bases. By 2010, it stored over a hundred billion, an increase of five orders of magnitude.

During the same period, the way researchers accessed sequence databases also underwent a profound transformation. The first GenBank contract required that the content of the database be released in print. The first release (1984), published jointly with EMBL, was printed as two volumes, the second (1984) as four, and the third (1986–87) took eight volumes.[69] It is telling that up to five years after *Time* magazine elected the computer as its "man of the year," the NIH still requested that the content of GenBank be printed on paper. Even though these heavy volumes were almost useless for any practical purposes, their very existence suggests that some still regarded biological databases primarily as repositories rather than as instruments for the production of knowledge. They probably thought GenBank should be to biology what the "Beilstein" was to chemistry.[70] The German organic chemist Friedrich Konrad Beilstein's *Handbuch der Organischen Chemie*, published since 1880, became the most used reference work in chemistry and was available only in print until 1994, when an electronic version was finally released. The use of these two data collections differed sharply. Whereas researchers looked up one entry at a time in the *Handbuch*, they routinely compared thousands of entries in GenBank, which could be done only electronically. Understandably, in 1987 the GenBank staff dropped the idea of providing the sequence database in printed format. It was a wise decision, as by 2010, each release would have required over twenty thousand printed volumes. The NIH's initial requirements show the extent to which the future growth of sequence data had been underestimated and its instrumental potential misunderstood.

From the beginning, the developers of sequence databases expected to handle all submissions and distributions electronically, if only to prevent the unavoidable typographic errors that crept in at every stage of transcription. Initially, GenBank sent computer tapes containing the entire database (for $65) through the mail. Researchers could return the tape with new sequences they had determined. But this system left "a lot of people out" because it was available only to researchers who had access to large computers that used this type of storage medium.[71] To improve the situation, from 1984 to 1992 the GenBank staff distributed "floppies," the computer disks used in increasingly popular "personal computers." Soon these low-capacity disks were unable to contain the entire database and thus proved inadequate for distribution. CD-ROMs were introduced in 1991 and became briefly popular but were soon abandoned for the same reason.

GenBank and EMBL staff placed great hopes in emerging computer networks as a solution to the problem of collecting and distributing data. ARPANET was being used by Los Alamos and BBN, in Boston, to exchange sequence data. The suggestion by Los Alamos that ARPANET could play a major role in the distribution of GenBank was, however, overly optimistic: Few laboratories involved in DNA sequencing or analysis had access to this network. Initially, Bolt, Beranek and Newman, which was in charge of the distribution of GenBank, offered access through telephone lines to its computer containing the database. But in 1984, only about five users accessed GenBank in that way.[72] IntelliGenetics, the Stanford-based startup company that had lost the first bid for GenBank to BBN, made GenBank and the EMBL library available to American researchers on their computer resource and to European researchers through a computer located in Paris. In 1982, thirty European research groups were already accessing the databases remotely.[73] In 1986, the database also became available from the NIH on the PROPHET system, which could also be accessed through telephone lines (slowly) or through the commercial Telenet network (expensively).[74]

To promote broader computer access among molecular biologists, the NIH's Department of Research Resources funded the BIONET project in 1984, run by IntelliGenetics.[75] BIONET aimed to "establish a community of molecular biologists who can communicate rapidly, effectively, and frequently with each other over a computer network."[76] The project was supervised by an advisory committee, most of whose members had been involved in some way in the development of GenBank: Joshua Lederberg (chairman), Richard Roberts, and Allan Maxam. Access to BIONET was restricted to academic and nonprofit users. Users could reach its fast computer (a DEC 2060) via a personal computer connected to a telephone line or a telecommunication network. They could remotely access data, such as GenBank or the EMBL sequence data library, run software to analyze the data, send electronic mail to other users, or post messages on electronic billboards. As Richard Roberts and molecular biologist Dieter Söll put it: BIONET "can provide a painless route through which local micro- and mini-computers can learn to talk to the rest of the world."[77] It was particularly suited to researchers with little experience in computing; within six months of its creation, about five hundred principal investigators had requested accounts. BIONET's rising popularity was fueled by the availability of a new network, BITNET, which was becoming widespread among academic institutions in the United States and Europe. As a researcher put it in 1987: "BITNET is how the scientific community is communicating."[78] It provided access that was much broader than ARPANET's and was much cheaper than commercial networks such as Telenet.

Sequence databases became widely accessible beginning in the second half of the 1980s thanks to the development of computer networks, but even more to the wider adoption of mini- and personal computers. GenBank advertised the circular flow of data from researchers using personal computers to GenBank and back to the researchers (figure 6.3). Even so, "wet-lab" researchers, who were providing the sequence data, took somewhat longer to adopt this new mode of communication than researchers who were analyzing it.[79] In 1984, Roberts and Söll assured the readers of *Nucleic Acids Research* that personal microcomputers were no longer "viewed as toys, but rather have assumed the importance of other, more familiar items of equipment in a molecular biology laboratory."[80] Two years later, they confirmed that "computers in all shapes and sizes" had become "an integral component of molecular biology laboratories."[81]

In 1987, the NIH renewed the GenBank contract with Los Alamos, but with a new partner, IntelliGenetics, which replaced Bolt, Beranek and Newman. With the growing success of BIONET, IntelliGenetics had demonstrated how it could effectively make sequence databases available to the entire scientific community.[82] This new mode of electronic communication allowed the producers, reviewers, publishers, curators, and users of sequences to share data almost instantaneously and at a minimal cost. Starting in 1993, users could access GenBank and the other sequence databases through the internet, which was becoming the standard way of interacting with the data bank.[83]

In 1992, the NCBI, which had been created four years earlier as part of the National Library of Medicine in Bethesda, Maryland, took over the operations of GenBank from Los Alamos.[84] The shift from a military to a biomedical institution reflected the changing fortunes of the physical and life sciences in a post–Cold War world and a loosening of some the links between physics and biology, which had been so important in the emergence of molecular biology.[85] But it also indicated that collections were no longer viewed as relics of an archaic past associated with naturalist pursuits or merely as specialized resources for a small community of biologists. They were coming into their own as essential tools in the production of knowledge in most of the

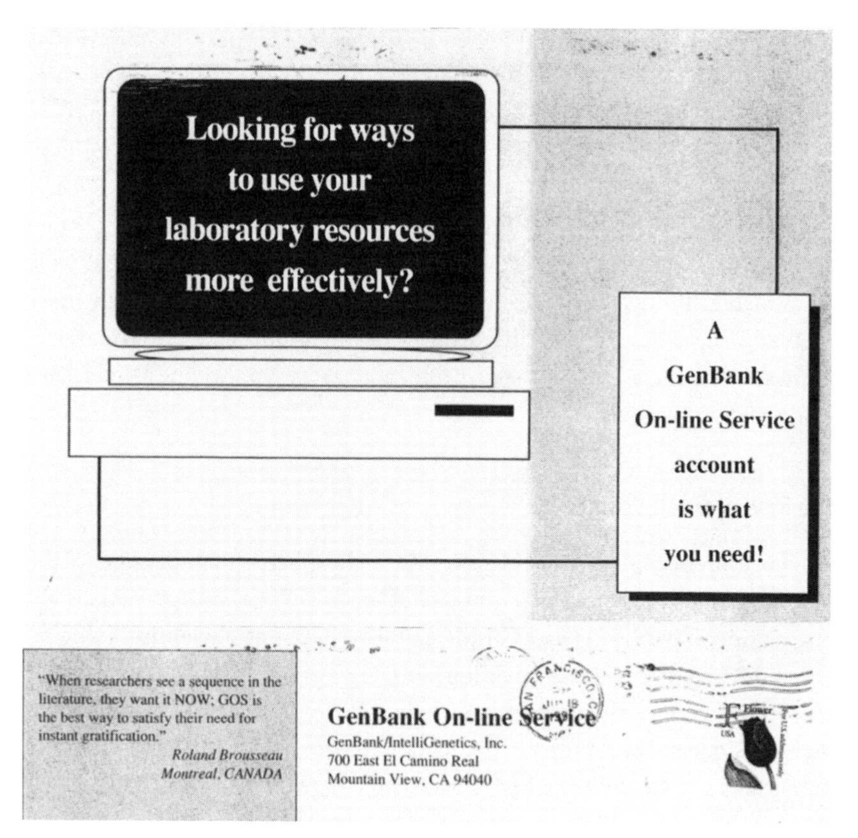

Fig. 6.3 GenBank postcard illustrating the circular flow of data between researchers producing data, the GenBank database, and researchers accessing data, July 1983. NBRF Archives. Printed with permission of Edward Feigenbaum.

experimental biomedical sciences. Under David J. Lipman's leadership, the NCBI developed new computational tools to search the database, including the algorithm BLAST.[86] It allowed researchers to make quick comparisons between a sequence and all the others in the database, leading to the discovery of possible homologues: molecules of common ancestry that usually carried out similar functions. Lipman also promoted the interconnection of a variety of molecular databases and bibliographic databases including PubMed, another product of the National Library of Medicine, which constituted the most comprehensive bibliographic databases in the life sciences.[87]

By 1995, more than one thousand computers were connecting daily to the databases offered by NCBI through the internet. Also that year, the EMBL

DNA sequence library moved to new quarters, the European Bioinformatics Institute (EBI). EBI was created in Hinxton on the site of the Sanger Center near Cambridge, UK, a major sequencing facility that was making a massive contribution to the "gene data deluge."[88] EBI brought together more than forty databases, including the Protein Information Resource (the successor of Dayhoff's *Atlas*), SwissProt (a manually curated database of protein sequences), the Protein Data Bank, and a range of software tools that were all available through the internet. As Graham Cameron noted with pride in 1995, "there is a huge awareness now of the importance of databases."[89] Things had come a long way—just fifteen years earlier, no one would have made such a statement.

The Rise of Open Science

The creation of the sequence data collections did more than just reflect the new moral economy of the experimental sciences (which had opposed the appropriation of published data) or the culture of computer scientists (who favored the free sharing of information); it also promoted a more general trend toward open access to scientific knowledge. The sequence databases began to serve as models for a number of new data collections that were similarly made public and free. In 1999, the EBI announced it would start a "publicly accessible repository" for "DNA microarray" data (data about gene expression). After just four years, scientists were predicting that the technology would soon lead to an "explosion in gene expression data,"[90] which could only be handled with a new database, considered "analogous to DDBJ/EMBL/GenBank."[91] Not only was it public and free, but it also adopted the policies of mandatory database submission that had been arranged with journal editors.[92] The same year, the NCBI established another new data collection with a focus on sequence variations (so-called single-nucleotide polymorphisms, or SNPs) "to supplement GenBank."[93] By 2010, more than 1,200 databases existed in the field of molecular studies alone—to the point that the journal *Nucleic Acids Research* created a database of databases. These collections might originally have been envisioned as a solution to the problem of data "explosion"; now their number began to "explode" too.

This proliferation of data collections had many significant consequences for the way scientific knowledge was produced (see the conclusion). It also played a major role in the general debates over data sharing in science. Beginning in the early 1980s, a growing number of scientists started to express concerns about the increasingly secretive practices of some of their colleagues and their unwillingness to share materials and data related to published articles. These changing attitudes can partly be explained by an encroaching "veil of commercialism" in the biomedical sciences, due to increasing opportunities to patent genes with lucrative potential and government pressure (especially in the United States) to do so.[94] But the lure of individual profit and the growing role of private research in the emerging biotechnology industry do not account for all the changes. Even without commercial prospects, there were plenty of reasons to be secretive within academic research. Nearly every year since 1962, Nobel Prizes have reflected the growing success of molecular biology at exposing fundamental mechanisms of life, triggering generous public funding and attracting an increasing number of young researchers to the field. Research on molecules quickly became increasingly crowded and "insanely competitive."[95] The degree of that competition seems to have varied greatly between fields; some small communities working on "new" model organisms, such as the worm *C. elegans,* seem to have been somewhat more cooperative, as the fly and the corn communities had been in their start-up phases (chapter 1).[96] Still, in most contexts—particularly human research—the temptation to withhold data was very high. This gave individual researchers more time to gain credit from new experiments based on a particular set of data. But as the molecular geneticist Philip Leder put it: "Credit is a bottomless pit—there is never enough for most people."[97] In 2002, a survey of almost two thousand geneticists in the United States concluded that nearly half had been denied "information, data, or materials regarding published research" in the last three years.[98]

Week after week in the 1990s, debates over the sharing of data and materials filled the pages of *Nature* and *Science.* By and large, the opinions expressed in public favored open access to data and research materials and condemned secretive practices of colleagues. But a few voices also called for limits. Molecular biologist Robert A. Weinberg argued that it was legitimate to withhold research materials for six to eighteen months or at least to negotiate co-authorship in exchange for material during that time period. Furthermore, he criticized calls for the sharing of raw data as misguided: without a thorough understanding of the context in which it had been produced, it could not easily be used by others.[99]

In addition to concerns about competition, many experimentalists were uncomfortable with the growing number of theoretical biologists analyzing the data they had produced. As biostatistician Marvin Zelen put it, those using data would sometimes rather brashly address those who had produced it: "Have computer—give me your data."[100] John Maddox, editor in chief of *Nature*, warned against a species of "sharpshooting theorist" who might unfairly exploit the data of experimentalists.[101]

The debates over data sharing became particularly acute as the Human Genome Project developed, because its successful outcome largely depended on the development of sequence databases. Open sharing was used to cement the international collaborations between groups that were carrying out the sequencing, to demarcate the public project from its private competitor, and to address various criticisms about the project as a whole that had arisen in the scientific community. As the idea to sequence the entire human genome gained momentum in the late 1980s, a number of scientists remained skeptical about its scientific merit. In 1987, Robert A. Weinberg mocked the project's promoters, saying he was "surprised that consenting adults have been caught in public talking about it. . . . It makes no sense."[102] One criticism was particularly damaging, that public money would be taken away from the research community as a whole for the benefit of just a few. Nobel Prize-winning molecular biologist Salvador Luria criticized it as a benefit to only a "small coterie of power-seeking enthusiasts."[103] In this context, the prompt and free availability of data became a crucial selling point, particularly the promise that the data (including sequences and physical maps of genes) would be deposited in a public database.

This public release not only helped secure support for the Human Genome Project from the broader scientific community; it also eased the dispute over the thorny issue of defining precise rules for the sharing of data and credit among the international partners involved in sequencing. After criticizing Japan's level of financial contributions to the project in 1989, James Watson, director of the NIH genome office, argued that the United States should restrict the dissemination of data in order "to reserve any commercial benefits to itself."[104] At a meeting of NIH and DOE representatives a few weeks later, Elke Jordan, who had been a leader in setting up the GenBank contract at the NIH and was now a deputy director of the genome office, insisted that any restrictions to the dissemination of data would be damaging to the international collaboration. She argued that all data should be deposited in GenBank for immediate access upon publication. She also suggested making deposition a condition for obtaining NIH grants.[105]

Over the next few years the issue kept resurfacing as an increasing numbers of centers became involved in the sequencing of model organism genomes, warming up for the Human Genome Project. In 1996, geneticist Michael Ashburner, co-director of the European Bioinformatics Institute, confessed that there was "lots of tension in the community" over the issue of data release.[106] That year, a European consortium completed the sequence of the first eukatyote genome, yeast, while announcing that it would not release its data for several months. This outraged a number of other genomic researchers, including John Sulston, who was heading the worm sequencing project at the Sanger Center. The Wellcome Trust, which funded the Sanger Center, convened a meeting in Bermuda in February 1996 to discuss the issue.[107] The participants established what came to be known as the "Bermuda Principles": "Raw" sequence data should be released immediately after its determination. The focus was on timing, but it affected sequencing centers in different ways. In Europe, making sequences public prohibited the subsequent filing of patents, whereas intellectual property laws in the United States allowed a grace period of twelve months.[108] Soon afterward, the NIH gave the Bermuda Principles political traction by adopting them and ensuring that its benefactors, the six centers in the United States involved in the Human Genome Project, agreed to release data "within a few days or weeks."[109] The NIH also discouraged its grantees from seeking patents on the sequences. By now Francis Collins had succeeded James Watson as director of the Human Genome Project at the NIH; he summarized by saying that the standards of openness reflected "the spirit and philosophy of the human genome project."[110]

But soon, in 1998, the situation became far more complex. Former NIH scientist J. Craig Venter founded Celera Genomics and basically made the completion of the human genome sequence a competition.[111] He boldly announced that his company would do it faster and at one-tenth of the price of the public consortium. His announcement threatened the political momentum of the public project. Why should science funding agencies shoulder the burden of sequencing the human genome if a private company could do it faster, without relying on taxpayer money? There was a good answer: only a public project could make the data immediately available to all free of charge and without restrictions.

The race to sequence the human genome turned into a bitter public controversy, with Collins and Venter exchanging criticisms over the respective merits of sequencing methods and the overall advance of their projects. By emphasizing openness, the advocates of the public project partially diverted general attention from these methodological criticisms.[112] In March 2000, they received political backing from the highest levels when US president

Bill Clinton and UK prime minister Tony Blair issued a statement arguing for "the rapid release of human genome data."[113] Three months later, Collins and Venter reached an agreement to simultaneously announce the completion of their respective sequences of the human genome.[114]

But by December 2000, a "storm" had erupted over the conditions under which Celera Genomics would release sequences. Venter had reached an agreement with *Science* to publish its results in the journal, but without *Science*'s usual requirement of submitting the sequence data to GenBank. Instead, Venter would make it available on the company's website. Individual researchers would be allowed to access it free, but only if they agreed not to redistribute it (echoing the practices Dayhoff had established with her database) or to use it for commercial purposes.[115] These restrictions outraged a number of researchers involved in the public project, including Michael Ashburner, who urged the community to boycott *Science*. The public consortium, which intended to publish its results in the same issue as Venter, decided to send them to *Nature*,[116] which announced it would deposit its sequence data "in the free public database GenBank."[117]

On February 15 and 16, 2001, *Nature* and *Science* published the initial results from the public and private drafts of the Human Genome Project. This disjointed publication only highlighted fundamentally different views of the principles of open access to scientific data. Of course, a company funded by private money was beyond the reach of the recommendations of the NIH, the National Academy of Science, and numerous other groups from the scientific community. It could do whatever was permitted by the law, including keeping the data secret. But what infuriated some genome scientists was the fact that Venter wanted both to restrict data access and at the same time to publish an analysis of the data in a scientific journal, and thus get "the academic kudos that goes with it," as Ashburner put it.[118]

Here again, journals stepped up as key gatekeepers of community norms. *Science* had usually been supportive of open access and published numerous editorial pieces on its merits. But its new editor in chief, Donald Kennedy, decided to make some exceptions for the human genome, arguing that at least the data was being made available—often not the case of other private sequencing efforts. Even so, the sequencing of the human genome brought the question of open access into public view and seemed to shift the consensus toward the immediate and unconditional release of data. And this position was increasingly being supported in public by scientific academies, funding agencies, and even governments.

Yet tensions over the open access to data collections keep resurfacing in the following years. As soon as the battle with Venter faded, the consensus among genome researchers around the theme of the immediate release of data became fragile and a new front opened. Once again, genome researchers feared being "scooped" by researchers doing solely theoretical work on data they had provided. In 2002, another episode in the long "string of clashes between those who collect and those who interpret data" took place. A researcher who had determined the genetic sequences of a protozoan made them publicly available, planning to publish conclusions about the organism's evolution later, only to be beaten to the punch ("scooped") by a data analyst who had accessed his data.[119] Such practices made some experimentalists consider the "computer wizards of bioinformatics" as "parasites" feeding on the work of others and taking credit for it.[120]

When the sequence of the mouse genome became available in 2001, staunch open-access supporter Francis Collins made a proposal to David Lipman, the head of the NCBI: the sequence should be deposited in GenBank, but under the condition that "no one should publish a global analysis" within six or possibly twelve months. Lipman flatly refused.[121] His decision was in line with the public statement made by the advisors of the International Nucleotide Sequence Database Collaboration (DDBJ, EMBL, and GenBank), who reaffirmed that "the success in building an immensely valuable, widely used public resource" had been made possible only by adhering to the strict policy that all the content of the databases would be freely accessible without any restrictions.[122]

The growing tensions between sequencers and bioinformaticians, including database managers, had to be resolved to move forward. The Wellcome Trust convened a meeting at Fort Lauderdale, Florida, in hopes of finding common ground. In a "late-night" session, a subgroup succeeded in reaching a consensus that reaffirmed the Bermuda Principles, namely, immediate release without restrictions, and a moral obligation on users to refrain from publishing analyses of the data before the producer had a chance to do so.[123] The view of database managers had finally prevailed. The Fort Lauderdale agreement gained support when the NIH adopted its principles as the basis for its new funding policies in 2003. It required all grantees to adopt an open access policy and explain how they would make data accessible.[124] In 2006, the Medical Research Council in the United Kingdom adopted a similar policy.

The continuing debates over data access and the growing consensus in favor of mandatory submission policies became even more widely relevant at the turn of the twenty-first century. Discussions that had begun around the sharing of scientific data were generally moving toward a broader sharing of scientific knowledge.[125] While sequence data was available free of charge from databases, published articles describing the sequences were usually available only through journal subscriptions, whose prices were skyrocketing in the 1990s.[126] Nobel Prize-winning molecular biologist Harold E. Varmus, who headed the NIH from 1993 to 1999, became one of the most vocal advocates for open access to scientific literature.[127] In 1999, he proposed an electronic archive of preprints for biomedical papers, modeled after arXiv, which Los Alamos had founded in 1991 for physical research.[128] The idea received a lukewarm reception from the scientific community. After discussing it with David Lipman, Varmus scaled the idea back to an online repository of *published* papers to be called PubMed Central (PMC) and integrated with the bibliographic database PubMed (also run by Lipman at the NCBI). It would make published articles available free six months after their publication in order not to encroach on the revenues of journal publishers.

Richard Roberts went on a campaign to support the project, christening it "the GenBank of the published literature."[129] In a 2001 editorial in *PNAS*, Roberts argued that "just as GenBank has proved invaluable to molecular biologists, PMC could serve an equally important role within the broader biological community." It would facilitate the dissemination of scientific knowledge, enable scientists to conduct searches of the entire content of articles, and thus increase scientific productivity. Roberts brushed away the idea that publishers would lose subscriptions; the articles would first be available on PubMed Central six months after their publication, and those who needed them would want them much sooner. Two months later, Roberts, Varmus, and eight other prominent scientists signed a petition published in *Science* that called for journals and scientists to participate in the project. They emphasized how a new repository of scientific knowledge could, just like sequence databases, become a powerful tool in the production of scientific knowledge: "Bringing all of the scientific literature together in a common format will encourage the development of new, more sophisticated, and valuable ways of using this information, much as GenBank has done for DNA sequences."[130] The petition called for a boycott of all scientific journals unwilling to participate. Unsurprisingly, a number of journals were hostile to the plan, including *Science*; right after publishing the letter by Roberts and his colleagues, it published a rebuttal entitled "Is a Government Archive the Best Option?," using "government" instead of "public" for a better effect, tapping into the scientists' fear of government regulation.[131] The NCBI, on the other hand, threw its weight behind the PubMed Central project, by playing up its "significant experience in the creation of online archives, exemplified by PubMed (MEDLINE) for biomedical abstracts and GenBank for nucleotide sequences."[132] GenBank was the living proof that a large open access database was feasible not only technically but also socially. The NCBI and the promoters of GenBank had demonstrated that an initially reluctant scientific community could be brought to collaborate and share some of their data through a public database. This was no small achievement, for many probably agreed with the popular joke that "Scientists would rather share their toothbrush than their data!"[133] Furthermore, after two decades of existence and hundreds of thousands of users, it had become clear that GenBank was useful for the scientific community and that a similar database for the published literature might prove just as valuable.

The balance of power between promoters of open access and journal publishers began to shift in October 2003 when Varmus, biochemist Patrick O. Brown, and computational biologist Michael Eisen launched the first of a series of open access journals under the heading *Public Library of Science* (*PLoS*).[134] *PLoS Biology* was similar to traditional journals in every way (its articles were peer-reviewed) except that it was available only online and was financed by author fees rather than by reader subscriptions. The articles were not copyrighted but protected by a "creative commons" license allowing anyone to reuse the content, including data, free. The promoters of open access made it clear that if they succeeded, journals would have no other choice but to go along with the movement, for example by collaborating with PubMed Central; otherwise they would see a loss of contributions from the scientific community.

In 2005, two years after it implemented its policy of mandatory data submission, the NIH began requesting that grantees submit the accepted version of their manuscripts to PubMed Central no later than twelve months after publication.[135] This new policy posed no problem for open access journals, such as *PLoS Biology*, but represented a serious change in policy, if not economics, for others. Even with the request in place, compliance remained as limited as it had been earlier with data. Only one in six grantees submitted their articles. As a response, the US Congress passed a law in 2007 making submission mandatory.[136] A group of large journal publishers hired a powerful PR firm to combat the proposal by claiming that "public access equals government censorship" (since by deciding what ought to be open access, government implicitly defined what was not), but to no avail.[137] The NIH's mandatory submission policy came into force on April 7, 2008. That same day, in Bethesda, the NIH celebrated the twenty-fifth anniversary of GenBank, a coincidence (or not) that pleased the director of the NCBI, the parent organization of GenBank and all bioinformatics efforts of the NIH.

Databases, Journals, and the Record of Science

In their first decade, sequence databases grew in size beyond the wildest expectations (or nightmares) of those who had planned them in the late 1970s. The increasing speed and diminishing costs of sequencing technologies had taken database planners greatly by surprise. The flood of sequences revealed another miscalculation, namely, the extent to which individual researchers would participate in the collecting effort. It was a lesson that could have been anticipated: Past experience with the Protein Data Bank or the *Atlas* should have made it clear that without offering proper rewards (such as authorship), researchers had many incentives not to make data public. The problem with sequence data was not as critical as with crystallographic data because the former were usually included in published papers, whereas the latter were often not made public at all. Nevertheless, getting researchers to share their sequences constituted a difficult challenge, especially in highly competitive research environments.

EMBL, as a powerhouse for European molecular biology, seems to have been more willing to take on journal editors than GenBank, whose staff did not enjoy the same authority in the field.[138] The ensuing negotiations between journal editors and database managers reflected a struggle over the question of who had the authority to validate scientific data. Journal editors argued that the peer review process was the only legitimate way to ensure the validity of scientific claims. In the case of sequences, however, reviewers were often not in a position to evaluate the data; they did not have access to the supporting empirical evidence or the data in an electronic format, which would have permitted running checks on computers. Databases, on the other hand, were in an ideal position to evaluate sequences, not from an empirical point of view, but computationally. Content could be examined for obvious errors and compared with other sequences in the databases. In effect, this gave database curators the peer review function that journals claimed as their own. Finally, most journal editors adopted mandatory submission policies and gave part of the critical examination of the data submitted by authors over to database curators. Data sharing became a condition for being granted authorship through publication in a journal. So the reward system succeeded in finding a sustainable solution to the problem of data collection, where an appeal to the communal ethos had failed.

The relationship between journals and the databases, which were now providing an important service, changed profoundly with the rise of large-scale genomics projects in the 1990s. Databases increasingly accepted sequences that would never see publication in scientific journals. While initially they had been intended to reflect only published data, databases quickly became far more comprehensive than the published literature. Up to that point the most complete repositories of scientific data had been large libraries that held entire runs of scientific journals. Now databases were slowly challenging their role in holding the definitive record of science.

As instruments for the production of knowledge, databases have surpassed all the expectations of their early advocates. The "generalist" databases, EMBL/GenBank/DDBJ, were soon joined by numerous specialized databases such as FlyBase, a database entirely devoted to the fruit fly *Drosophila melanogaster*.[139] They incorporated sequence data that had been copied from the "generalist" databases, as well as numerous other kinds of biological information. The rise of these new collections was made possible by the fact that the "generalist" databases (unlike Dayhoff's *Atlas*) placed no restrictions on the redistribution of sequences.

Sequence databases not only transformed the practices of experimental science but also led to the emergence of a new kind of scientific practice, "bioinformatics," centered on the computational analysis of sequences. Alongside developments in computer technology, the creation of sequence databases was the single most important factor in the rise of this new scientific specialty. The comprehensive collection of standardized biological data in electronic format opened the door for the development of sophisticated algorithms to analyze it. In addition to studying nature *in vivo* and *in vitro*, research *in silico* was becoming a legitimate way of producing knowledge about nature. But the rise of the new specialty did not fail to create tensions between the experimentalists who produce data and those who were "merely" analyzing it. By refraining from analyzing the data themselves, the way Dayhoff had done, database managers remained neutral in the debate.

Biologists played a particularly active role in the promotion of open access, building on their experience with GenBank and other open databases, more than chemists and physicists, for example. The American Chemical

Society has also been a leader of the opposition to open-access publishing, a stand that led to the resignation of molecular biologist Richard Roberts, one of its Nobel Prize-winning members.[140] Databases in chemistry, such the CAS Registry, a collection of chemical substance information compiled from the published literature, has been available only at a very high cost for scientific institutions. Physicists were less pressed to fight for open access, having created in 1991 the arXiv, an electronic database of article preprints in the physical sciences, which practically reduced the role of journals as gatekeepers of scientific knowledge.

The most surprising aspect of the history of sequence databases is that they have remained free and open, with no restrictions on the access to and distribution of the data they contain. This could hardly have been predicted at a time of increasing commercialization among researchers and powerful journal publishers. As a result, sequence databases provided a powerful model for numerous other data collections. And they have made an important contribution to the extension of the open access movement to scientific literature as a whole. While using sequence databases, scientists became accustomed to easy and free access to data and less willing to pay for access to other forms of content published in scientific journals. The rise of open databases has transformed how knowledge is produced in the sciences, and by contributing to the rise of open access to the scientific literature, it has transformed who can access knowledge and thus contribute to the production of knowledge. The significance of a history of collections in the experimental sciences lies exactly here: collections have deeply changed the epistemic practices and the moral and political economy of science.

Conclusion

In this book I have attempted to make sense of the role of today's "big data" in the production of biological knowledge by placing current practices into a long historical tradition of scientific collections. Instead of focusing on the recent proliferation of databases and marveling at the amount of data they contain, I have assigned them to the broader analytical category of "collection," which encompasses all human-made organized sets of items, whether material or abstract.[1] This methodological decision permitted me to connect scientific institutions that at first glance appear very different, such as zoological, paleontological, or medical museums, botanical gardens, stock collections, blood banks, and databases. These institutions have usually been discussed separately because they were considered to "belong" to the naturalist, the medical, or the experimental traditions, to museums or to laboratories. While the differences between these institutions are very real, they should not distract us from their deep commonalities. Collections seem to "belong" to the naturalist and museum traditions mainly because their importance in the experimentalist tradition has largely been overlooked, as the examples given in this book should make clear.

The historical connections between different institutions hosting the collections examined in this book have been very strong at times, at others rather weak. The people organizing these different collections have generally worked independently from one another, at least until the end of the twentieth century. Now collections of things, from books to bones, are increasingly becoming digitized and handled like other databases, by professional database managers, curators, and information scientists. The strongest historical continuity among these collections is at the level of practices, i.e., how the items were collected, curated, computed, compared, and classified. The epistemological practices of people working with collections exhibit such similarities that one can learn much by engaging them in dialog. And a number of actors, especially in the study of evolution, learned important lessons as they moved from the study of collections of specimens to collections of data.[2] To apply the essential biological distinction between homology (structural similarity due to common descent) and analogy (structural similarity due to a common context), the collections discussed in this book are mostly analogous, even though the epistemic practices can, to some extent, be considered homologous.[3]

When we reexamine the story of sequence databases, it becomes clear that their significance lies not only in the possibility of investigating "biology *without doing lab experiments*,"[4] or in the revitalization of theoretical biology under the banner of computational biology or bioinformatics, but in the creation of a hybrid epistemic practice drawn from two ways of knowing in the history of the life sciences: comparing and experimenting.

The End of Model Organisms?

What are the consequences of this new perspective for our understanding of contemporary biomedical research? The production of biomedical knowledge has entered a new regime, with distinct epistemic, material, and social characteristics. Defining this regime as "big data biology" says close to nothing about what makes it so different from those of previous eras. As discussed in this book, one of the most distinctive features of the experimental life and biomedical sciences in the twentieth century has been its extensive reliance on model organisms. The vast majority of our knowledge about the processes of life and disease has been produced through the study of a handful of species.[5] In 2016, the National Institutes of Health listed just thirteen model organisms as standards of biomedical research.[6] For researchers, model organisms provided an experimental system and served as the basis for a social community, often defining their professional identity ("I am a drosophilist" was an acceptable answer to the question "What's your field?"). This is understandable because the introduction of a new model organism was considered a significant professional achievement. As a developmental biologist put it: "In the economy of modern competitive science, there can be few career outcomes more satisfactory for a bioscience professor than the successful introduction of a 'new' model organism."[7]

Yet the dominance of model organisms at the heart of the experimental life sciences is slowly fading. It is being replaced by a renewed focus on the diversity of biological systems, which was once the hallmark of natural history and more generally of the comparative way of knowing. The emphasis on biological differences and diversity, a subject of pride for naturalists and of ridicule for many others, has been as central to natural history and other fields as model organisms have been to experimentalism. The naturalist's passion for the diversity of forms certainly has many origins, including affective ones, but it was rarely mere "stamp collecting."[8] It served a specific epistemic purpose: the systematic comparison of a wide diversity of cases, both within and between species. This method could expose regularities and support general claims about the structures and functions of the components of living organisms, usually by identifying features to be considered as homologous with a clear function in at least one species. By contrast, experimentalists working on a single but well-defined "exemplary" organism extended their results to make claims of wider validity—or as the molecular biologists Jacques Monod and François Jacob, cited in the introduction, famously put it, what is "true of *E. coli* must also be true of Elephants."

The "comparative" and the "exemplary" approaches rest on fundamentally different assumptions about the world and about how local claims can be made universal. Ernst Mayr's contrast between the "comparative method" and the "experimental method," laid out in his monumental *The Growth of Biological Thought*, captures this conceptual distinction. Ernst Mayr's intellectual and disciplinary agenda was to affirm the legitimacy of comparative methods in science and counter the view that "experiment is . . . the method of science." According to Mayr, this view was dominant in the historiography of science written by "physical scientists" and "historians of the physical sciences." But these authors "display[ed] an extraordinary ignorance when discussing methods other than the experimental one" and thus overlooked the importance of the "comparative method" in other sciences.[9]

For Mayr, the issue was not simply historiographic, nor did it concern only the value of biology versus the physical sciences. He aimed to confront deep tensions that had been growing within biology since the beginnings of the twentieth century and had grown particularly acute in the 1960s, especially between "organismic biology" (as it came to be called) and the new field of molecular biology.[10] The entomologist and evolutionary biologist E. O. Wilson, Mayr's protégé at Harvard as a student and later a faculty colleague, described the fights between the proponents of the two approaches vividly; they took place in faculty meetings and spilled out into the corridors of the Biological Laboratories building. For Wilson, these amounted to "Molecular Wars."[11] Although he was tenured the same year (1958) as James Watson, whom Wilson called "the Caligula of biology," organismic biology was losing power across the board under the assaults of molecular biologists.[12] Wilson later described Watson as "the most unpleasant human being [he] had ever met."[13] Thus, when Mayr attempted to portray the "comparative method" as a legitimate method for the production of knowledge, his main goal was to rehabilitate "organismic biology" (essentially evolutionary biology, systematics, ecology, and ethology) in an academic landscape that was becoming increasingly dominated by experimental biology (particularly molecular biology).

By associating the "comparative method" with organismic biology and the "experimental method" with functional biology (for which molecular biology was the prime example), Mayr made the same assumption as those who thought "the difference between physical and biological research [is] a difference of methodology," equating research fields and methodologies. Auguste Comte had made a similar distinction a century earlier, contrasting the "comparative method of the biologists" with the "experimental method of the physicist."[14] Mayr rightly argued that both methods were important for biology—but only when applied to their respective fields, organismic biology and functional biology.

The argument of this book is that these two ways of knowing have been essential not only in particular subfields, but in all of biology. To be sure, organismic biology has relied more on the comparative than the experimental way of knowing, at least until the 1970s, and functional biology has been deeply experimental all along. But the key historical argument of this book is that even the crowning achievements of molecular biology—deciphering the genetic code, understanding the relationship between the structure and function of proteins, and elucidating the organization and workings of entire genomes—were not only the products of experimental virtuosity, but also the fruit of comparative approaches involving many species that were not model organisms.[15] Comparative approaches, as now applied to the analysis of data collections, have become essential for contemporary biomedical research, and the combination of the experimental and the comparative can be considered one of its most distinctive features.

The deployment of comparative approaches has brought biodiversity back into the laboratory, from which it was mostly excluded in the twentieth century. Organisms, molecules, and data from a wide variety of species are being studied experimentally to produce scientific knowledge. But how does the range of biodiversity examined in laboratory studies compare with that of museum studies? In 2019, the American Museum of Natural History and the British Museum (Natural History), two of the largest natural history museums in the world, were home to slightly over 500,000 species.[16] That same year, GenBank, the largest collection of molecular data, hosted DNA sequences of over 430,000 species (not counting viruses). If this rate of growth continues, by 2025 there will be more species represented in GenBank than in the largest natural history museums.[17] Data obtained from these many species are routinely used in laboratories, as discussed in chapter 5; for example when researchers determine the sequence of a gene in a given species, they typically "blast" it (a verb derived from "BLAST," the sequence comparison algorithm) against the sequences of all other species represented in a database such as GenBank. Every "match," or strong similarity, will suggest potential homologies and similar functions (the same is true for protein sequence comparisons).

To illustrate this approach, consider an organism that is quite common in museums and most unusual in laboratories: the elephant. In 1980, a group of researchers that included molecular anthropologist Maurice Goodman determined the amino acid sequence of the Asian elephant myoglobin in order to clarify its phylogenetic position. When the Asian elephant sequence was compared with that of the sperm whale, they found several differences that reflected the evolutionary distance between these two species. One surprising difference, of potentially functional consequence, was at a single position in the molecule where one amino acid was replaced by another one, even though the original one was believed to be crucial for the binding of oxygen and thus the main function of the molecule.[18] But elephant myoglobin worked just fine, like myoglobin of other species.

Fifteen years later, another group of researchers succeeded in determining the three-dimensional structure of Asian elephant myoglobin to place this unusual residue in a functional context and shed light on the mystery.

Myoglobin is arguably "one of the best characterized proteins and therefore can be used as a reference for comparative analysis," they argued.[19] What they found was yet another amino acid difference, far away from the main oxygen-binding site, but which the folding of the molecule brought into proximity with the first site. The substitution was able to compensate, to a degree, for the first sequence difference at this location. The authors thus concluded that "obviously, nature is able to achieve solutions for a given biological problem by different approaches."[20] All the experimental work carried out to gain a better understanding of the structure and function of myoglobin was guided by comparisons between various species, including the elephant.

Another way to look at the explosion of biodiversity in laboratory studies is to follow the professional trajectories of individual scientists. Take the molecular and developmental biologist Denis Duboule. Starting in the mid-1980s he made a number of contributions to the understanding of the role of Hox genes in animal development. All of his early studies were carried out on mice, with an occasional comparison with the fruit fly *Drosophila* or the African toad *Xenopus*, all standard model organisms.[21] Starting in the 1990s, his focus shifted owing to the surprising discovery that the spatial order of the Hox genes on chromosomes could be matched to the spatial and the temporal order of their expression along the body axis of the embryo. At that point he began including references to birds and fish, in addition to mice and *Drosophila*.[22] A decade later, one of his papers on the same issue listed more than twenty different species, including mosquitos (*Anopheles*), beetles (*Trilobium*), silkworms (*Bombix*), round worms (*Chaetopterus*), flat worms (*Schistosoma*), squid (*Euprymna*), fish (*Fugu*), sea anemone (*Nematostella*), and many others.[23] And his more recent studies on the role of variation in Hox genes expression in the emergence of vertebrate body plans included data from the corn snake (*Pantherophis guttatus*), turtles (*Trachemys scripta*), tuatara (*Sphenodon punctatus*), the green anole lizard (*Anolis carolinensis*), the gecko (*Gekko ulikovskii*), and the slow-worm (*Anguis fragilis*)—not your typical laboratory model organism. [24]

A recent collaborative study by Duboule and another laboratory on the role of Hox genes in determining the body plan of vertebrates brings us back to elephants. The researchers compared the sequence of a genomic region to which an important Hox protein binds in an even wider range of species. Snakes (python, boa, corn snake) exhibited a different sequence than most other vertebrates, but similar to elephants, hyrax, and manatees, despite the rather obvious differences in snake and elephant morphologies. But this masks an important similarity: both snakes and elephants have unusually long rib cages, made of at least nineteen vertebrae. The similarity in the sequences found in these two species, compared with other vertebrates, yielded important insights into the evolution of vertebrate body plans as well as into the molecular mechanisms underlying vertebrate development.[25] In a 2009 interview, after outlining the value of model organism research, Duboule acknowledged: "The epistemological question is whether or not one can fully understand one given mechanism without considering its realm of variations, i.e. without looking at many other species."[26] His personal research practices seem to provide an eloquent answer to this rhetorical question.

The museum provided fertile ground for the comparative perspective to flourish in natural history (just as the anatomical collection did for comparative anatomy and the embryological collection for comparative embryology). In the nineteenth century, the comparative work of Georges Cuvier and Louis Agassiz was possible only given their access to this special knowledge infrastructure, the Muséum National d'Histoire Naturelle in Paris and the Museum of Comparative Zoology in Cambridge, Massachusetts, respectively. Experimentalists followed a similar path with the creation of collections of mutant organisms, tissues, molecules, and eventually data, all discussed in this book, and essential in bringing the comparative perspective so productively into the experimental life sciences.

The laboratory revolution in the late nineteenth century had brought experimentalism progressively to the forefront of life science research, overshadowing the sciences, including natural history, where the comparative way of knowing was most present. The comparative approach was progressively rediscovered in experimental research, but after an eclipse of almost a century, becoming part again of the avant-garde of biological research.[27] This new periodization of the history of the life sciences stands in sharp contrast to narratives focused on the decline of natural history and the triumph of the reductionist agenda ("biology goes molecular") or those centered on the revolutionary nature of *in silico* biology and the emergence of biology as an "information science."[28] My narrative, stressing continuity, does not imply that the life sciences have simply returned to their origins in eighteenth- and nineteenth-century natural history, and the reason for this is not to be found in the many differences between bone collections and databases, which are too numerous (and often trivial) to mention. The true novelty of the new regime for the production of knowledge is that it is a hybrid of both experimental and comparative perspectives. After identifying that a new gene of unknown function has a sequence similar to that of a well-characterized gene or protein in a database, a researcher will then move to experimental studies to examine whether this function can be demonstrated *in vivo* or *in vitro*. Comparisons and experimentation have become two moments in the production of knowledge. In the experimental "moment," model organisms (and particularly humans) play a central role; in the comparative "moment," all available data about individual and species diversity is taken to shed light on the process under scrutiny.

If museums rather than laboratories have been a prime location of natural historical practice, this was not the result of some historical connection between disciplines and institutions, but because museums have been tools for collecting, and laboratories for isolating—museums assemble, laboratories disassemble. The material nature of scientific objects constrained the physical shapes of the institutions in which they were studied. The objects of natural history collections were often voluminous, heavy, cumbersome, and almost impossible to reproduce. Housing them required centralized institutions such as museums. Collections of physical objects almost never moved, if only for reasons of preservation, and generally remained permanently attached to an institution unless they were sold as a whole.[29] Individual objects such as specimens, on the other hand, were temporarily lent to other museums, so that researchers could compare them with the collections housed there. More typically, the researcher traveled to another museum to examine another set of specimens, particularly for type specimens, which were often barred from leaving the museum.[30]

But herein lies one significant difference between the new databases and earlier museum collections: in the twenty-first century, data collections are freely accessible, at least in the life sciences, from anywhere, including within the laboratories, through any computer (or even a smartphone) with internet access. What has changed with electronic digital data collections is the extent to which the content circulates, the range of people who have access to the collections, and the comprehensiveness of the comparisons. The physical locations of these data collections are often separate from the institutions of the laboratories, typically in large "data centers" that host servers, like those at the NCBI in Washington, DC, or at the EBI in Hinxton, UK. But since data can be here and there at the same time,[31] their physical location becomes irrelevant for users, who can access entire collections from the laboratory, allowing for seamless combinations of experiments and comparisons of any type the scientist can imagine.

The New Politics of Knowledge

Comparative and experimental work differed not only epistemically, but also socially, particularly in regard to the moral economies upon which they rested. Throughout this book we have seen how collection-based research required extensive forms of cooperation among people whose status was often quite diverse—professionals and amateurs, researchers and curators—and who lived within different moral economies, specifically with regard to issues of credit and authorship, in ways that created tensions or even clashes between them (see especially chapter 3 and 5). Early collections of objects in the experimental life sciences such as stock collections (chapter 1) emerged within tightly knit communities of researchers whose communitarian ethos placed a value on sharing objects and data and regulated claims for individual credit.[32] Single-author papers masked the fact that the results were based on a much wider community of people who contributed indirectly to the research, for example through the gathering of the mutants that were added to the stock collection upon which further research was based. The situation was no different in natural history, where single-author monographs depended on the extensive and often anonymous work of amateur collectors. But whereas the amateur (and often female) status of these contributors made it possible for professional naturalists to render their work almost invisible, this option was not available within the dominant culture of assigning credit among professional experimentalist (except in the case of communities of researchers centered on model organisms).

But as applications of comparative approaches to experimental data became more common in the second half of the twentieth century, researchers struggled to find ways to combine the dominant individualist moral economy of the experimental life sciences with the need for more communitarian practices, in cases such as open data sharing. Particular arrangements have been worked out in certain disciplines, tying authorship to data sharing (chapter 5), thus making data available without challenging the traditional form of knowledge appropriation through authorship. The larger context for these arrangements, however, is the rise of open access (chapter 6). Although the ideal of making (scientific) knowledge freely accessible and reusable by anyone has a very long history—it was practiced by movements as different as the avant-garde situationists in the 1950s and the computer hackers in the 1970s—it was only in the 1980s that a significant number of initiatives began to regulate access to data and published literature in a new way.[33] In the sciences, alongside the free software movement,[34] it was mainly the growing importance of databases and the kind of research they made possible that prompted the development of open data policies. The success of these policies for scientific data, so evident in the case of genomic databases, served as a model to foster a more general open access to the scientific literature. In the early twenty-first century, it has become a powerful norm, enforced by science funding agencies, professional societies, and scientific journals. The NIH Public Access Policy, requiring all the publications resulting from NIH-funded research to be rendered open access within six months, went into effect on April 7, 2008. Six years later, the NIH and the Wellcome Trust in the United Kingdom actually took action to enforce their mandatory open-access policies, withholding grant payments from researchers who had not made their publications available.[35]

"Open science" constitutes a renewed attempt to find a balance between the interests of public disclosure and intellectual ownership in the scientific community. It has strongly tilted the balance in favor of public disclosure; some have argued that this has taken place at the expense of individual intellectual ownership and benefited private corporations, favored by a neoliberal governmental agenda.[36] Furthermore, because the sciences mainly reward individuals for what they can claim to be their own intellectual contributions, open access could in fact discourage researchers from producing knowledge, because it prevents them from getting full credit for it.[37] Two recent sets of initiatives have begun to address this problem. First, database managers have attempted to use the model of incentives created by journal editors to encourage submissions: the granting of authorship. The Protein Data Bank made it possible to cite "an entry without a published reference," by attributing a Digital Object Identifier to each entry as for a journal article.[38] A data entry in the Protein Data Bank could then be listed in an author's publications and thus contribute to his or her intellectual credit. Journal editors have responded to databases' challenges of their exclusive rights to grant data authorship by launching new electronic journals solely for the publication of data. Nature Publishing Group started *Scientific Data* in 2014 for this purpose.[39] Now databases and data journals both grant authorship, allowing researchers to claim the professional rewards attached to publication. The ability of this model to entice data sharing is mitigated by the fact that the credit associated with publications depends on the estimated "impact" (a measure of a journal's influence and selectivity) of the journal. Thus, the incentive provided by authorship in "data journals" or directly in databases will depend on the "impact" that these venues are able to achieve. Given that they will hold large amounts of data that is never reused, their calculated "impact factor" will likely remain low, and the credit associated with publishing in these journals or databases will be as well.

A second model based on data citations was developed to overcome these limitations. Another type of scientific reward, beyond a scientist's publication record, has been based on the number of citations of an author's work. Since 2005, the number of citations has been used to calculate the *h*-index, a measurement of a scientist's productivity and impact that has now become a standard part of a scientist's resume and that is often required by science funding agencies and academic search committees.[40] But if data is considered a common property that does not require citing the researcher who produced it, then there is little individual incentive to produce data, let alone disclose it. For this reason, database managers, journal editors, and funding agencies began to encourage authors to cite individual data entries, including the name of the researchers who had deposited the data, just as they would for a published paper. Efforts by the Committee on Data for Science and Technology (CODATA), the US National Academy of Sciences, and various other groups of scientists led to the Joint Declaration of Data Citation Principles in 2013, emphasizing the importance of setting specific minimal standards for data citations.[41] At the same time, the media multinational Thomson Reuters, which maintains the bibliographic record and citation index Web of Science, launched its Data Citation Index, making it possible to easily quantify how often a particular data set is cited in the scientific literature.[42]

These two sets of initiatives, which promote data authorship and data citations, aimed to increase the individual rewards attached to the disclosure of data, thus encouraging researchers to make their data "open access." It is too early to say whether these initiatives will have the expected results, but what is clear is that data collections will continue to be a marketplace for the negotiation of the political economy of knowledge. What seems already historically significant is the degree to which there has been a retreat from the idealistic attempts of the 1960s and 1970s to transform the moral economy of experimental science. Instead, those promoting "open science" all rely on existing reward systems based on the way authorship is granted by community-based journals (or databases) and the citations of published work by members of the scientific community.

Here, history provides some guidance to solve the current challenges of implementing open data and open access policies. First, as this book makes clear, ideas about credit, authorship, and ownership are understood very differently among various (sub)specialties because they cultivate different moral economies. One-size-fits-all solutions to the problem of "data sharing" are thus unlikely to be effective. Disciplinary repositories are thus much preferable to institutional repositories. Subfield repositories, such as model organism databases, are even more likely to attend precisely to the norms and values of the communities of researchers producing and using data, thus ensuring the cooperation of data producers and making sure the data collected is actually reusable.

Second, as the examples of the Protein Data Bank and GenBank show, deposition policies based solely on the goodwill of individual scientists and collective rewards, as well as mandatory deposition policies without enforcement mechanisms, are only moderately effective. Yet this is still what many academic institutions and science funding agencies propose, especially in Europe. In the case of the databases examined here, only when data deposition became linked to the individual interests of researchers (e.g., through access to authorship) did the compliance become almost universal. In the case of open access to the scientific literature, the University of Liège 2007 open-access mandate followed precisely that path. Only the publications deposited in a timely manner in the university's Open Repository and Bibliography (ORBi) were taken into consideration for internal evaluations, including job appointments and promotions.[43] By linking publication deposition to the individual interest of researchers, this mandate became almost instantaneously effective and widely recognized as a success. Research funding agencies withholding grant renewals from grantees not complying with mandatory open-access policies have similarly linked compliance to individual self-interest and proved equally effective.[44]

Scholarly literature concerning the rise of "open science" has focused on policies elaborated by governments and science funding agencies, often overlooking the role of journal editors and database managers. The author of *Reinventing Discovery*, for example, claimed, "The granting agencies are the de facto governance mechanism in the republic of science, and have great power to compel change, more power even than superstar scientists such as Nobel prizewinners."[45] Others have described the "open science revolution" as an essentially spontaneous revolution "from below," where individual researchers at some point became voluntarily committed to "open science" and shared data as something that served the best interests of the scientific community.[46] But as this book makes clear, there was no revolution at all, or at most only a very conservative one, and journal editors played a key role. The "open science revolution" might have changed how much data was made available publicly, but not the reasons why individual researchers shared data. By and large, researchers have shared data when it served their own interests to do so, as defined by the reward system that had already been established in the experimental sciences. Far from upsetting the current moral economy of science, the rise of "open science" illustrates how much moral economies remained entrenched in scientific practices. For this reason, researchers have even suggested that the term "data sharing," with its communitarian overtones, be abandoned and replaced by "data publication," a term that is perfectly in line with the individualistic ethos prevalent in the experimental sciences.[47]

The rise of open access data collections has led to the emergence of the autonomous field of data analysis. Although producers of (experimental) data are often the main users, in the life sciences researchers who solely analyze data without producing their own (under the banner of "bioinformatics") have gained increasing autonomy since the late twentieth century (chapter 6).[48] In the 1960s, statisticians, computer scientists, and other mathematically inclined researchers made experimental data their prime object of study (chapter 3), a move that became professionalized in the 1990s through the creation of journals, associations, and institutions devoted to "bioinformatics" or "computational biology." As a well-recognized profession and set of methods, bioinformatics has produced indispensable tools for experimentalists as well as become a scientific field of its own, although it still lacks some of the luster of experimental work. No Nobel Prize has yet been awarded to research based exclusively on the *in silico* analysis of experimental data, although that may change in the future, and numerous Nobel Prizes in physiology or medicine have already rewarded research that did rely, in part, on bioinformatics methods.

Bioinformatics analysis has already produced major insights into the working of living organisms. In a follow-up to the Human Genome Project, the NIH launched the project ENCODE (Encyclopedia of DNA Elements), aiming to identify the functions of all the elements of the human genome. In 2012, the ENCODE consortium simultaneously published thirty papers by over one thousand researchers to announce their results, which relied on the analysis of vast amounts of data produced by the genome-sequencing projects, as well as on a small set of standardized biochemical assays (for example, to identify sequences that proteins directly attach themselves to).[49] These studies have led to claims that approximately 80 percent of the human genome has some biochemical function (a number in stark contrast to the widely held proposition that only 1.5 percent of the genome encodes proteins), that 90 percent of the variations in the genomes lie outside protein-coding regions, and that 95 percent of the genome lies in proximity to a sequence with a regulatory function. These results received great attention in the scientific community (and beyond) and generated controversies of their own. Taken together these results emphasized the importance of networks of regulation and not only the presence or absence of a set of elements (often proteins) whose expression was being regulated.[50] Bioinformatics analysis began to highlight the fact that the roles of regulatory networks were far more complex than simply serving as on/off switches and could not be conceptualized without massive computing power.[51] These results could not have come solely from an experimental study of individual model organisms. They came to light only by bringing together experimental and comparative ways of knowing. The ENCODE project represents a good example of a type of massive, collaborative experimental and bioinformatic effort that has become increasingly common in "big data biology." The fact that individual authors tend to disappear in these collaborative projects, even though some still stand out, does not diminish the significance of bioinformatics research as an essential means of producing knowledge from data collections.

The growing use of data collections has also led to the professionalization of those who take care of them. In the 1960s, researchers were unsure how to address Margaret Dayhoff, who was managing a major collection of protein sequences. She was variously referred to as an "editor," a "librarian," or a "compiler" (chapter 3). Although "curator" had long been recognized as a title for a scientist in charge of a natural history collection, it had no equivalent in the experimental life sciences. Yet by the end of the twentieth century, the terms "database curator," "annotator," or "manager" had become common titles in job advertisements published in scientific journals. In 2009, the newly created International Society of Biocuration and Oxford University Press launched *Database: The Journal of Biological Databases and Curation*, marking another step toward the professionalization of this field.[52] However, there remains much unease regarding the status that should be accorded these scientists. As one of them commented: "[database annotators] dread the immortal cocktail party question 'So, what do you do?'"[53]

The growing professionalization of bioinformatics and database management should not distract from the bigger picture of an increasing public participation in big-data science. Even as they were struggling for professional recognition, data curators were quick to realize that they could never handle the data deluge on their own. In 2006, after reading a report in *Nature* showing that the reliability of Wikipedia was on a par with that of the *Encyclopedia Britannica*, a graduate student wrote a letter to *Nature* asking for a "wiki on gene function" that would use "the collective brain power of biologists around the world."[54] Two years later, 256 scientists wrote a letter to *Science* asking the managers of GenBank at the NCBI to open annotation to the public as a way of "preserving accuracy in GenBank."[55] NCBI director David Lipman refused, arguing that since GenBank was an archive, its records should not be open to modification by anybody other than their authors or the curators; otherwise, the result would be "chaos."[56] That did not prevent researchers from starting Gene Wiki, embedded in the Wikipedia encyclopedia.[57] Automated procedures created Wikipedia articles for each human gene, added all relevant information that could be culled from various databases, and opened the rest for annotation by the scientific community (or anyone else). WikiGenes, a similar database, was launched the same year, but the project attributed individual authorship to each contributor, in contrast to Wikipedia (and thus Gene Wiki) and to earlier community databases, which initially lacked a mechanism to elicit participation.[58] The attribution of authorship created a potential incentive to contribute and a way to evaluate the credibility of authors and the reliability of their contributions. More recently, the managers of Gene Wiki realized that "the incentives to contribute are not well-aligned with traditional academic rewards" and as a result "many qualified scientists hesitate to devote their limited time to contributing."[59] This prompted them to team up with the editors of the journal *GENE* to invite authors to publish a peer-reviewed synthesis of knowledge pertaining to a given gene in the journal and simultaneously contribute to the corresponding entry in Wikipedia. As the editors explained, "This design creates two versions that have distinct functions—one article of record that can be cited and treated as an authoritative snapshot of the field, and one 'living article' that will continue to evolve as new biological insights are revealed."[60]

These are examples of many initiatives that illustrate how profoundly the rise of research based on data collections is transforming the moral and political economy of science, while maintaining features of previous norms regarding authorship and credit. It seems uncontroversial that a Wikipedia article on the "Big Apple" (or on "apple pie") might benefit from the contributions of people with many different points of view. Far more consequential is the idea that scientific matters of fact, which were long protected as areas of professional expertise, should be open for debate to anyone with a connection to the internet and (usually) a bit of familiarity with at least one scientific domain. This deep epistemic and political shift had many causes, but in part it was a result of the emergence of "big data" that was itself a consequence of the return of collection-based research. The "crowd," and "crowdsourcing," was increasingly becoming a solution to the problem of "big data" curation, at least in some fields, although professional curators surely maintained some misgivings about the value of "amateur" participation. A discourse about public participation as a universal solution to all big data problems has become fashionable, with populist overtones, masking the great diversity of situations in which amateur participation might, or might not, prove helpful. When data analysis can be broken down into simple tasks or cognitive processes dependent on unique human skills, such as pattern recognition, then public participation seems to hold great potential. On the other hand, when data analysis requires skills that specialized professional curators need several years to acquire, it seems doubtful that public participation will have a significant impact on database curation.[61]

But the broader impact of open access data collections goes far beyond the sciences. The fact that scientific knowledge is increasingly available to anyone, including in the Global South, is changing who actually accesses and reuses knowledge. As the historical examples discussed in this book so clearly illustrate, those who developed biological collections were eager to make them as widely available as possible. At first, the users were essentially those who produced the material and data for the collection. But soon, the community of users expanded far beyond the producers, with far-reaching consequences for the production of knowledge. As the Wikipedia example illustrates, open access to knowledge can lead to powerful collective actions that produce more knowledge. Although Wikipedia articles concerning specific pieces of scientific data, such as those centered on genes, are most likely edited by professional scientists, lay people are increasingly contributing to the production of scientific knowledge by analyzing scientific data—often from home. These activities involve professionals and amateurs and are known as "citizen science," "participatory research," or "crowdsourcing." They are shifting the boundaries between the sciences and the public. In the online astronomical data analysis project Galaxy Zoo, citizens classify photographic images of galaxies; in Eye Wire, they map data on networks of neurons; in Old Weather, they decipher weather data obtained from maritime logbooks. These projects have already resulted in numerous publications, often in high-profile journals. They follow every standard of a scientific publication, except that they include collective and distributed authors, such as "the EyeWirers" in a publication in *Nature*.[62] Open access data collections are making this new kind of public participation possible in which lay people are not only providing data through their observations of nature, but also analyzing data. Citizens have long contributed to the construction of data collections—as observers of comets and migrating birds—and have even occasionally engaged in their own research projects.[63] But today, open access collections available through the internet are fostering the emergence of countless new initiatives to involve the public in data analysis, with the aim of producing new scientific knowledge.[64]

Historically, the involvement of amateurs in the collection of observations and specimens was a major reason for the overflow of data and objects in natural history collections. In today's experimental sciences, amateurs are contributing less to the production of data than to its analysis, and could thus become part of the solution to the "data deluge." At the same time, these new forms of public participation are challenging the boundaries between professionals and experts that were so impenetrable in the experimental sciences throughout most of the twentieth century. Therefore, the "big data" revolution in the life sciences and beyond should be considered not simply a matter of quantity or of technological change, but the product of a deep epistemic, social, and political transformation that once again engaged the public as an active player in scientific research, at least in some fields. Taking into account the history of collecting, comparing, and computing data helps us look beyond the hype and think more deeply about the true novelty of the current regime of biomedical knowledge production.

The rise of modern science in the seventeenth century represented a deep transformation not only in *how* knowledge about the natural world should be produced, but also in *who* should be considered a legitimate producer of knowledge.[65] Initially open to a wide range of practitioners—provided they were gentlemen—the production of scientific knowledge became progressively restricted to professional scientists. This movement was accentuated at the end of the nineteenth century, when the experimental tradition came to dominate the life sciences and most other fields of science. As an editorialist put it in 1902, "The era of the amateur scientist is passing; science must now be advanced by the professional expert."[66] In the twentieth century, this separation between "professional experts," on the one side, and the "amateur scientists" and "lay public," on the other, became wider than perhaps at any previous time in history. Thus, if a new regime of producing biomedical knowledge, which combines collecting and experimenting, is also enlarging, once again, the range of people who can participate in the production of scientific knowledge, then an even deeper historical transformation is indeed under way.

Archives Consulted

APS Archives Archives of the American Philosophical Society, Philadelphia, Simpson Papers, Box 9, "Boyden, Alan A."; Demerec Papers, Box 2, "Beadle"; Box 3, "Boyden"; Box 7, "*Drosophila Information Service*"; Box 8, "Drosophila Stock Center"; Genetics Society of America Papers, Box 17, "1958 Report of the AIBS Committee on Maintenance of Genetic Stocks"; Box 18, "1959 Committee on Maintenance of Genetic Stocks"; Box 19, "1960 Committee on Maintenance of Genetic Stocks"; Box 25, "1965 Committee on Maintenance of Genetic Stocks"; Box 30, "Committee on Maintenance of Genetic Stocks, 1971"

ASM Archives American Society of Microbiology Archives, University of Maryland, Baltimore, Special Collections, American Type Culture Collection, Boxes 1–8; Jeff Karr Collection; Section 12—Other Orgs, IB folders 1–10 ATCC 1935–1973

EBI Archives Archives of the European Molecular Biology Institute, Hinxton, UK, [unsorted].

EMBL Archives European Molecular Biology Laboratory Archives, Heidelberg, Germany.

EMBO Archives European Molecular Biology Organization Archives, Heidelberg, Germany.

HUG Archives Harvard University Archives, Cambridge, MA, HUG (FP) 14.7, Boxes 2, 8, 10; HUG (FP) 74.7, Boxes 3–5, 8, 11, 16–26, 29, 34–35, folder 92; HUG (FP) 74.10, Box 1

MIT Archives	History of Recent Science & Technology Dibner Institute Project, History of Bioinformatics documents, formerly hosted at MIT
NBRF Archives	Archives of the National Biomedical Research Foundation, Washington, DC, [unsorted]
NCBI Archives	National Center for Biotechnology Information Archives, Bethesda, MD, GenBank Collection, [unsorted]
PDB Archives	Protein Data Bank Archives, Rutgers University, New Brunswick, NJ, [unsorted]
Peabody Archives	Archives of the Peabody Museum of Natural History, Yale University, New Haven, CT, Sibley Collection
RAC Archives	Rockefeller Archives Center, Sleepy Hollow, NY, Series 200.D, Box 136, Folders 1679–84; Box 137, Folder 137; Box 172, Folder 863
Rutgers Archives	Special Collections and University Archives, Office of the President, (Robert C. Clothier), Rutgers University Libraries, New Brunswick, NJ, RG 04/A14, Box 84, Folder 1, "Boyden, Alan A., Zoology Department, 1931–1951"; Box 1, Folder 6, "Boyden, 1949–1969"; Box 41, Folder 11, Faculty Correspondence: "Boyden, Dr. Alan A. (Zoology/Serological Museum), 1953–1962"; Box 101, Folder 4, "Serological Museum, 1950–1952"; Box 120, Folder 21, "Speeches: Serological Museum"; Faculty Biographical Information, "Alan A. Boyden"

Bibliography

Abbott, Alison. "Bioinformatics Institute Plans Public Database for Gene Expression Data." *Nature* 398, no. 6729 (Apr 22 1999): 646.

Abelson, Philip H. "Amino Acid Sequence in Proteins." *Science* 160 (1968): 951.

Abir-Am, Pnina. "The Politics of Macromolecules: Molecular Biologists, Biochemists, and Rhetoric." *Osiris* 7 (1992): 164–91.

———. "Themes, Genres and Orders of Legitimation in the Consolidation of New Scientific Disciplines: Deconstructing the Historiography of Molecular Biology." *History of Science* 23 (1985): 73–117.

Acharya, K. Ravi, David I. Stuart, David C. Phillips, and Harold A. Scheraga. "A Critical Evaluation of the Predicted and X-Ray Structures of Alpha-Lactalbumin." *Journal of Protein Chemistry* 9, no. 5 (Oct 1990): 549–63.

Acher, Roger, Jacqueline Chauvet, and Marie-Thérèse Chauvet. "Phylogeny of the Neurohypophysial Hormones." *Nature* 216, no. 5119 (Dec 9 1967): 1037–38.

Adelman, Benjamin. "Blood Museum Traces Animal Cousins." *Science Digest* 30 (1951): 6–8.

Agassiz, Louis. *Methods of Study in Natural History*. Boston: Fields, Osgood, 1869.

Ahlquist, Jon E. "Charles G. Sibley: A Commentary on 30 Years of Collaboration." *Auk* 116, no. 3 (Jul 1999): 856–60.

Alberti, Samuel J. M. M. *Morbid Curiosities: Medical Museums in Nineteenth-Century Britain*. Oxford: Oxford University Press, 2011.

Alberti, Samuel J. M. M., and Elizabeth Hallam, eds. *Medical Museums: Past, Present, Future.* London: Royal College of Surgeons of England, 2013.

Aldhous, Peter. "Managing the Genome Data Deluge." *Science* 262, no. 5133 (Oct 22 1993): 502–3.

Allen, David E. *The Naturalist in Britain: A Social History*. London: A. Lane, 1976.

Allen, Garland E. *Life Science in the Twentieth Century*. Cambridge: Cambridge University Press, 1978.

———. *Thomas Hunt Morgan: The Man and His Science*. Princeton: Princeton University Press, 1978.

Allen, J. F. "Bioinformatics and Discovery: Induction Beckons Again." *Bioessays* 23, no. 1 (Jan 2001): 104–7.

———. "*In Silico Veritas.* Data-Mining and Automated Discovery: The Truth Is in There." *EMBO Reports* 2, no. 7 (Jul 2001): 542–44.

Alston, Ralph E., and Billie Lee Turner. *Biochemical Systematics*. Englewood Cliffs, NJ: Prentice-Hall, 1963.

Altman, Philip L., and Kenneth D. Fisher. *Guidelines for Development of Biology Data Banks.* Bethesda, MD: Life Sciences Research Office, Federation of American Societies for Experimental Biology, 1981.

Altschul, Stephen F., Warren Gish, Webb Miller, Eugene W. Myers, and David J. Lipman. "Basic Local Alignment Search Tool." *Journal of Molecular Biology* 215, no. 3 (Oct 5 1990): 403–10.

"The American Type Culture Collection." *Science* 79, no. 2050 (1934): 336.

Andersen, Jackey S., EMBL, and GenBank, eds. *Nucleotide Sequences, 1984: A Compilation from the GenBank and EMBL Data Libraries*. Oxford and Washington, DC: IRL Press, 1984.

Anderson, Warwick. *The Collectors of Lost Souls: Turning Kuru Scientists into Whitemen.* Baltimore: Johns Hopkins University Press, 2008.

Anfinsen, Christian B. *The Molecular Basis of Evolution*. New York: Wiley, 1959.

Anfinsen, Christian B., Stig E. Aqvist, Juanita P. Cooke, and Börje Jonsson. "A Comparative Study of the Structures of Bovine and Ovine Pancreatic Ribonucleases." *Journal of Biological Chemistry* 234, no. 5 (May 1959): 1118–23.

Ankeny, Rachel A., and Sabina Leonelli. "What's So Special about Model Organisms?" *Studies in History and Philosophy of Science Part C* 42, no. 2 (Jun 2011): 313–23.

Ankeny, Rachel. A., Sabina Leonelli, Nicole C. Nelson, and Edmund Ramsden. "Making Organisms Model Human Behavior: Situated Models in North-American Alcohol Research, since 1950." *Science in Context* 27, no. 3 (Sep 2014): 485–509.

Annual Report. New York: American Museum of Natural History, 1911.

Annual Report. New York: American Museum of Natural History, 1912.

Annual Report. New York: American Museum of Natural History, 1915.

Annual Report. New York: American Museum of Natural History, 1923.

Annual Report. New York: American Museum of Natural History, 1929.

Appel, Toby A. *The Cuvier-Geoffroy Debate: French Biology in the Decades before Darwin.* New York: Oxford University Press, 1987.

———. *Shaping Biology: The National Science Foundation and American Biological Research, 1945–1975*. Baltimore, MD: Johns Hopkins University Press, 2000.

Aspray, William. "Command and Control, Documentation, and Library Science: The Origins of Information Science at the University of Pittsburgh." *IEEE Annals of the History of Computing* 21, no. 4 (Oct–Dec 1999): 4–20.

Aspray, William, and Bernard O. Williams. "Arming American Scientists: NSF and the Provision of Scientific Computing Facilities for Universities, 1950–1973." *IEEE Annals of the History of Computing* 16, no. 4 (Winter 1994): 60–74.

Atencio, Edwin J., ed. *Nucleotide Sequences 1986/1987: A Compilation from the GenBank and EMBL Data Libraries*. Orlando: Academic Press, 1987.

Bachmann, Barbara J. "Pedigrees of Some Mutant Strains of Escherichia Coli K-12." *Bacteriological Reviews* 36, no. 4 (Dec 1972): 525–57.

———. "This Week's Citation Classic." *Current Contents, Life Sciences* 8, (Feb 22 1982): 22.

Bachmann, Barbara J., and K. Brooks Low. "Linkage Map of Escherichia Coli K-12, Edition 6." *Microbiological Reviews* 44, no. 1 (Mar 1980): 1–56.

Bahar, Sonya. "Ribbon Diagrams and Protein Taxonomy: A Profile of Jane S. Richardson." *Biological Physicist* 4, no. 3 (2004): 5–8.

Bahl, Om P., and Emil L. Smith. "Amino Acid Sequence of Rattlesnake Heart Cytochrome C." *Journal of Biological Chemistry* 240, no. 9 (Sep 1965): 3585–93.

Bajorath, Jürgen, Ronald Stenkamp, and Alejandro Aruffo. "Knowledge-Based Model-Building of Proteins: Concepts and Examples." *Protein Science* 2, no. 11 (Nov 1993): 1798–1810.

Baldwin, Ernest. *An Introduction to Comparative Biochemistry*. Cambridge: Cambridge University Press, 1937.

Baldwin, Melinda Clare. "Credibility, Peer Review, and *Nature*, 1945–1990." *Notes and Records of the Royal Society of London* 69, no. 3 (Sep 20 2015): 337–52.

———. *Making Nature: The History of a Scientific Journal*. Chicago: University of Chicago Press, 2015.

Ball, Catherine A., Alvis Brazma, Helen Causton, Steve Chervitz, Ron Edgar, Pascal Hingamp, John C. Matese, et al. "Submission of Microarray Data to Public Repositories." *PLoS Biology* 2, no. 9 (Sep 2004): E317.

Barinaga, Marcia. "The Missing Crystallography Data." *Science* 245, no. 4923 (Sep 15 1989): 1179–81.

Barker, Winona C., and Margaret O. Dayhoff. "Viral *Src* Gene Products Are Related to the Catalytic Chain of Mammalian cAMP-Dependent Protein Kinase." *Proceedings of the National Academy of Sciences* 79, no. 9 (May 1982): 2836–39.

Barker, Winona C., Lynne K. Ketcham, and Margaret O. Dayhoff. "A Comprehensive Examination of Protein Sequences for Evidence of Internal Gene Duplication." *Journal of Molecular Evolution* 10, no. 4 (Feb 21 1978): 265–81.

Barrow, Mark V. *A Passion for Birds: American Ornithology after Audubon*. Princeton: Princeton University Press, 1998.

———. "The Specimen Dealer: Entrepreneurial Natural History in America's Gilded Age." *Journal of the History of Biology* 33 (2000): 493–534.

Barry, C. David, and Anthony C. T. North. "The Use of a Computer-Controlled Display System in the Study of Molecular Conformations." *Cold Spring Harbor Symposia on Quantitative Biology* 36 (1972): 577–84.

Basilio, Carlos, Albert J. Wahba, Peter Lengyel, Joseph F. Speyer, and Severo Ochoa. "Synthetic Polynucleotides and the Amino Acid Code. V," *Proceedings of the National Academy of Sciences* 48 (Apr 15 1962): 613–16.

Bell, Gordon, Tony Hey, and Alex Szalay. "Computer Science: Beyond the Data Deluge." *Science* 323, no. 5919 (Mar 6 2009): 1297–98.

Benson, Dennis, Mark Boguski, David J. Lipman, and James Ostell. "The National Center for Biotechnology Information." *Genomics* 6, no. 3 (Mar 1990): 389–91.

Benson, Dennis, David J. Lipman, and James Ostell. "GenBank." *Nucleic Acids Research* 21, no. 13 (Jul 1 1993): 2963–65.

Benson, Etienne. "A Centrifuge of Calculation: Spinning Off Data and Enthusiasts in the US Bird Banding Program, 1920–1940," *Osiris* 32 (2017): 286–306.

Benson, Keith R. "From Museum Research to Laboratory Research: The Transformation of Natural History into Academic Biology." In *The American Development of Biology*, ed. Ronald Rainger, Keith R. Benson, and Jane Maienschein, 49–83. Philadelphia: University of Pennsylvania Press, 1988.

———. "Laboratories on the New England Shore: The 'Somewhat Different Direction' of American Marine Biology." *New England Quarterly* 61, no. 1 (1988): 55–78.

Benson, Keith R., Ronald Rainger, and Jane Maienschein, eds. *The Expansion of American Biology*. New Brunswick, NJ: Rutgers University Press, 1987.

Berman, Helen M. "X-Ray Crystallography of Biological Molecules." In *Spectroscopy in Biology and Chemistry*, ed. Sow-Hsin Chen and Sidney Yip, 145–75. New York: Academic Press, 1974.

Berman, Helen M., George A. Jeffrey, and Robert D. Rosenstein. "Crystal Structures of α' and β Forms of D-Mannitol." *Acta Crystallographica Section B Structural Crystallography and Crystal Chemistry* B 24 (1968).

Berman, Helen M. Gerard J. Kleywegt, Haruki Nakamura, and John L. Markley. "How Community Has Shaped the Protein Data Bank." *Structure* 21, no. 9 (Sep 3 2013): 1485–91.

Bernhard, Sidney A., D. F. Bradley, and William L. Duda. "Automatic Determination of Amino Acid Sequences." *IBM Journal of Research and Development* 7, no. 3 (1963): 246–51.

Bernstein, Frances C., Thomas F. Koetzle, Graheme J. B. Williams, Edgar F. Meyer, Michael D. Brice, John R. Rodgers, Olga Kennard, Takehiko Shimanouchi, and Mitsuo Tasumi. "Protein Data Bank: Computer-Based Archival File for Macromolecular Structures." *European Journal of Biochemistry* 80, no. 2 (1977): 319–24.

Bernstein, Herbert J., Lawrence C. Andrews, Helen M. Berman, Frances C. Bernstein, G. H. Campbell, H. L. Carrell, H. B. Chiang, et al. "Second Annual AEC Scientific Computer Information Exchange Meeting." *Proceedings of the Technical Program* 148 (1974): 158.

Biagioli, Mario, and Peter Galison, eds. *Scientific Authorship: Credit and Intellectual Property in Science*. New York: Routledge, 2003.

Bidartondo, Martin I. "Preserving Accuracy in GenBank." *Science* 319, no. 5870 (Mar 21 2008): 1616.

Bilofsky, Howard S., Christian Burks, James W. Fickett, Walter B. Goad, Frances I. Lewitter, Wayne P. Rindone, C. David Swindell, and Chang-Shung Tung. "The GenBank Genetic Sequence Databank." *Nucleic Acids Research* 14, no. 1 (Jan 10 1986): 1–4.

Bisig, Daniel A., Ernesto E. Di Iorio, Kay Diederichs, Kaspar H. Winterhalter, and Klaus Piontek. "Crystal-Structure of Asian Elephant (*Elephas-Maximus*) Cyano-Met-Myoglobin at 1.78- Å Resolution. *Journal of Biological Chemistry* 270, no. 35 (Sep 1 1995): 20754–62.

Blackwelder, Richard E. *Taxonomy: A Text and Reference Book*. New York: Wiley, 1967.

Blair, Ann. *The Theater of Nature: Jean Bodin and Renaissance Science*. Princeton: Princeton University Press, 1997.

———. *Too Much to Know: Managing Scholarly Information before the Modern Age*. New Haven: Yale University Press, 2010.

Blom, Philipp. *To Have and to Hold: An Intimate History of Collectors and Collecting*. Woodstock, NY: Overlook Press, 2003.

Blombäck, Birger. "Travels with Fibrinogen." *Journal of Thrombosis and Haemostasis* 4, no. 8 (Aug 2006): 1653–60.

Blombäck, Birger, Margareta Blombäck, and Nils Jakob Grondahl. "Studies on Fibrinopeptides from Mammals." *Acta Chemica Scandinavica* 19 (1965): 1789–91.

Blombäck, Margareta. "Thrombosis and Haemostasis Research: Stimulating, Hard Work and Fun." *Journal of Thrombosis and Haemostasis* 98, no. 1 (Jul 2007): 8–15.

"Blood Crystals Aid Detectives." *Washington Post*, Sep 1 1912, M1.

"Blood Proteins to Be Kept in New, Special Museum." *Science News Letter*, Jun 12 1948, 377.

"Blood Will Really Tell." *Washington Post*, Sep 28 1913, MS3.

Bloom, Floyd E. "Policy Change." *Science* 281, no. 5374 (1998): 175.

Blumenthal, D. David, Eric G. Campbell, Manjusha Gokhale, Recai Yucel, Brian Clarridge, Stephen Hilgartner, and Neil A. Holtzman. "Data Withholding in Genetics and the Other Life Sciences: Prevalences and Predictors." *Academic Medicine* 81, no. 2 (Feb 2006): 137–45.

Bolker, J. A. "Exemplary and Surrogate Models: Two Modes of Representation in Biology." *Perspectives in Biology and Medicine* 52, no. 4 (Autumn 2009): 485–99.

Bourne, Philip. "Will a Biological Database Be Different from a Biological Journal?" *PLoS Computational Biology* 1, no. 3 (Aug 2005): 179–81.

Boyd, William C. "Systematics, Evolution, and Anthropology in the Light of Immunology." *Quarterly Review of Biology* 24, no. 2 (1949): 102–8.

Boyden, Alan A. "Collecting Serological Samples." *Atoll Research Bulletin* 17 (1953): 96–99.

———. "Fifty Years of Systematic Serology." *Systematic Zoology* 2, no. 1 (1953): 19–30.

———. "The 'Flying Lemur' Comes to the Museum." *Serological Museum Bulletin* 4 (1950): 4–5.

———. "Introductory Remarks." In *Serological and Biochemical Comparisons of Proteins*, ed. William H. Cole, 1–2. New Brunswick: Rutgers University Press, 1958.

———. "Major Objectives." *Serological Museum Bulletin* 4 (1950): 1.

———. "Our Contacts Grow." *Serological Museum Bulletin* 3 (1949): 1.

———. "The Precipitin Reaction in the Study of Animal Relationships." *Biological Bulletin* 50, no. 2 (Feb 1926): 73–107.

———. "Precipitin Tests as a Basis for a Quantitative Phylogeny." *Proceeding of the Society for Experimental Biology* 29 (1932): 955–57.

———. "Precipitins and Phylogeny in Animals." *American Naturalist* 68 (1934): 516–36.

———. "Serology and Animal Systematics." *American Naturalist* 77 (1943): 234–55.

———. "Systematic Serology: A Critical Appreciation." *Physiological Zoölogy* 15, no. 2 (1942): 109–45.

———. "Zoological Collecting Expeditions and the Salvage of Animal Bloods for Comparative Serology." *Science* 118, no. 3054 (Jul 10 1953): 57–58.

Boyden, Alan A., and Joseph G. Baier Jr. "A Rapid Quantitative Precipitin Technic." *Journal of Immunology* 17, no. 1 (Jul 1929): 29–37.

Boyden, Alan A., and Ralph J. DeFalco. "Report on the Use of the Photoreflectometer in Serological Comparisons." *Physiological Zoölogy* 16, no. 3 (1943): 229–41.

Boyden, Alan A., and G. Kingsley Noble. "The Relationships of Some Common Amphibia as Determined by Serological Study." *American Museum Novitates* 606 (1933): 1–24.

Brandt, Christina. *Metapher und Experiment: Von der Virusforschung zum genetischen Code.* Göttingen: Wallstein, 2004.

Brazma, Alvis, Helen Parkinson, Ugis Sarkans, Mohammadreza Shojatalab, Jaak Vilo, Niran Abeygunawardena, Ele Holloway, et al. "Arrayexpress: A Public Repository for Microarray Gene Expression Data at the EBI." *Nucleic Acids Research* 31, no. 1 (Jan 1 2003): 68–71.

Brecht, Christine and Sybilla Nikolow. "Displaying the Invisible: *Volkskrankheiten* on Exhibition in Imperial Germany," *Studies in History and Philosophy of Biological and Biomedical Sciences* 31, no. 4 (2000): 511–30.

Brembs, Björn, Katherine Button, and Marcus Munafò. "Deep Impact: Unintended Consequences of Journal Rank." *Frontiers in Human Neuroscience* 7, no. 291 (2013): 1–12.

Brenner, Sydney. "On the Impossibility of All Overlapping Triplet Codes in Information Transfer from Nucleic Acid to Proteins." *Proceedings of the National Academy of Sciences* 43, no. 8 (Aug 15 1957): 687–94.

Brockway, Lucile. *Science and Colonial Expansion: The Role of the British Royal Botanic Gardens.* New Haven: Yale University Press, 2002.

Brown, H., Frederick Sanger, and Ruth Kitai. "The Structure of Pig and Sheep Insulins." *Biochemical Journal* 60, nos. 1–4 (1955): 556–65.

Brown, Patrick O., and David Botstein. "Exploring the New World of the Genome with DNA Microarrays." *Nature Genetics* 21, no. 1 suppl. (Jan 1999): 33–37.

Browne, Wynne J., Anthony C North, David C. Phillips, Keith Brew, Thomas C. Vanaman, and Robert L. Hill. "A Possible Three-Dimensional Structure of Bovine Lactalbumin Based on That of Hen's Egg-White Lysozyme." *Journal of Molecular Biology* 42, no. 1 (May 28 1969): 65–86.

Brunak, Søren, Antoine Danchin, Masahira Hattori, Haruki Nakamura, Kazuo Shinozaki, Tara Matise, and Daphne Preuss. "Nucleotide Sequence Database Policies." *Science* 298, no. 5597 (Nov 15 2002): 1333.

Brush, Alan H. "Charles Gald Sibley." *Biographical Memoirs of the National Academy of Sciences* 83 (2003): 216–39.

Brutlag, Douglas L., Jan Clayton, Peter Friedland, and Laurence H. Kedes. "SEQ: A Nucleotide Sequence Analysis and Recombination System." *Nucleic Acids Research* 10, no. 1 (Jan 11 1982): 279–94.

Buchanan, Robert E. "History and Development of the American Type Culture Collection." *Quarterly Review of Biology* 41, no. 2 (1966): 101–4.

Buettner-Janusch, John, and Robert L. Hill. "Molecules and Monkeys." *Science* 147 (Feb 19 1965): 836–42.

Buffon, Georges Louis Leclerc. *Histoire naturelle, générale et particulière, avec la description du Cabinet du Roy*. Paris, 1749.

Bungener, Patrick, and Marino Buscaglia. "Cytology and Mendelism: Early Connection between Michael F. Guyer's Contribution." *History and Philosophy of the Life Sciences* 25, no. 1 (2003): 27–50.

Burge, Sarah, Teresa K. Attwood, Alex Bateman, Tanya Z. Berardini, Michael Cherry, Claire O'Donovan, Ioannis Xenarios, and Pascale Gaudet. "Biocurators and Biocuration: Surveying the 21st Century Challenges." *Database* (2012): bar059.

Burian, Richard M. "How the Choice of Experimental Organism Matters: Epistemological Reflections on an Aspect of Biological Practice." *Journal of the History of Biology* 26, no. 2 (1993): 351–67.

Burkhardt, Kyle, Bohdan Schneider, and Jeramia Ory. "A Biocurator Perspective: Annotation at the Research Collaboratory for Structural Bioinformatics Protein Data Bank." *PLoS Computational Biology* 2, no. 10 (Oct 2006): 1186–89.

Burkhardt, Richard W., Jr. "The Leopard in the Garden: Life in Close Quarters at the Museum d'Histoire *Naturelle*." *Isis* 98, no. 4 (Dec 2007): 675–94.

———. "Naturalists' Practices and *Nature*'s Empire: Paris and the Platypus, 1815–1833." *Pacific Science* 55 (2001): 327–41.

Burks, Christian. "Biotechnology and the Human Genome Innovations and Impact." In *Basic Life Sciences 46*, ed. Avril D. Woodhead and Benjamin J. Barnhart, 51–56. Boston, MA: Springer US, 1988.

Burks, Christian, James W. Fickett, and Walter B. Goad. "GenBank." *Science* 235 (1987): 267–68.

Burks, Christian, James W. Fickett, Walter B. Goad, Minoru Kanehisa, Frances I. Lewitter, Wayne P. Rindone, C. David Swindell, Chang-Shung Tung, and Howard S. Bilofsky. "The GenBank Nucleic Acid Sequence Database." *Computer Applications in the Biosciences* 1, no. 4 (Dec 1985): 225–33.

Bussard, Alain E. "Data Proliferation: A Challenge for Science and for Codata." In *Biomolecular Data: A Resource in Transition*, ed. Rita Colwell, 11–15. Oxford: Oxford University Press, 1989.

Cain, Joseph Allen. "Common Problems and Cooperative Solutions: Organizational Activity in Evolutionary Studies, 1936–1947." *Isis* 84, no. 1 (Mar 1993): 1–25.

———. "Launching the Society of Systematic Zoology in 1947." In *Milestones in Systematics*, ed. David M. Williams and Peter L. Forey, 19–48. Boca Raton: CRC Press, 2004.

Cairns, John, Gunther S. Stent, and James D. Watson, eds. *Phage and the Origins of Molecular Biology*. New York: Cold Spring Harbor Press, 1966.

Cameron, Graham. "The Impact of Electronic Publishing on the Academic Community." In *Electronic Databases and the Scientific Record*. London: Portland Press, 1997.

Campbell, Philip. "New Policy for Structure Data." *Nature* 394, no. 6689 (Jul 9 1998): 105.

Campbell-Kelly, Martin, and William Aspray. *Computer: A History of the Information Machine*. Boulder: Westview Press, 2004.

Cassatt, James C., and Jane L. Peterson. "GenBank Information." *Science* 238, no. 4831 (Nov 27 1987): 1215.

Castelao-Lawless, Teresa. "Phenomenotechnique in Historical Perspective: Its Origins and Implications for Philosophy of Science." *Philosophy of Science* 62, no. 1 (Mar 1995): 44–59.

Castleman, Paul. "Medical Application of Computers at BBN." *IEEE Annals of the History of Computing* 28, no. 1 (Jan–Mar 2006): 6–16.

Castleman, Paul A., Channing H. Russell, Frederick N. Webb, Charlotte A. Hollister, J. R. Siegel, S. R. Zdoxik, and David M. Fram. "The Implementation of the PROPHET System." *AFIPS Conference Proceedings* 43 (Jan–Mar 1974): 457–68.

"Charles-Edward Amory Winslow, February 4, 1887–January 8, 1957." *American Journal of Public Health and the Nation's Health* 47, no. 2 (Feb 1957): 153–67.

Charmantier, Isabelle, and Staffan Muller-Wille. "Carl Linnaeus's Botanical Paper Slips (1767–1773)." *Intellectual History Review* 24, no. 2 (Apr 3 2014): 215–38.

———. "Natural History and Information Overload: The Case of Linnaeus." *Studies in History and Philosophy of Biological and Biomedical Sciences* 43, no. 1 (2012): 4–15.

Chester, K. Starr. "A Critique of Plant Serology. Part I. The Nature and Utilization of Phytoserological Procedures." *Quarterly Review of Biology* 12, no. 1 (1937): 19–46.

———. "A Critique of Plant Serology. Part II. Application of Serology to the Classification of Plants and the Identification of Plant Products." *Quarterly Review of Biology* 12, no. 2 (1937): 165–90.

———. "A Critique of Plant Serology. Part III. Phytoserology in Medicine and General Biology. Bibliography." *Quarterly Review of Biology* 12, no. 3 (1937): 294–321.

Chothia, Cyrus. "One Thousand Families for the Molecular Biologist." *Nature* 357, no. 6379 (Jun 18 1992): 543–44.

———. "Structural Invariants in Protein Folding." *Nature* 254, no. 5498 (Mar 27 1975): 304–8.

Chothia, Cyrus, and Arthur M. Lesk. "Canonical Structures for the Hypervariable Regions of Immunoglobulins." *Journal of Molecular Biology* 196, no. 4 (Aug 20 1987): 901–17.

———. "The Relation between the Divergence of Sequence and Structure in Proteins." *EMBO Journal* 5, no. 4 (Apr 1986): 823–26.

Churchill, Frederick B. "In Search of the New Biology: An Epilogue." *Journal of the History of Biology* 14, no. 1 (Spring 1981): 177–91.

———. "Life before Model Systems: General Zoology at August Weismann's Institute." *American Zoologist* 37, no. 3 (Jun 1997): 260–68.

Cinkosky, Michael J., James W. Fickett, Paul Gilna, and Christian Burks. "Electronic Data Publishing and GenBank," *Science* 252, no. 5010 (May 31 1991): 1273–77.

Clark, William A., and Dorothy H. Geary. "The Story of the American Type Culture Collection: Its History and Development (1899–1973)." *Advances in Applied Microbiology* 17 (1974): 295–309.

Clarke, Adele E., and Joan H. Fujimura, eds. *The Right Tool for the Job: At Work in Twentieth-Century Life Sciences*. Princeton: Princeton University Press, 1992.

Clause, Bonnie T. "The Wistar Rat as a Right Choice: Establishing Mammalian Standards and the Ideal of a Standardized Mammal." *Journal of the History of Biology* 26, no. 2 (Sum 1993): 329–49.

Cobb, Matthew. "The Prehistory of Biology Preprints: A Forgotten Experiment from the 1960s." *PeerJ Preprints* 5:e3174v1 (2017).

Coe, Edward H., Jr. "East, Emerson, and the Birth of Maize Genetics." In *Handbook of Maize: Genetics and Genomics*, ed. J. L. Bennetzen and S. Hake, 3–15. New York: Springer-Verlag, 2009.

———. "The Origins of Maize Genetics." *Nature Reviews Genetics* 2, no. 11 (Nov 2001): 898–905.

Cohen, Elias, and Gunnar B Stickler. "Absence of Albuminlike Serum Proteins in Turtles." *Science* 127 (1958): 1392.

Cohen, Jon. "The Culture of Credit." *Science* 268, no. 5218 (Jun 23 1995): 1706–11.

Cole, Douglas. *Captured Heritage: The Scramble for Northwest Coast Artifacts*. Norman: University of Oklahoma Press, 1995.

Coleman, William R. *Biology in the Nineteenth Century: Problems of Form, Function and Transformation*. Cambridge: Cambridge University Press, 1971.

Collins, Douglas M., F. Albert Cotton, Edward E. Hazen Jr., Edgar F. Meyer, and Carl N. Morimoto. "Protein Crystal Structures: Quicker, Cheaper Approaches." *Science* 190, no. 4219 (Dec 12 1975): 1047–53.

Collins, Francis. "In the Crossfire: Collins on Genomes, Patents, and 'Rivalry.' Interview by Eliot Marshall, Elizabeth Pennisi, and Leslie Roberts." *Science* 287, no. 5462 (Mar 31 2000): 2396–98.

Comfort, Nathaniel C. *The Science of Human Perfection: How Genes Became the Heart of American Medicine*. New Haven: Yale University Press, 2012.

Commission on Journals. "Deposition of Macromolecular Atomic Coordinates and Structure Factors with the Protein Data Bank." *Acta Crystallographica* B37 (1981): 1161–62.

———. "Deposition of Macromolecular Atomic Coordinates and Structure Factors with the Protein Data Bank: Modified Policy." *Acta Crystallographica* B38 (1982): 1050.

"Computing in the University." *Datamation* 8 (1962): 27–30.

Comte, Auguste. *Cours de philosophie positive*. Paris: Bachelier, 1830.

Contreras, Jorge L. "Bermuda's Legacy: Policy, Patents, and the Design of the Genome Commons." *Minnesota Journal of Law, Science and Technology* 12, no. 1 (2011): 61–125.

Cook, Harold J. *Matters of Exchange: Commerce, Medicine, and Science in the Dutch Golden Age*. New Haven: Yale University Press, 2007.

Cooper, Caren. *Citizen Science: How Ordinary People Are Changing the Face of Discovery*. New York: Overlook Press, 2016.

Corbin, Kendall W., and Alan H. Brush. "In Memoriam: Charles Gald Sibley, 1917–1998." *Auk* 116, no. 3 (1999): 806–14.

Costello, Mark J. "Motivating Online Publication of Data." *Bioscience* 59, no. 5 (May 2009): 418–27.

Cox, Francis E. G. "George Henry Falkiner Nuttall and the Origins of Parasitology and *Parasitology*." *Parasitology* 136, no. 12 (Mar 30 2009): 1389–94.

Craven, Bryan. "George A. Jeffrey (1915–2000)." *Acta Crystallographica Section B—Structural Science* 56 (Aug 2000): 545–46.

Creager, Angela N. H. *The Life of a Virus: Tobacco Mosaic Virus as an Experimental Model, 1930–1965*. Chicago: University of Chicago Press, 2002.

———. "The Paradox of the Phage Group: Essay Review." *Journal of the History of Biology* 43, no. 1 (Feb 2010): 183–93.

Creager, Angela N. H., Elizabeth Lunbeck, and M. Norton Wise. *Science without Laws: Model Systems, Cases, Exemplary Narratives*. Durham: Duke University Press, 2007.

Crease, Robert P. *Making Physics. A Biography of Brookhaven National Laboratory, 1946–1972*. Chicago: The University of Chicago Press, 1999.

Crick, Francis H. C., and John C. Kendrew. "X-Ray Analysis and Protein Structure." *Advances in Protein Chemistry* 12 (1957): 133–214.

Crippen, Gordon M. "The Tree Structural Organization of Proteins." *Journal of Molecular Biology* 126, no. 3 (Dec 15 1978): 315–32.

Crowther, James G. *Famous American Men of Science*. Harmondsworth, UK, 1944.

"Crystallography Protein Data Bank." *Journal of Molecular Biology* 78 (1971): 587.

Cumley, Russell W. "Comparison of Serologic and Taxonomic Relationships of Drosophila Species." *Journal of the New York Entomological Society* 48, no. 3 (1940): 265–74.

Cunningham, Andrew. *The Anatomist Anatomis'd: An Experimental Discipline in Enlightenment Europe*. Farnham, Surrey, UK: Ashgate, 2010.

Curry, Helen A. "From Working Collections to the World Germplasm Project: Agricultural Modernization and Genetic Conservation at the Rockefeller Foundation." *History and Philosophy of the Life Sciences* 39, no. 2 (Jun 2017): 5.

Curry, Helen A., Nicholas Jardine, James A. Secord, and Emma C. Spary, eds. *Worlds of Natural History*. Cambridge: Cambridge University Press, 2018.

Cuvier, Georges, *Le règne animal distribué d'après son organisation, pour servir de base à l'histoire naturelle des animaux et d'introduction à l'anatomie comparée*. Paris: Deterville, 1817.

Darnton, Robert. "An Early Information Society: News and the Media in Eighteenth-Century Paris." *American Historical Review* 105, no. 1 (2000): 1–35.

Daston, Lorraine. "The Empire of Observation, 1600–1800." In *Histories of Scientific Observation*, ed. Lorraine Daston and Elizabeth Lunbeck, 81–113. Chicago: University of Chicago Press, 2011.

———. *Science in the Archives: Pasts, Presents, Futures*. Chicago: University of Chicago Press, 2017.

———. "Type Specimens and Scientific Memory." *Critical Inquiry* 31 (2004): 153–82.

Daston, Lorraine, and Peter Galison. “The Image of Objectivity.” *Representations* 40 (1992): 81–128.
———. *Objectivity*. New York: Zone Books, 2007.
Daston, Lorraine, and Elizabeth Lunbeck, eds. *Histories of Scientific Observation*. Chicago: University of Chicago Press, 2011.
Daston, Lorraine, and Katharine Park, eds. *Wonders and the Order of Nature, 1150–1750*. Cambridge, MA: Zone Books, 1998.
Dawid, Igor B. “Editorial Submission of Sequences.” *Proceedings of the National Academy of Sciences* 86 (1989): 407.
Dayhoff, Margaret O. *Atlas of Protein Sequence and Structure*. Silver Spring, MD: National Biomedical Research Foundation, 1969.
———. *Atlas of Protein Sequence and Structure. Vol. 5. Suppl. 2.* Washington, DC: National Biomedical Research Foundation, 1976.
———. “Computer Analysis of Protein Evolution.” *Scientific American* 221 (1969): 86–95.
———. “Computer Analysis of Protein Sequences.” *Federation Proceedings* 33, no. 12 (Dec 1974): 2314–16.
———. “Computer Search for Active Site Configurations.” *Journal of the American Chemical Society* 86, no. 11 (1964).
Dayhoff, Margaret O., Richard V. Eck, Marie A. Chang, and Minnie R. Sochard. *Atlas of Protein Sequence and Structure*. Silver Spring, MD: National Biomedical Research Foundation, 1965.
Dayhoff, Margaret O., Richard V. Eck, Ellis R. Lippincott, and Carl Sagan. “Venus: Atmospheric Evolution.” *Science* 155, no. 3762 (Feb 3 1967): 556–58.
Dayhoff, Margaret O., and Robert S. Ledley. “Comprotein: A Computer Program to Aid Primary Protein Structure Determination.” In *Proceedings of the Fall Joint Computer Conference*, 262–74. Santa Monica: American Federation of Information Processing Societies, 1962.
Dayhoff, Margaret O., Ellis. R. Lippincott, and Richard V. Eck. “Thermodynamic Equilibria in Prebiological Atmospheres.” *Science* 146, no. 1461 (Dec 11 1964): 1461–64.
Dayhoff, Margaret O., Peter J. McLaughlin, Winona C. Barker, and Lois T. Hunt. “Evolution of Sequences within Protein Superfamilies.” *Naturwissenschaften* 62, no. 4 (1975): 154–61.
Dayhoff, Margaret O., Gertrude E. Perlmann, and Duncan A. MacInnes. “The Partial Specific Volumes, in Aqueous Solution, of Three Proteins.” *Journal of the American Chemical Society* 74, no. 10 (1952): 2515–17.
Dayhoff, Margaret O., Robert M. Schwartz, H. R. Chen, Lois T. Hunt, Winona C. Barker, and Bruce C. Orcutt. “Nucleic Acid Sequence Bank.” *Science* 209, no. 4462 (Sep 12 1980): 1182.
Dayhoff, Margaret O., Robert M. Schwartz, H. R. Chen, Lois T. Hunt, Bruce C. Orcutt, and Winona C. Barker. “Banking DNA Sequences.” *Nature* 286 (1980): 326.
de Bon, Raf. “Between the Laboratory and the Deep Blue Sea: Space Issues in the Marine Stations of Naples and Wimereux.” *Social Studies of Science* 39, no. 2 (2009): 199–227.
———. *Stations in the Field: A History of Place-Based Animal Research, 1870–1930*. Chicago: University of Chicago Press, 2015.

de Chadarevian, Soraya. *Designs for Life: Molecular Biology after World War II*. Cambridge: Cambridge University Press, 2002.

———. "Following Molecules: Haemoglobin between the Clinic and the Laboratory." In *Molecularizing Biology and Medicine: New Practices and Alliances, 1910s–1970s*, ed. Soraya de Chadarevian and Harmke Kamminga, 171–201. Amsterdam: Harwood Academic Publishers, 1998.

———. "Models and the Making of Molecular Biology." In *Models: The Third Dimension of Science*, ed. Soraya de Chadarevian and Nick Hopwood, 339–68. Stanford: Stanford University Press, 2004.

———. "Of Worms and Programmes: *Caenorhabditis Elegans* and the Study of Development." *Studies in History and Philosophy of Biological and Biomedical Sciences* 29, no. 1 (1998): 81–105.

———. "Protein Sequencing and the Making of Molecular Genetics." *Trends in Biochemical Sciences* 24 (1999): 203–6.

de Chadarevian, Soraya, and Jean-Paul Gaudillière. "The Tools of the Discipline: Biochemists and Molecular Biologists." *Journal of the History of Biology* 29, no. 3 (1996): 327–30.

de Chadarevian, Soraya, and Harmke Kamminga, eds. *Molecularizing Biology and Medicine: New Practices and Alliances, 1910s–1970s*. Amsterdam: Harwood Academic Publishers, 1998.

de Solla Price, Derek J. *Little Science, Big Science*. New York: Columbia University Press, 1963.

Delbrück, Max. "Experiments with Bacterial Viruses (Bacteriophages)." *Harvey Lectures* 41 (1946): 161–87.

Delfanti, Alessandro. *Biohackers: The Politics of Open Science*. London: Pluto Press, 2013.

Demerec, Milislav. "Department of Genetics." *Carnegie Institution of Washington Year Book* 47 (1948): 139–43.

"Democracy and the Recognition of Science." *Popular Science*, Mar 1902, 477–78.

Dene, Howard, Morris Goodman, and Alejo E. Romero-Herrera. "The Amino Acid Sequence of Elephant (*Elephas Maximus*) Myoglobin and the Phylogeny of Proboscidea." *Proceedings of the Royal Society B—Biological Sciences* 207, no. 1166 (Feb 13 1980): 111–27.

Dessauer, Herbert C., and Wade Fox. "Characteristic Electrophoretic Patterns of Plasma Proteins of Orders of Amphibia and Reptilia." *Science* 124, no. 3214 (Aug 3 1956): 225–26.

Dessauer, Herbert C., and Mark S. Hafner. *Collections of Frozen Tissues: Value, Management, Field and Laboratory Procedures, and Directory of Existing Collections*. Lawrence, KS: Association of Systematics Collections, 1984.

Deutsch, Harold F., and Martha B. Goodloe. "An Electrophoretic Survey of Various Animal Plasmas." *Journal of Biological Chemistry* 161 (1945): 1–20.

Diamond, Louis K. "The Story of Our Blood Groups." In *Blood, Pure and Eloquent*, ed. Maxwell M. Wintrobe, 691–717. New York: McGraw-Hill Book Company, 1980.

Dick, Steven J., and James Edgar Strick. *The Living Universe: NASA and the Development of Astrobiology*. New Brunswick, NJ: Rutgers University Press, 2004.

Dickson, David. "Stanford Ready to Fight for Patent." *Nature* 292, no. 5824 (Aug 13 1981): 573.

Dietrich, Michael R. "Paradox and Persuasion: Negotiating the Place of Molecular Evolution within Evolutionary Biology." *Journal of the History of Biology* 31 (1998): 85–111.

Dietrich, Michael R., and Brandi H. Tambasco. "Beyond the Boss and the Boys: Women and the Division of Labor in Drosophila Genetics in the United States, 1934–1970." *Journal of the History of Biology* 40, no. 3 (2007): 509–28.

Dietrich, Michael R., Rachel A. Ankeny, and Patrick M. Chen. "Publication Trends in Model Organism Research." *Genetics* 198, no. 3 (Nov 2014): 787–94.

Di-Poi, Nicolas, Juan I. Montoya-Burgos, Hilary Miller, Olivier Pourquie, Michel C. Milinkovitch, and Denis Duboule. "Changes in Hox Genes' Structure and Function during the Evolution of the Squamate Body Plan." *Nature* 464, no. 7285 (Mar 4 2010): 99–103.

Dobzhansky, Theodosius. "Taxonomy, Molecular Biology, and the Peck Order." *Evolution* 15, no. 2 (1961): 263–64.

Dobzhansky, Theodosius. "Biology, Molecular and Organismic." *American Zoologist* 4 (Nov 1964): 443–52.

Doggett, Philip E., and Frederick R. Blattner. "Personal Access to Sequence Databases on Personal Computers." *Nucleic Acids Research* 14, no. 1 (Jan 10 1986): 611–19.

Doolittle, Russell F. "Characterization of Lamprey Fibrinopeptides." *Biochemical Journal* 94 (Mar 1965): 742–50.

———. *Of Urfs and Orfs: A Primer on How to Analyze Derived Amino Acid Sequences*. Mill Valley, CA: University Science Books, 1986.

———. "On the Trail of Protein Sequences." *Bioinformatics* 16, no. 1 (Jan 2000): 24–33.

Doolittle, Russell F., and Birger Blombäck. "Amino-Acid Sequence Investigations of Fibrinopeptides from Various Mammals: Evolutionary Implications." *Nature* 202 (1964): 147–52.

Doolittle, Russell F., Seymour J. Singer, and Henry Metzger. "Evolution of Immunoglobulin Polypeptide Chains: Carboxy-Terminal of an IgM Heavy Chain." *Science* 154, no. 756 (Dec 23 1966): 1561–62.

Duboule, Denis. "The Hox Complex: An Interview with Denis Duboule. Interviewed by Michael K. Richardson," *International Journal of Developmental Biology* 53, nos. 5–6 (2009): 717–23.

———. "The Rise and Fall of Hox Gene Clusters." *Development* 134, no. 14 (Jul 2007): 2549–60.

———. "Temporal Colinearity and the Phylotypic Progression: A Basis for the Stability of a Vertebrate Bauplan and the Evolution of Morphologies through Heterochrony." *Development Supplement* (1994): 135–42.

Duboule, Denis, Agnès Baron, Philippe Mahl, and Brigitte Galliot. "A New Homeo-Box Is Present in Overlapping Cosmid Clones Which Define the Mouse Hox-1 Locus." *EMBO Journal* 5, no. 8 (Aug 1986): 1973–80.

Dunn, Leslie C. "William Ernest Castle." *Biographical Memoirs of the National Academy of Science of the United States of America* 38 (1965): 33–80.

Dutfield, Graham. *Intellectual Property Rights and the Life Science Industries: A Twentieth Century History*. Hackensack, NJ: World Scientific, 2009.

Dyer, Owen. "Publishers Hire PR Heavyweight to Defend Themselves against Open Access." *British Medical Journal* 334, no. 7587 (Feb 3 2007): 227.

Eck, Richard V. "Cryptogrammic Detection of a Pattern in Amino Acid 'Alleles': Its Use in Tracing the Evolution of Proteins." *Proceedings of the 17th Annual Conference on Engineering in Medicine and Biology* 6 (1964): 115.

———. "Non-randomness in Amino-Acid 'Alleles.'" *Nature* 191 (Sep 23 1961): 1284–85.

———. "The Protein Cryptogram: I Non-random Occurrence of Amino Acid 'Alleles.'" *Journal of Theoretical Biology* 2 (1962): 139–51.

Eck, Richard V., and Margaret O. Dayhoff. *Atlas of Protein Sequence and Structure*. Silver Spring, MD: National Biomedical Research Foundation, 1966.

———. "Evolution of the Structure of Ferredoxin Based on Living Relics of Primitive Amino Acid Sequences." *Science* 152, no. 3720 (Apr 15 1966): 363–66.

The Editors. "Deposition of Nucleotide Sequence Data in the Data Banks." *Nucleic Acids Research* 15, no. 18 (1987).

———. "Instruction to Authors." *Nucleic Acids Research* 11, no. 24 (1983).

Edman, Pehr, and Geoffrey Begg. "A Protein Sequenator." FEBS Journal 1, no. 1 (March 1967): 80–91.

The Editors. "Is a Government Archive the Best Option?" *Science* 291, no. 5512 (Mar 23 2001): 2318–19.

"Edward Dayhoff; Contributor to Nobel-Winning Work on Atom." *Washington Post*, Jan 11 2007.

"Egg White Clue to Evolution." *Hartford Courant*, Jan 11 1968, 43C.

Elson, D., and E. Chargaff. "Evidence of Common Regularities in the Composition of Pentose Nucleic Acids." *Biochimica et Biophysica Acta* 17, no. 3 (Jul 1955): 367–76.

Emerson, Rollins A. "The Present Status of Maize Genetics." *Proceedings of the Sixth International Congress of Genetics* 1 (1932): 141–52.

Emerson, Rollins A., George W. Beadle, and Allan C. Fraser. "A Summary of Linkage Studies in Maize." *Cornell University Agricultural Experiment Station Memoirs* 180 (1935).

ENCODE Project Consortium. "An Integrated Encyclopedia of DNA Elements in the Human Genome." *Nature* 489, no. 7414 (Sep 6 2012): 57–74.

Endersby, Jim. *A Guinea Pig's History of Biology*. Cambridge: Harvard University Press, 2007.

———. *Imperial Nature: Joseph Hooker and the Practices of Victorian Science*. Chicago: University of Chicago Press, 2008.

Engelward, Bevin P., and Richard J. Roberts. "Open Access to Research Is in the Public Interest." *PLoS Biology* 5, no. 2 (2007): e48.

Epstein, Steven. *Impure Science: AIDS, Activism, and the Politics of Knowledge*. Berkeley: University of California Press, 1996.

Erhardt, Albert. "Die Verwandtschaftsbestimmungen Mittels der Immunitätsreaktionen in der Zoologie und ihr Wert für phylogenetische Untersuchungen." *Ergebnisse und Fortschritte der Zoologie* 7 (1931): 279–377.

Ewing, Anna H. "Blood Will Tell." *Popular Science* 93, no. 6 (1918): 72–73.

Fabian, Ann. *The Skull Collectors: Race, Science, and America's Unburied Dead*. Chicago: University of Chicago Press, 2010.

Fan, Fa-ti. *British Naturalists in Qing China: Science, Empire, and Cultural Encounter*. Cambridge: Harvard University Press, 2003.

Farber, Paul L. "Development of Taxidermy and History of Ornithology." *Isis* 68, no. 244 (1977): 550–66.

Feduccia, Alan. "Reviewed Work: A Comparative Study of the Egg-White Proteins of Passerine Birds by Sibley, Charles G.; A Comparative Study of the Egg-White Proteins of Non-passerine Birds by Sibley, Charles G.; Jon E. Ahlquist." *Auk* 90, no. 4 (1973): 919–21.

Feigenbaum, Edward A., and Edward H. Shortliffe. "Sumex Annual Report—Year 09." 1982.

Felsenstein, Joseph. *Inferring Phylogenies*. Sunderland, MA: Sinauer Associates, 2004.

Ferguson, Andrew. *Biochemical Systematics and Evolution*. New York: Wiley, 1980.

Ferretti, Fred. "Fining of Bird Scholar Stirs Colleagues." *New York Times*, Jul 13 1973, 55.

Ferry, Georgina. *Max Perutz and the Secret of Life*. New York: Cold Spring Harbor Laboratory Press, 2007.

Fickett, James W. "Recognition of Protein Coding Regions in DNA Sequences." *Nucleic Acids Research* 10, no. 17 (Sep 11 1982): 5303–18.

Findlen, Paula. "The Economy of Scientific Exchange in Early Modern Italy." In *Patronage and Institutions: Science, Technology, and Medicine at the European Court, 1500–1750*, ed. Bruce T. Moran, 5–24. Rochester, NY: Boydell Press, 1991.

———. *Possessing Nature: Museums, Collecting, and Scientific Culture in Early Modern Italy*. Berkeley: University of California Press, 1994.

Fischer, Ernst P., and Carol Lipson. *Thinking About Science: Max Delbrück and the Origins of Molecular Biology*. New York: W. W. Norton, 1988.

Fitch, Walter M. "Evidence Suggesting a Partial, Internal Duplication in the Ancestral Gene for Heme-Containing Globins." *Journal of Molecular Biology* 16, no. 1 (Mar 1966): 17–27.

———. "The Probable Sequence of Nucleotides in Some Codons." *Proceedings of the National Academy of Sciences* 52 (Aug 1964): 298–305.

———. "The Relation between Frequencies of Amino Acids and Ordered Trinucleotides." *Journal of Molecular Biology* 16, no. 1 (Mar 1966): 1–8.

Fitch, Walter M., and Emanuel Margoliash. "Construction of Phylogenetic Trees." *Science* 155, no. 760 (Jan 20 1967): 279–84.

Fitzgerald, Deborah Kay. *The Business of Breeding: Hybrid Corn in Illinois, 1890–1940*. Ithaca, NY: Cornell University Press, 1990.

Florkin, Marcel. *Biochemical Evolution*. New York: Academic Press, 1949.

———. *L'évolution biochimique*. Paris: Masson, 1944.

Fontes da Costa, Palmira. "The Culture of Curiosity at the Royal Society in the First Half of the Eighteenth Century." *Notes and Records of the Royal Society* 56, no. 2 (May 2002): 147–66.

Forgan, Sophie. "The Architecture of Display: Museums, Universities and Objects in Nineteenth-Century Britain." *History of Science* 32, no. 96 (Jun 1994): 139–62.

Forman, Paul. "Social Niche and Self-Image of the American Physicist." In *The Restructuring of Physical Sciences in Europe and the United States, 1945–1960*, ed. Michelangelo De Maria, Mario Grilli, and Fabio Sebastiani, 96–104. Singapore: World Scientific Publishing, 1989.

"For the Birds." *New York Times*, Jul 17 1974, 36.

Fortun, Michael. "Projecting Speed Genomics." In *The Practices of Human Genetics*, ed. M. Fortun and E. Mendelsohn, 25–87. Dordrecht: Kluwer, 1999.

Francoeur, Eric. "Cyrus Levinthal, the Kluge and the Origins of Interactive Molecular Graphics." *Endeavour* 26, no. 4 (Dec 2002): 127–31.

Francoeur, Eric, and Jérôme Segal. "From Model Kits to Interactive Graphics." In *Models: The Third Dimension of Science*, ed. Soraya de Chadarevian and Nick Hopwood, 402–29. Stanford: Stanford University Press, 2004.

Frankel, Felice, and Rosalind Reid. "Big Data: Distilling Meaning from Data." *Nature* 455, no. 7209 (Sep 4 2008): 30.

Frayling, Christopher. *Mad, Bad and Dangerous?: The Scientist and the Cinema*. London: Reaktion, 2005.

French, Alfred D. "George Alan Jeffrey." *Advances in Carbohydrate Chemistry and Biochemistry* 57 (2001): 1–9.

Friedland, Peter, and Laurence H. Kedes. "Discovering the Secrets of DNA." *Computer* 18, no. 11 (1985): 49–69.

Friedmann, Herbert C. "From 'Butyribacterium' to 'E. Coli': An Essay on Unity in Biochemistry." *Perspectives in Biology and Medicine* 47, no. 1 (Win 2004): 47–66.

Friedmann, Theodore, Russell F. Doolittle, and Gernot Walter. "Amino-Acid Sequence Homology between Polyoma and SV40 Tumor Antigens Deduced from Nucleotide-Sequences." *Nature* 274, no. 5668 (1978): 291–93.

Fulton, John F. "C.-E. A. Winslow, Leader in Public Health." *Science* 125, no. 3260 (Jun 21 1957): 1236.

Gachelin, Gabriel. *Les organismes modèles dans la recherche médicale*. Paris: Presses universitaires de France, 2006.

Galison, Peter, and Lorraine Daston. "Scientific Coordination as Ethos and Epistemology." In *Instruments in Art and Science: On the Architectonics of Cultural Boundaries in the Seventeenth Century*, ed. Helmar Schramm, Ludger Schwarte, and Jan Lazardzig, 296–333. Berlin: Walter de Gruyter, 2008.

Gammon, Clive. "The Case of the Absent Eggs," *Sports Illustrated* 40, no. 25 (Jun 24 1974): 26–33.

Gamow, George. "Possible Relation between Desoxyribonucleic Acid and Protein Structure." *Nature* 173 (1954): 318.

Gamow, George, and Nicholas Metropolis. "Numerology of Polypeptide Chains." *Science* 120, no. 3124 (1954): 779–80.

Gamow, George, Alexander Rich, and Martynas Yčas. "The Problem of Information Transfer from the Nucleic Acids to Proteins." *Advances in Biological and Medical Physics* 4 (1956): 23–68.

Gamow, George, and Martynas Yčas. "Statistical Correlation of Protein and Ribonucleic Acid Composition." *Proceedings of the National Academy of Sciences* 41, no. 12 (Dec 15 1955): 1011–19.

Garcia-Sancho, Miguel. *Biology, Computing, and the History of Molecular Sequencing*. New York: Palgrave Macmillan, 2012.

———. "From Metaphor to Practices: The Introduction of 'Information Engineers' into the First DNA Sequence Database." *History and Philosophy of the Life Sciences* 33, no. 1 (2011): 71–104.

———. "A New Insight into Sanger's Development of Sequencing: From Proteins to DNA, 1943–1977." *Journal of the History of Biology* 43 (2010): 265–323.

Gardey, Delphine. *Ecrire, calculer, classer: Comment une révolution de papier a transformé les sociétés contemporaines, 1800–1940.* Paris: Découverte, 2008.

Gaudillière, Jean-Paul. "Paris-New York Roundtrip: Transatlantic Crossings and the Reconstruction of the Biological Sciences in Post-War France." *Studies in the History and Philosophy of Biological and Biomedical Sciences* 33C (2002): 389–417.

Gaudillière, Jean-Paul, and Hans-Jörg Rheinberger. *From Molecular Genetics to Genomics: The Mapping Cultures of Twentieth-Century Genetics.* London: Routledge, 2004.

Geison, Gerald L., and Manfred D. Laubichler. "The Varied Lives of Organisms: Variation in the Historiography of the Biological Sciences." *Studies in History and Philosophy of Biological and Biomedical Sciences* 32, no. 1 (2001): 1–29.

Geron, Tomio. "Can the U.N. Use Big Data to Respond to Global Disasters." *Forbes/Tech,* 2011.

Gerstein, Mark B., Anshul Kundaje, Manoj Hariharan, Stephen G. Landt, Koon Kiu Yan, Chao Cheng, Xinmeng Jasmine Mu, et al. "Architecture of the Human Regulatory Network Derived from ENCODE Data." *Nature* 489, no. 7414 (Sep 6 2012): 91–100.

Gieryn, Thomas F. "Laboratory Design for Post-Fordist Science." *Isis* 99, no. 4 (Dec 2008): 796–802.

———. "Three Truth-Spots." *Journal of the History of the Behavioral Sciences* 38, no. 2 (Spr 2002): 113–32.

Giles, Jim. "Internet Encyclopaedias Go Head to Head." *Nature* 438, no. 7070 (Dec 15 2005): 900–901.

———. "PR's 'Pit Bull' Takes on Open Access." *Nature* 445, no. 7126 (Jan 25 2007): 347.

Gill, Frank B., and Frederick H. Sheldon. "The Birds Reclassified." *Science* 252, no. 5008 (May 17 1991): 1003–5.

Gillispie, Charles C. "Scientific Aspects of the French Egyptian Expedition, 1798–1801." *Proceedings of the American Philosophical Society* 133, no. 4 (1989): 447–74.

Gingeras, Thomas R., and Richard J. Roberts. "Steps toward Computer Analysis of Nucleotide Sequences." *Science* 209, no. 4463 (Sep 19 1980): 1322–28.

Glusker, Jenny. "Lost Data." *Accounts of Chemical Research* 18, no. 4 (1985): 95.

Goad, Walter B. "GenBank." *Los Alamos Science* 9 (1983): 52–63.

Gooday, Graeme. "Placing or Replacing the Laboratory in the History of Science?" *Isis* 99, no. 4 (Dec 2008): 783–95.

Goodman, Morris, and G. William Moore. "Immunodiffusion Systematics of the Primates. I. The Catarrhini." *Systematic Biology* 20, no. 1 (1971): 19–62.

Gordin, Michael D. "Beilstein Unbound: The Pedagogical Unraveling of a Man and His *Handbuch*." In *Pedagogy and the Practice of Science: Historical and Contemporary Perspectives,* ed. David Kaiser, 11–39. Cambridge: MIT Press, 2005.

Gormley, Melinda. "Geneticist L. C. Dunn: Politics, Activism, and Community." PhD diss., Oregon State University, 2006.

Gradmann, Christoph, and Elborg Forster. *Laboratory Disease: Robert Koch's Medical Bacteriology.* Baltimore: Johns Hopkins University Press, 2009.

Graham-Smith, George S. "George Henry Falkiner Nuttall (5 July 1862–16 December 1937)." *Journal of Hygiene* 38, no. 2 (1938): 129–40.

Graham-Smith, George S., and David Keilin. "George Henry Falkiner Nuttall, 1862–1937." *Obituary Notice of Fellows of the Royal Society* 2, no. 7 (1939): 493–99.

Graham-Smith, George S., and Frederick Sanger. "The Biological or Precipitin Test for Blood Considered Mainly from Its Medico-Legal Aspect." *Journal of Hygiene* 3, no. 2 (Apr 1903): 258–91.

Groeben, Christiane. "The Stazione Zoologica: A Clearing House for Marine Organisms." In *Oceanographic History: The Pacific and Beyond*, ed. Keith Rodney Benson and Philip F. Rehbock, 537–47. Seattle: University of Washington Press, 2002.

Group on Data Citation Standards and Practices CODATA-ICSTI. "Out of Cite, out of Mind: The Current State of Practice, Policy, and Technology for the Citation of Data." *Data Science Journal* 12 (2013): CIDCR1-CIDCR75.

Guerreiro, Isabel, Andreia Nunes, Joost M. Woltering, Ana Casaca, Ana Novoa, Tania Vinagre, Margaret E. Hunter, Denis Duboule, and Moisiés Mallo. "Role of a Polymorphism in a Hox/Pax-Responsive Enhancer in the Evolution of the Vertebrate Spine." *Proceedings of the National Academy of Sciences* 110, no. 26 (Jun 25 2013): 10682–86.

Guyer, Michael F. "Blood Reactions of Man and Animals." *Scientific Monthly* 21, no. 10 (1925): 145–46.

———. "Immune Sera and Certain Biological Problems." *American Naturalist* 55, no. 637 (1921): 97–115.

Hacking, Ian. *Representing and Intervening*. Cambridge: Cambridge University Press, 1983.

———. "The Self-Vindication of the Laboratory Sciences." In *Science as Practice and Culture*, ed. Andrew Pickering, 29–64. Chicago: University of Chicago Press, 1992.

Hadley, Caroline, and David T. Jones. "A Systematic Comparison of Protein Structure Classifications: SCOP, CATH and FSSP." *Structure* 7, no. 9 (Sep 15 1999): 1099–1112.

Hagen, Joel B. "Experimental Taxonomy, 1920–1950: The Impact of Cytology, Ecology, and Genetics on the Ideas of Biological Classification." PhD diss., Oregon State University, 1984.

———. "Experimentalists and Naturalists in Twentieth-Century Botany: Experimental Taxonomy, 1920–1950." *Journal of the History of Biology* 17, no. 2 (1984): 249–70.

———. "The Introduction of Computers into Systematic Research in the United States during the 1960s." *Studies in the History and Philosophy of Biological and Biomedical Sciences* 32, no. 2 (2001): 291–314.

———. "Naturalists, Molecular Biology, and the Challenge of Molecular Evolution." *Journal of the History of Biology* 32 (1999): 321–41.

———. "The Origins of Bioinformatics." *Nature Reviews* 1 (2000): 231–36.

———. "The Statistical Frame of Mind in Systematic Biology from *Quantitative Zoology* to *Biometry*." *Journal of the History of Biology* 36, no. 2 (Sum 2003): 353–84.

———. "Waiting for Sequences: Morris Goodman, Immunodiffusion Experiments, and the Origins of Molecular Anthropology." *Journal of the History of Biology* 43, no. 4 (Dec 2010): 697–725.

Hagner, Michael. *Geniale Gehirne: Zur Geschichte der Elitegehirnforschung*. Göttingen: Wallstein, 2004.

Hagner, Michael. 2018. "Open Access, Data Capitalism and Academic Publishing." *Swiss Medical Weekly* 148: w14600.

Hagstrom, Warren. "Gift Giving as an Organizing Principle in Science." In *Science in Context: Readings in the Sociology of Science*, ed. Barry Barnes and David Edge, 21–34. Cambridge: MIT Press, 1982.

Hall, R. M. S. "Development of the United Kingdom Data Program." *Journal of Chemical Documentation* 7, no. 1 (1967).

Hallam, Elizabeth. *Anatomy Museum: Death and the Body Displayed*. London: Reaktion Books, 2016.

Hamilton, Andrew, and Quentin D. Wheeler. "Taxonomy and Why History of Science Matters for Science: A Case Study." *Isis* 99, no. 2 (Jun 2008): 331–40.

Hamilton, Walter C. "The Revolution in Crystallography." *Science* 169, no. 941 (Jul 10 1970): 133–41.

Hamilton, Walter C., Michel N. Frey, Ljubo Golic, Jacques J. Verbist, Thomas F. Koetzle, and Mogens S. Lehmann. "Stereoscopic Atlas of Amino-Acid Structures." *Materials Research Bulletin* 7, no. 11 (1972).

Hanson, Elizabeth. *Animal Attractions: Nature on Display in American Zoos*. Princeton: Princeton University Press, 2002.

Haraway, Donna. "Teddy Bear Patriarchy: Taxidermy in the Garden of Eden, New York City, 1908–1936." *Social Text* 11 (1984–85): 20–64.

Harding, Anne. "Blast: How 90,000 Lines of Code Helped Spark the Bioinfomatics Explosion." *Scientist* 19, no. 16 (2005): 21–26.

Harris, J. Ieuan, Michael A. Naughton, and Frederick Sanger. "Species Differences in Insulin." *Archives of Biochemistry and Biophysics* 65, no. 1 (Nov 1956): 427–38.

Heesen, Anke te, and Emma C. Spary, eds. *Sammeln als Wissen: Das Sammeln und seine wissenschaftsgeschichtliche Bedeutung*. Göttingen: Wallstein, 2001.

Heidelberger, Michael, and Karl Landsteiner. "On the Antigenic Properties of Hemoglobin." *Journal of Experimental Medicine* 38, no. 5 (Oct 31 1923): 561–71.

Heijne, Gunnar von. *Sequence Analysis in Molecular Biology: Treasure Trove or Trivial Pursuit*. San Diego: Academic Press, 1987.

Henahan, John F. "Dr. Doolittle: Making Big Changes in Small Steps." *Chemical and Engineering News*, Feb 9 1970, 22–32.

Hey, Anthony John G., Stuart Tansley, and Kristin Tolle. *The Fourth Paradigm: Data-Intensive Scientific Discovery*. Redmond, WA: Microsoft Research, 2009.

Hillis, David M., Craig Moritz, and Barbara K. Mable. *Molecular Systematics*. Sunderland, MA: Sinauer Associates, 1996.

Hiscock, Ira V. "Charles-Edward Amory Winslow, February 4, 1877–January 8, 1957." *Journal of Bacteriology* 73, no. 3 (Mar 1957): 295–96.

Hodgson, John. "A Certain Lack of Coordination." *Trends in Biotechnology* 5, no. 3 (Mar 1987): 59–60.

Hodson, Simon. "Data-Sharing Culture Has Changed." *Research Information*, Dec 2009–Jan 2010.

Hoffmann, Robert. "A Wiki for the Life Sciences Where Authorship Matters." *Nature Genetics* 40, no. 9 (Sep 2008): 1047–51.

Holley, Robert W., Jean Apgar, George A. Everett, James T. Madison, Mark Marquisee, Susan H. Merrill, John Robert Penswick, and Ada Zamir. "Structure of a Ribonucleic Acid." *Science* 147 (Mar 19 1965): 1462–65.

Holm, Liisa, and Chris Sander. "The FSSP Database of Structurally Aligned Protein Fold Families." *Nucleic Acids Research* 22, no. 17 (Sep 1994): 3600–3609.

———. "Parser for Protein Folding Units." *Proteins* 19, no. 3 (Jul 1994): 256–68.

Holmes, Frederic L. *Claude Bernard and Animal Chemistry: The Emergence of a Scientist.* Cambridge: Harvard University Press, 1974.

———. *Reconceiving the Gene: Seymour Benzer's Adventures in Phage Genetics.* New Haven: Yale University Press, 2006.

Hope, Janet. *Biobazaar: The Open Source Revolution and Biotechnology.* Cambridge: Harvard University Press, 2008.

Hopwood, Nick. "Embryology." In *The Cambridge History of Science: The Modern Biological and Earth Sciences,* ed. Peter J. Bowler and John V. Pickstone, 285–315. Cambridge: Cambridge University Press, 2009.

———. *Haeckel's Embryos: Images, Evolution, and Fraud.* Chicago: University of Chicago Press, 2015.

———. "Producing Development: The Anatomy of Human Embryos and the Norms of Wilhelm His." *Bulletin of the History of Medicine* 74, no. 1 (Spr 2000): 29–79.

Houde, Peter. "Review: *Phylogeny and Classification of Birds: A Study in Molecular Evolution,* by Charles G. Sibley and Jon E. Ahlquist." *Quarterly Review of Biology* 67, no. 1 (1992): 62–63.

Hoyer, Bill H., Brian J. McCarthy, and Ellis T. Bolton. "A Molecular Approach in the Systematics of Higher Organisms: DNA Interactions Provide a Basis for Detecting Common Polynucleotide Sequences among Diverse Organisms." *Science* 144 (May 22 1964): 959–67.

Hughes, Sally Smith. *Genentech: The Beginnings of Biotech.* Chicago: University of Chicago Press, 2011.

———. "Making Dollars out of DNA: The First Major Patent in Biotechnology and the Commercialization of Molecular Biology, 1974–1980." *Isis* 92, no. 3 (2001): 541–75.

Hull, David L. "Openness and Secrecy in Science: Their Origins and Limitations." *Science, Technology, and Human Values* 10, no. 2 (1985): 4–13.

———. *Science as a Process.* Chicago: University of Chicago Press, 1988.

Hunt, John A., and Vernon M Ingram. "The Chemical Effects of Gene Mutations in Some Abnormal Human Haemoglobins." In *Symposium on Protein Structure,* ed. Albert Neuberger, 148–51. New York: Wiley, 1958.

Huss, Jon W., III, Camilo Orozco, James Goodale, Chunlei Wu, Serge Batalov, Tim J. Vickers, Faramarz Valafar, and Andrew I. Su. "A Gene Wiki for Community Annotation of Gene Function." *PLoS Biology* 6, no. 7 (Jul 8 2008): e175.

Huxley, Julian, ed. *The New Systematics.* Oxford: Clarendon Press, 1940.

Ibers, James A. "Walter Clark Hamilton, 1931–1973." *Acta Crystallographica Section A—Crystal Physics, Diffraction, Theoretical and General Crystallography* 29, no. 4 (1973): 483–84.

Ingram, Vernon M. "Sickle-Cell Anemia Hemoglobin: The Molecular Biology of the First 'Molecular Disease'—the Crucial Importance of Serendipity." *Genetics* 167, no. 1 (May 2004): 1–7.

"The International Exhibition at Rome." *Lancet* 143, no. 3686 (1894): 1030–31.
Jackson, Catherine M. "The Laboratory." In *A Companion to the History of Science*, ed. Bernard V. Lightman, 296–309. West Sussex, UK: John Wiley & Sons, 2016.
Jacob, François. *La logique du vivant: Une histoire de l'hérédité*. Paris: Gallimard, 1970.
Jardine, Nicholas, James A. Secord, and Emma C. Spary, eds. *Cultures of Natural History*. London; New York etc.: Cambridge University Press, 1996.
Jeffrey, George A., Edith Mootz, and Dietrich Mootz. "A Knowledge Availability Survey of the Crystal Structural Data for Pyrimidine Derivatives." *Acta Crystallographica* 19, no. 4 (Oct 10 1965): 691–92.
Johnsgard, Paul A. "In Memoriam: Charles G. Sibley." *Nebraska Bird Review* 66, no. 2 (1998): 68–69.
Johnson, Kristin. "Ernst Mayr, Karl Jordan, and the History of Systematics." *History of Science* 43 (2005): 1–35.
———. "Natural History as Stamp Collecting: A Brief History." *Archives of Natural History* 34, no. 2 (2007): 244–58.
———. *Ordering Life: Karl Jordan and the Naturalist Tradition*. Baltimore: Johns Hopkins University Press, 2012.
Jones, Donald F. "Edward Murray East, 1879–1938." *Biographical Memoirs of the National Academy of Science of the United States of America* 23 (1944): 215–42.
Jordan, Elke. "DNA Database." *Science* 218 (1982): 108.
Judd, Walter S. *Plant Systematics: A Phylogenetic Approach*. Sunderland, MA: Sinauer Associates, 2008.
Jue, Rodney A., Neal W. Woodbury, and Russell F. Doolittle. "Sequence Homologies among E. Coli Ribosomal Proteins: Evidence for Evolutionarily Related Groupings and Internal Duplications." *Journal of Molecular Evolution* 15, no. 2 (1980): 129–48.
Jukes, Thomas H. "Beta Lactoglobulins and the Amino Acid Code." *Biochemical and Biophysical Research Communications* 7, no. 4 (1962): 281–83.
———. "Possible Base Sequences in the Amino Acid Code." *Biochemical and Biophysical Research Communications* 7, no. 6 (1962): 497–502.
———. "Relations between Mutations and Base Sequences in the Amino Acid Code." *Proceedings of the National Academy of Sciences* 48, no. 10 (1962): 1809–15.
———. "Some Recent Advances in Studies of the Transcription of the Genetic Message." *Advances in Biological and Medical Physics* 9 (1963): 1–41.
Kabat, Elvin A. "The Problem with GenBank." In *Biomolecular Data: A Resource in Transition*, ed. Rita Colwell, 127–28. Oxford: Oxford University Press, 1989.
Kahn, Patricia. "Human Genome Project: Sequencers Split over Data Release." *Science* 271, no. 5257 (Mar 29 1996): 1798–99.
Kahn, Patricia, and David Hazledine. "NAR's New Requirement for Data Submission to the EMBL Data Library: Information for Authors." *Nucleic Acids Research* 16, no. 10 (May 25 1988): I–IV.
Kaiser, David. *How the Hippies Saved Physics: Science, Counterculture, and the Quantum Revival*. New York: W. W. Norton, 2011.
Kaiser, David, and Patrick McCray. *Groovy Science: Knowledge, Innovation, and American Counterculture*. Chicago: University of Chicago Press, 2016.

Kaiser, Jocelyn. "Chemists Want NIH to Curtail Database." *Science* 308, no. 5723 (May 6 2005): 774.

Kamminga, Harmke, and Mark W. Weatherall. "The Making of a Biochemist, I: Frederick Gowland Hopkins' Construction of Dynamic Biochemistry." *Medical History* 40, no. 3 (Jul 1996): 269–92.

Kass, Lee B., and Christophe Bonneuil. "Mapping and Seeing: Barbara Mcclintock and the Linking of Genetics and Cytology in Maize Genetics, 1928–35." In *Classical Genetic Research and Its Legacy: The Mapping Cultures of Twentieth-Century Genetics*, ed. Hans-Jörg Rheinberger and Jean-Paul Gaudillière, 91–118. London: Routledge, 2004.

Kass, Lee B., Christophe Bonneuil, and Edward H. Coe Jr. "Cornfests, Cornfabs and Cooperation: The Origins and Beginnings of the Maize Genetics Cooperation News Letter." *Genetics* 169, no. 4 (Apr 2005): 1787–97.

Kay, Lily E. *The Molecular Vision of Life: Caltech, the Rockefeller Foundation and the Rise of the New Biology*. New York: Oxford University Press, 1993.

———. *Who Wrote the Book of Life: A History of the Genetic Code*. Stanford: Sanford University Press, 2000.

Keilin, David. "Preparation of Pure Cytochrome C from Heart Muscle and Some of Its Properties." *Proceedings of the Royal Society of London B—Biological Sciences* 122 (1937): 298–308.

Keilin, David, and Yin-Lai Wang. "Stability of Haemoglobin and of Certain Endoerythrocytic Enzymes in Vitro." *Biochemical Journal* 41, no. 4 (1947): 491–500.

Keller, Evelyn Fox. *Secrets of Life, Secrets of Death: Essays on Language, Gender, and Science*. New York: Routledge, 1992.

Kelty, Christopher M. "This Is Not an Article: Model Organism Newsletters and the Question of 'Open Science.'" *Biosocieties* 7, no. 2 (Jun 2012): 140–68.

———. *Two Bits: The Cultural Significance of Free Software*. Durham: Duke University Press, 2008.

Kendrew, John C., Richard E. Dickerson, Bror E. Strandberg, Roger G. Hart, David R. Davies, David C Phillips, and V. C. Shore. "Structure of Myoglobin: A Three-Dimensional Fourier Synthesis at 2 A. Resolution." *Nature* 185, no. 4711 (Feb 13 1960): 422–27.

Kennard, Olga, William G. Town, and David G. Watson. "Cambridge Crystallographic Data Center. 1. Bibliographic File." *Journal of Chemical Documentation* 12, no. 1 (1972): 14–19.

Kennedy, Michael. *Philanthropy and Science in New York City: The American Museum of Natural History, 1868–1968*. 1968. Microform.

Kevles, Daniel J. *The Baltimore Case: A Trial of Politics, Science, and Character*. New York: W. W. Norton, 1998.

———. "Big Science and Big Politics in the United States: Reflections on the Death of the SSC and the Life of the Human Genome Project." *Historical Studies in the Physical and Biological Sciences* 27 (1997): 269–97.

———. "*Diamond v. Chakrabarty* and Beyond: The Political Economy of Patenting Life." In *Private Science: Biotechnology and the Rise of the Molecular Sciences*, ed. Arnold Thackray, 65–79. Philadelphia: University of Pennsylvania Press, 1998.

———. "Of Mice and Money: The Story of the World's First Animal Patent." *Daedalus* 131, no. 2 (Spr 2002): 78–88.

Kim, Jinseop S., Matthew J. Greene, Aleksandar Zlateski, Kisuk Lee, Mark Richardson, Srinivas C. Turaga, Michael Purcaro, et al. "Space-Time Wiring Specificity Supports Direction Selectivity in the Retina." *Nature* 509, no. 7500 (May 15 2014): 331–36.

Kimmelman, Barbara A. "Organisms and Interests: R. A. Emerson's Claims for the Unique Contributions of Agricultural Genetics." In *The Right Tool for the Job*, ed. A. E. Clarke and J. H. Fujimura, 198–232. Princeton: Princeton University Press, 1992.

Kingsland, Sharon. "Robert E. Kohler, *Landscapes and Labscapes: Exploring the Lab-Field Border in Biology*." *Isis* 95 (Sep 2004): 509–10.

Kingston, William. "Streptomycin, Schatz v. Waksman, and the Balance of Credit for Discovery." *Journal of the History of Medicine and Allied Sciences* 59, no. 3 (2004): 441–62.

Kirk, R. G. W. "A Brave New Animal for a Brave New World: The British Laboratory Animals Bureau and the Constitution of International Standards of Laboratory Animal Production and Use, circa 1947–1968." *Isis* 101, no. 1 (Mar 2010): 62–94.

———. "'Wanted—Standard Guinea Pigs': Standardisation and the Experimental Animal Market in Britain ca. 1919–1947." *Studies in the History and Philosophy of Biological and Biomedical Sciences* 39 (Mar 2008): 280–91.

Kitchin, Rob. *The Data Revolution: Big Data, Open Data, Data Infrastructures and Their Consequences*. Los Angeles: SAGE Publications, 2014.

Klein, Ursula. "The Laboratory Challenge: Some Revisions of the Standard View of Early Modern Experimentation." *Isis* 99, no. 4 (Dec 2008): 769–82.

Kloppenburg, Jack Ralph. *First the Seed: The Political Economy of Plant Biotechnology, 1492–2000*. Cambridge: Cambridge University Press, 1988.

Kneale, Geoff G., and Martin J. Bishop. "Nucleic Acid and Protein Sequence Databases." *Computer Applications in the Biosciences* 1, no. 1 (1985): 11–17.

Knell, Simon J. *The Culture of English Geology, 1815–1851: A Science Revealed through Its Collecting*. Aldershot, UK: Ashgate, 2000.

Knoeff, Rina, and Robert Zwijnenberg. *The Fate of Anatomical Collections*. Farnham, Surrey, UK: Ashgate, 2015.

Kocur, Miloslav. "History of the Král Collection." In *100 Years of Culture Collections*, ed. Lindsay I. Sly, Teiji Iijima and Barbara E. Kirsop, 4–12. Osaka: Institute for Fermentation, 1990.

Koetzle, Thomas F. "Benefits of Databases." *Nature* 342, no. 6246 (Nov 9 1989): 114.

Koetzle, Thomas F., Lawrence C. Andrews, Frances C. Bernstein, and Herbert J. Bernstein. "CRYSNET: A Network for Crystallographic Computing." In *Computer Networking and Chemistry*, ed. Peter Lykos, 1–8. Washington, DC: American Chemical Society, 1975.

Kofoid, Charles Atwood. *The Biological Stations of Europe*. Washington, DC: Government Printing Office, 1910.

Kohler, Robert E. "Finders, Keepers: Collecting Sciences and Collecting Practice." *History of Science* 45, no. 4 (2007): 428–54.

———. *From Medical Chemistry to Biochemistry. The Making of a Biomedical Discipline*. Cambridge: Cambridge University Press, 1982.

———. "Lab History Reflections." *Isis* 99, no. 4 (Dec 2008): 761–68.

———. "Labscapes: Naturalizing the Lab." *History of Science* 40, no. 130 (Dec 2002): 473–501.

———. *Landscapes and Labscapes: Exploring the Lab-Field Border in Biology*. Chicago: University of Chicago Press, 2002.

———. *Lords of the Fly: Drosophila Genetics and the Experimental Life*. Chicago: University of Chicago Press, 1994.

———. "Place and Practice in Field Biology." *History of Science* 40, no. 128, pt. 2 (Jun 2002): 189–210.

Kohlstedt, Sally Gregory. "Museums on Campus: A Tradition of Inquiry and Teaching." In *The American Development of Biology*, ed. Ronald Rainger, Keith R Benson, and Jane Maienschein, 15–47. Philadelphia: University of Pennsylvania Press, 1988.

Korn, Laurence Jay, and Cary Queen. "Analysis of Biological Sequences on Small Computers." *DNA* 3, no. 6 (1984): 421–36.

Kraft, Alison, and Samuel J. M. M. Alberti. "'Equal Though Different': Laboratories, Museums and the Institutional Development of Biology in Late-Victorian Northern England." *Studies in History and Philosophy of Biological and Biomedical Sciences* 34, no. 2 (2003): 203–36.

Krige, John. "The Birth of EMBO and the Difficult Road to EMBL." *Studies in the History and Philosophy of Biological and Biomedical Sciences* 33 (2002): 547–64.

"The Laboratory in Modern Science." *Science* 3 (1884): 172–74.

Lamanna, Carl. "Microbiology, Museums, and the American Type Culture Collection." *Quarterly Review of Biology* 41, no. 2 (1966): 95–97.

Landauer, Walter. "Shall We Lose or Keep Our Plant and Animal Stocks." *Science* 101, no. 2629 (May 18 1945): 497–99.

Landecker, Hannah. "Sending Cells Around: How to Exchange Biological Matter." In *The Moment of Conversion: Exchange Networks in Modern Biomedical Science*. Department of Anthropology, History, and Social Medicine, UC San Francisco, Apr 4–5 2003.

Landsman, David, Robert Gentleman, Janet Kelso, and B. F. Francis Ouellette. "Database: A New Forum for Biological Databases and Curation." *Database (Oxford)* (2009): bap002.

Landsteiner, Karl. "Cell Antigens and Individual Specificity." *Journal of Immunology* 15, no. 6 (Nov 1928): 589–600.

———. *The Specificity of Serological Reactions*. Springfield, IL: C. C. Thomas, 1936.

Landsteiner, Karl, Lewis G. Longsworth, and James van der Scheer. "Electrophoresis Experiments with Egg Albumins and Hemoglobins." *Science* 88, no. 2273 (Jul 22 1938): 83–85.

Landsteiner, Karl, and C. Philip Miller Jr. "Serological Observations on the Relationship of the Bloods of Man and the Anthropoid Apes." *Science* 61, no. 1584 (May 8 1925): 492–93.

Laporte, Léo F. *George Gaylord Simpson: Paleontologist and Evolutionist*. New York: Columbia University Press, 2000.

Latour, Bruno. "Give Me a Laboratory and I Will Raise the World." In *Science Observed*, ed. Karin Knorr-Cetina and Michael Mulkay, 258–75. New York: Routledge, 1983.

———. *Pandora's Hope: Essays on the Reality of Science Studies*. Cambridge: Harvard University Press, 1999.

Laubichler, Manfred Dietrich, and Jane Maienschein. *From Embryology to Evo-Devo: A History of Developmental Evolution*. Cambridge: MIT Press, 2007.

Lederberg, Joshua. "Digital Communications and the Conduct of Science; New Literacy." *Proceedings of the IEEE* 66, no. 11 (1978): 1314–19.

Lederer, Susan E. *Flesh and Blood: Organ Transplantation and Blood Transfusion in Twentieth-Century America*. Oxford: Oxford University Press, 2008.

———. *Frankenstein: Penetrating the Secrets of Nature*. New Brunswick, NJ: Rutgers University Press, 2002.

Ledley, Robert S. "Digital Computational Methods in Symbolic Logic, with Examples in Biochemistry." *Proceedings of the National Academy of Sciences* 41, no. 7 (1955): 498–511.

———. "Digital Electronic Computers in Biomedical Sciences." *Science* 130 (1959): 1225–34.

———. "Letters to the Editor." *Science* 131, no. 3399 (Feb 19 1960): 474–564.

———. *The Use of Computers in Biology and Medicine*. New York: McGraw-Hill, 1965.

Ledley, Robert S., and Lee B. Lusted. "Probability, Logic and Medical Diagnosis." *Science* 130, no. 3380 (Oct 9 1959): 892–930.

———. "Reasoning Foundations of Medical Diagnosis: Symbolic Logic, Probability, and Value Theory Aid Our Understanding of How Physicians Reason." *Science* 130, no. 3366 (Jul 3 1959): 9–21.

Lehmann, Mogens S., Thomas F. Koetzle, and Walter C Hamilton. "Precision Neutron Diffraction Structure Determination of Protein and Nucleic Acid Components. I. The Crystal and Molecular Structure of the Amino Acid L-Alanine." *Journal of the American Chemical Society* 94, no. 8 (Apr 19 1972): 2657–60.

Lengyel, Peter, Joseph F. Speyer, Carlos Basilio, and Severo Ochoa. "Synthetic Polynucleotides and the Amino Acid Code, III." *Proceedings of the National Academy of Sciences* 48 (Feb 1962): 282–84.

Lengyel, Peter, Joseph F. Speyer, and Severo Ochoa. "Synthetic Polynucleotides and the Amino Acid Code." *Proceedings of the National Academy of Sciences* 47 (Dec 15 1961): 1936–42.

Lenoir, Timothy. "Shaping Biomedicine as an Information Science." In *Proceedings of the 1998 Conference on the History and Heritage of Science Information Systems*, ed. Mary Ellen Bowden, Trudi Bellardo Hahn, and Robert V. Williams, 27–45. Medford, NJ: Information Today, 1999.

Leone, Charles A. "Comparative Serology of Some Branchyuran Crustacea and Studies in Hemocyanin Correspondence." *Biological Bulletin* 97, no. 3 (Dec 1949): 273–86.

———. *Taxonomic Biochemistry and Serology*. New York: Ronald Press, 1964.

Leonelli, Sabina. "*Arabidopsis*, the Botanical *Drosophila*: From Mouse Cress to Model Organism." *Endeavour* 31, no. 1 (Mar 2007): 34–38.

———. *Data-centric Biology: A Philosophical Study*. Chicago: University of Chicago Press, 2016.

———. "Introduction: Making Sense of Data-Driven Research in the Biological and Biomedical Sciences." *Studies in History and Philosophy of Science Part C* 43, no. 1 (Mar 2012): 1–3.

———. “When Humans Are the Exception: Cross-Species Databases at the Interface of Biological and Clinical Research.” *Social Studies of Science* 42, no. 2 (Apr 2012): 214–36.

Leonelli, Sabina, and Rachel A. Ankeny. “Re-thinking Organisms: The Impact of Databases on Model Organism Biology.” *Studies in History and Philosophy of Science Part C* 43, no. 1 (Mar 2012): 29–36.

———. “What Makes a Model Organism?” *Endeavour* 37, no. 4 (Dec 2013): 209–12.

Leonelli, Sabina, Daniel Spichtinger, and Barbara Prainsack. “Sticks and Carrots: Encouraging Open Science at Its Source.” *Geo* 2, no. 1 (Jun 30 2015): 12–16.

Leslie, Stuart W. *The Cold War and American Science: The Military-Industrial-Academic Complex at MIT and Stanford*. New York: Columbia University Press, 1993.

Levine, N. “The Nature of the Glut: Information Overload in Postwar America.” *History of the Human Sciences* 30, no. 1 (Feb 2017): 32–49.

Levinthal, Cyrus. “Molecular Model-Building by Computer.” *Scientific American* 214 (1966): 41–52.

Levinthal, Elliott C., Raymond E. Carhart, Suzanne M. Johnson, and Joshua Lederberg. “When Computers ‘Talk’ to Computers.” *Industrial Research* 17, no. 12 (1975): 35–42.

Levitt, Michael, and Cyrus Chothia. “Structural Patterns in Globular Proteins.” *Nature* 261, no. 5561 (Jun 17 1976): 552–58.

Levitt, Michael, and Jonathan Greer. “Automatic Identification of Secondary Structure in Globular Proteins.” *Journal of Molecular Biology* 114, no. 2 (Aug 5 1977): 181–239.

Lewin, Roger. “Conflict over DNA Clock Results.” *Science* 241, no. 4873 (Sep 23 1988): 1598–1600.

———. “DNA Clock Conflict Continues.” *Science* 241, no. 4874 (Sep 30 1988): 1756–59.

———. “Long-Awaited Decision on DNA Database.” *Science* 217 (1982): 817–18.

———. “National Networks for Molecular Biologists.” *Science* 223, no. 4643 (Mar 30 1984): 1379–80.

———. “Proposal to Sequence the Human Genome Stirs Debate.” *Science* 232, no. 4758 (Jun 27 1986): 1598–1600.

Light, Jennifer S. “When Computers Were Women.” *Technology and Culture* 40, no. 3 (1999): 455–83.

Lim, Wendell A. “Frederic M. Richards, 1925–2009, Obituary.” *Nature Structural & Molecular Biology* 16, no. 3 (Mar 2009): 230–32.

Lindee, M. Susan. *Moments of Truth in Genetic Medicine*. Baltimore: Johns Hopkins University Press, 2005.

Lindsay, Robert K. *Applications of Artificial Intelligence for Organic Chemistry: The Dendral Project*. New York: McGraw-Hill, 1980.

Lindsey, J. Russell. “Historical Foundations.” In *The Laboratory Rat*, ed. Henry J. Baker, J. Russell Lindsey, and Steven H. Weisbroth, 1–36. San Diego: Academic Press, 1979.

Linton, Derek S. *Emil von Behring: Infectious Disease, Immunology, Serum Therapy*. Philadelphia: American Philosophical Society, 2005.

Loeb, Jacques. “Is Species Specificity a Mendelian Character?” *Science* 45 (Jan–Jun 1917): 191–93.

Loeb, Leo. “Scientific Books.” *Science* 33 (Jan–Jun 1911): 147–50.

Loison, Laurent. "Controverses sur la méthode dans les sciences du vivant: physiologie, zoologie, botanique (1865–1931)." In *Claude Bernard: La méthode de la physiologie*, ed. François Deschenaux, Jean-Jacques Kupiec, and Michel Morange, 63–82. Paris: Editions Rue d'Ulm, 2013.

———. *Qu'est-ce que le Néolamarckisme? Les biologistes français et la question de l'évolution des espèces*. Paris: Vuibert, 2010.

Longo, Dan L., and Jeffrey M. Drazen. "Data Sharing." *New England Journal of Medicine* 374, no. 3 (Jan 21 2016): 276–77.

Luria, Salvador E. "Human Genome Program." *Science* 246 (1989): 873–74.

MacArthur, Malcolm W., and Janet M. Thornton. "Influence of Proline Residues on Protein Conformation." *Journal of Molecular Biology* 218, no. 2 (Mar 20 1991): 397–412.

MacLeod, Colin M. "Some Thoughts on Microbial Taxonomy." *Quarterly Review of Biology* 41, no. 2 (1966): 98–100.

Maddox, John. "Making Authors Toe the Line." *Nature* 342 (1989): 855.

———. "Making Good Databanks Better." *Nature* 341, no. 6240 (Sep 28 1989): 277.

Maienschein, Jane. "Shifting Assumptions in American Biology: Embryology, 1890–1910." *Journal of the History of Biology* 14, no. 1 (Spring 1981): 89–113.

———. *Transforming Traditions in American Biology, 1880–1915*. Baltimore: Johns Hopkins University Press, 1991.

Malakoff, David. "Scientific Publishing: Opening the Books on Open Access." *Science* 302, no. 5645 (Oct 24 2003): 550–54.

Mares, Michael A. "Natural Science Collections: America's Irreplaceable Resource." *Bioscience* 59, no. 7 (Jul–Aug 2009): 544–45.

Marks, Jonathan. "The Legacy of Serological Studies in American Physical Anthropology." *History and Philosophy of the Life Sciences* 18 (1996): 345–62.

Marshall, Eliot. "Data Sharing: DNA Sequencer Protests Being Scooped with His Own Data." *Science* 295, no. 5558 (Feb 15 2002): 1206–7.

———. "Genome Researchers Take the Pledge." *Science* 272, no. 5261 (Apr 26 1996): 477–78.

———. "Genome Sequencing: Claim and Counterclaim on the Human Genome." *Science* 288, no. 5464 (Apr 14 2000): 242–43.

———. "Genome Sequencing: Clinton and Blair Back Rapid Release of Data." *Science* 287, no. 5460 (Mar 17 2000): 1903.

———. "Human Genome: Rival Genome Sequencers Celebrate a Milestone Together." *Science* 288, no. 5475 (Jun 30 2000): 2294–95.

———. "The Human Genome: Sharing the Glory, Not the Credit." *Science* 291, no. 5507 (Feb 16 2001): 1189–93.

———. "Human Genome: Storm Erupts over Terms for Publishing Celera's Sequence." *Science* 290, no. 5499 (Dec 15 2000): 2042–43.

———. "Sequencers Call for Faster Data Release." *Science* 276, no. 5316 (May 23 1997): 1189–90.

———. "Transfer of Protein Data Bank Sparks Concern." *Science* 281, no. 5383 (Sep 11 1998): 1584–85.

———. "Varmus Circulates Proposal for NIH-Backed Online Venture." *Science* 284, no. 5415 (Apr 30 1999): 718.

———. "Varmus Defends E-Biomed Proposal, Prepares to Push Ahead." *Science* 284, no. 5423 (Jun 25 1999): 2062–63.

Marshall, Eliot, and Elizabeth Pennisi. "NIH Launches the Final Push to Sequence the Genome." *Science* 272, no. 5259 (Apr 12 1996): 188–89.

Marshall, Garland R., Jeremy G. Vinter, and Hans-Dieter Holtje. "The Veil of Commercialism." *Journal of Computer-Aided Molecular Design* 2, no. 1 (Apr 1988): 1–2.

Marx, Jean L. "Bionet Bites the Dust." *Science* 245, no. 4914 (Jul 14 1989): 126.

Maxam, Allan M., and Walter Gilbert. "A New Method for Sequencing DNA." *Proceedings of the National Academy of Sciences* 74, no. 2 (Feb 1977): 560–64.

Mayer-Schönberger, Viktor, and Kenneth Cukier. *Big Data: A Revolution That Will Transform How We Live, Work, and Think*. Boston: Houghton Mifflin Harcourt, 2013.

Mayr, Ernst. *The Growth of Biological Thought: Diversity, Evolution, and Inheritance*. Cambridge, MA: Belknap Press, 1982.

———. "A New Classification of the Living Birds of the World." *Auk* 106 (1989): 508–12.

———. *Systematics and the Origin of Species from the Viewpoint of a Zoologist*. New York: Columbia University Press, 1942.

Mayr, Ernst, E. Gorton Linsley, and Robert Leslie Usinger. *Methods and Principles of Systematic Zoology*. New York: McGraw-Hill, 1953.

Mazumder, Raja, Darren A. Natale, Jessica A. Julio, Lai-Su S. Yeh, and Cathy H. Wu. "Community Annotation in Biology." *Biology Direct* 5 (2010): 12.

McCarthy, Brian J., and Ellis T. Bolton. "An Approach to the Measurement of Genetic Relatedness among Organisms." *Proceedings of the National Academy of Sciences* 50 (Jul 1963): 156–64.

McClintock, Barbara. "Chromosome Morphology in Zea Mays." *Science* 69, no. 1798 (Jun 14 1929): 629.

McGourty, Christine. "Human Genome Project: Dealing with the Data." *Nature* 342, no. 6246 (Nov 9 1989): 108.

McGuigan, Glenn S. "Publishing Perils in Academe." *Journal of Business and Finance Librarianship* 10, no. 1 (2004): 13–26.

McKusick, Victor A. *Mendelian Inheritance in Man: Catalogs of Autosomal Dominant, Autosomal Recessive, and X-Linked Phenotypes*. Baltimore: Johns Hopkins University Press, 1966.

McLaughlin, Peter J., and Margaret O. Dayhoff. "Eukaryote Evolution: A View Based on Cytochrome C Sequence Data." *Journal of Molecular Evolution* 2, nos. 2–3 (1973): 99–116.

Medical Research Council (London). *Mathematics and Computer Science in Biology and Medicine: Proceedings of Conference Held by Medical Research Council in Association with the Health Departments: Oxford, July 1964*. London: Her Majesty's Stationery Office, 1965.

Merriam, C. Hart. "Biology in Our Colleges: A Plea for a Broader and More Liberal Biology." *Science* 21 (1893): 352–55.

Meyer, Edgar F., Jr. "The First Years of the Protein Data Bank." *Protein Science* 6, no. 7 (Jul 1997): 1591–97.

———. "Interactive Computer Display for the Three-Dimensional Study of Macromolecular Structures." *Nature* 232, no. 5308 (Jul 23 1971): 255–57.

———. "Three-Dimensional Graphical Models of Molecules and a Time-Slicing Computer." *Journal of Applied Crystallography* 3 (1970): 392.

———. "Towards an Automatic, Three-Dimensional Display of Structural Data." *Journal of Chemical Documentation* 10, no. 2 (1970): 85.

Meyer, Edgar F., Jr., C. N. Morimoto, J. Villarreal, H. M. Berman, H. L. Carrell, R. K. Stodola, T. F. Koetzle, et al. "CRYSNET, a Crystallographic Computing Network with Interactive Graphics Display." *Federation Proceedings* 33, no. 12 (Dec 1974): 2402–5.

Mez, Carl. "Die Bedeutung der Sero-Diagnostik für die stammesgeschichtliche Forschung." *Botanisches Archiv* 16 (1926): 1–23.

Milam, Erika L. "The Equally Wonderful Field: Ernst Mayr and Organismic Biology." *Historical Studies in the Natural Sciences* (2010): 279–317.

Miles, Wyndham D. *A History of the National Library of Medicine: The Nation's Treasury of Medical Knowledge*. Washington, DC: US Dept. of Health and Human Services, 1982.

Miller, Stanley L., and Harold C. Urey. "Organic Compound Synthesis on the Primitive Earth." *Science* 130, no. 3370 (Jul 31 1959): 245–51.

Mindell, David P. "Review: DNA-DNA Hybridization and Avian Phylogeny." *Systematic Biology* 41, no. 1 (1992): 126–34.

Mirowski, Philip. "The Future(s) of Open Science." *Social Studies of Science* 48, no. 2 (2018): 171–203.

Mitman, Gregg. "Robert E. Kohler, *Landscapes and Labscapes: Exploring the Lab-Field Border in Biology*." *Journal of the History of Biology* 36 (Sep 1 2003): 599–629.

Mitman, Gregg, and Richard W. Burkhardt Jr. "Struggling for Identity: The Study of Animal Behavior in America, 1930–1950." In *The American Expansion of Biology*, ed. Keith R. Benson, Ronald Rainger, and Jane Maienschein, 164–94. New Brunswick: Rutgers University Press, 1987.

Mitman, Gregg, and Anne Fausto-Sterling. "Whatever Happened to Planaria? C. M. Child and the Physiology of Inheritance." In *The Right Tool for the Job*, ed. A. E. Clarke and J. H. Fujimura, 172–97. Princeton: Princeton University Press, 1992.

Monod, Jacques, and François Jacob. "General Conclusions: Teleonomic Mechanisms in Cellular Metabolism, Growth, and Differentiation." *Cold Spring Harbor Symposia on Quantitative Biology* 21 (1961): 389–401.

Moore, Kelly. *Disrupting Science: Social Movements, American Scientists, and the Politics of the Military, 1945–1975*. Princeton: Princeton University Press, 2008.

Moore, Stanford, Darrel H. Spackman, and William H. Stein. "Automatic Recording Apparatus for Use in the Chromatography of Amino Acids." *Federation Proceedings* 17, no. 4 (Dec 1958): 1107–15.

Morange, Michel. *A History of Molecular Biology*. Cambridge: Harvard University Press, 2000.

"More Bang for Your Byte." *Scientific Data* 1 (2014): 140010.

Morgan, Diana. "American Type Culture Collection Seeks to Expand Research Effort." *Scientist* 4, no. 16 (Aug 20 1990): 1–7.

Morgan, Thomas Hunt. *Regeneration*. New York: Macmillan, 1901.

———. "Review of Experimental Morphology." *Science* 9 (1899): 648–50.

Mount, David W. "Computer Analysis of Sequence, Structure and Function of Biological Macromolecules." *BioTechniques* 3, no. 2 (1985): 102–12.

Muka, Samantha K. "Working at Water's Edge: Life Sciences at American Marine Stations, 1880–1930." PhD diss., University of Pennsylvania, 2014.

Müller-Wille, Staffan. "Carl Von Linnés Herbarschrank. Zur epistemischen Funktion eines Sammlungsmöbels." In *Sammeln als Wissen: Das Sammeln und seine wissenschaftsgeschichtliche Bedeutung*, ed. Anke te Heesen and Emma C. Spary. Göttingen: Wallstein, 2001.

Müller-Wille, Staffan, and Hans-Jörg Rheinberger. *A Cultural History of Heredity*. Chicago: University of Chicago Press, 2012.

Murzin, Alexey G., Steven E. Brenner, Tim Hubbard, and Cyrus Chothia. "SCOP: A Structural Classification of Proteins Database for the Investigation of Sequences and Structures." *Journal of Molecular Biology* 247, no. 4 (Apr 7 1995): 536–40.

Myers, Natasha. "Molecular Embodiments and the Body-Work of Modeling in Protein Crystallography." *Social Studies of Science* 38, no. 2 (Apr 2008): 163–99.

National Commission on Terrorist Attacks upon the United States. *The 9/11 Commission Report: Final Report of the National Commission on Terrorist Attacks upon the United States*. New York: Norton, 2004.

National Research Council (US). *Systematic Biology: Proceedings of an International Conference*. Washington, DC: National Academy of Sciences, 1969.

National Science Foundation. *Twentieth Annual Report of the National Science Foundation*. Washington, DC: National Science Foundation, 1970.

Needham, James G. "Methods of Securing Better Co-operation between Government and Laboratory Zoologists in the Solution of Problems of General or National Importance." *Science* 49 (1919): 455–58.

"New and Unique Kind of Museum." *Science* 107, no. 2774 (Feb 27 1948): 217.

Newell, Allen, and Robert F. Sproull. "Computer-Networks: Prospects for Scientists." *Science* 215, no. 4534 (1982): 843–52.

Nielsen, Michael. *Reinventing Discovery: The New Era of Networked Science*. Princeton: Princeton University Press, 2012.

November, Joseph A. *Biomedical Computing: Digitizing Life in the United States*. Baltimore: Johns Hopkins University Press, 2012.

———. *"Digitizing Life: The Introduction of Computers to Biology and Medicine."* PhD diss., Princeton University, 2006.

Nuttall, George H. F. *Blood Immunity and Blood Relationship: A Demonstration of Certain Blood-Relationships amongst Animals by Means of the Precipitin Test for Blood*. Cambridge: Cambridge University Press, 1904.

———. "A Further Note on the Biological Test for Blood and Its Importance in Zoological Classification." *British Medical Journal* 2 (1901): 669.

———. "The New Biological Test for Blood in Relation to Zoological Classification." *Proceedings of the Royal Society of London B—Biological Sciences* 69, no. 453 (Dec 1901): 150–53.

———. "On the Formation of Specific Anti-bodies in the Blood, Following upon Treatment with the Sera of Different Animals." *American Naturalist* 35, no. 419 (1901): 927–32.

———. "Progress Report upon the Biological Test for Blood as Applied to Over 500 Bloods from Various Sources, Together with a Preliminary Note upon a Method for Measuring the Degree of Reaction." *British Medical Journal* 1 (1902): 825–27.

Nuttall, George H. F., and Edgar M. Dinkelspiel. "Experiments upon the New Specific Test for Blood (Preliminary Note)." *British Medical Journal* 1 (Jan–Jun 1901): 1141.

———. "On the Formation of Specific Anti-bodies in the Blood Following upon Treatment with the Sera of Different Animals, Together with Their Use in Legal Medicine." *Journal of Hygiene* 1, no. 3 (1901): 367–87.

Nyhart, Lynn K. *Biology Takes Form: Animal Morphology and the German Universities, 1800–1900*. Chicago: University of Chicago Press, 1995.

———. *Modern Nature: The Rise of the Biological Perspective in Germany*. Chicago: University of Chicago Press, 2009.

———. "Natural History and the 'New' Biology." In *Cultures of Natural History*, ed. Nicholas Jardine, James A. Secord, and Emma C. Spary, 426–43. London: Cambridge University Press, 1996.

Oakley, Margaret B., and George E. Kimball. "Punched Card Calculation of Resonance Energies." *Journal of Chemical Physics* 17, no. 8 (1949): 706–17.

Ogilvie, Brian W. "The Many Books of *Nature*: Renaissance Naturalists and Information Overload." *Journal of the History of Ideas* 64, no. 1 (2003): 29–40.

———. *The Science of Describing: Natural History in Renaissance Europe*. Chicago: University of Chicago Press, 2006.

Olby, Robert. *The Path to the Double Helix*. New York: Dover, 1994 [1974].

O'Malley, Maureen A., Kevin C. Elliott, Chris Haufe, and Richard M. Burian. "Philosophies of Funding." *Cell* 138, no. 4 (Aug 21 2009): 611–15.

Ophir, David, Seymour Rankowitz, Barry J. Shepherd, and Robert J. Spinrad. "BRAD: Brookhaven Raster Display." *Communications of the ACM* 11, no. 6 (1968): 415–16.

Orengo, Christine A., T. P. Flores, W. R. Taylor, and Janet M. Thornton. "Identification and Classification of Protein Fold Families." *Protein Engineering* 6, no. 5 (Jul 1993): 485–500.

Orengo, Christine A., Alex D. Michie, Susan Jones, David T. Jones, Mark B. Swindells, and Janet M. Thornton. "CATH—a Hierarchic Classification of Protein Domain Structures." *Structure* 5, no. 8 (Aug 15 1997): 1093–1108.

Packer, Laurence, Jason Gibbs, Cory Sheffield, and Hanner Robert. "DNA Barcoding and the Mediocrity of Morphology." *Molecular Ecology Resources* 9 suppl. s1 (May 2009): 42–50.

Page, Roderic D. M., and Edward C. Holmes. *Molecular Evolution: A Phylogenetic Approach*. Oxford: Blackwell Science, 1998.

Paléus, Sven, and Hans Tuppy. "A Hemopeptide from a Tryptic Hydrolysate of *Rhodospirillum-rubrum* Cytochrome C." *Acta Chemica Scandinavica* 13, no. 4 (1959): 641–46.

Paul, Diane B. "Mendel in America: Theory and Practice, 1900–1919." In *The American Development of Biology*, ed. Ronald Rainger, Keith R Benson, and Jane Maienschein, 281–310. Philadelphia: University of Pennsylvania Press, 1988.

Paul, Harry W. *From Knowledge to Power: The Rise of the Science Empire in France, 1860–1939*. Cambridge: Cambridge University Press, 1985.

Pauly, Philip J. "*Landscapes and Labscapes: Exploring the Lab-Field Border in Biology*, by Robert E. Kohler." *Quarterly Review of Biology* 78 (Jun 2003): 214–15.

———. "Summer Resort and Scientific Discipline: Woods Hole and the Structure of American Biology." In *The American Development of Biology*, ed. Ronald Rainger, Keith R Benson, and Jane Maienschein, 121–50. Philadelphia: University of Pennsylvania Press, 1988.

Payne, Robert B. "Charles Gald Sibley (1917–1998)—Obituary." *Ibis* 140, no. 4 (Oct 1998): 697–99.

Pearl, Raymond. "Trends of Modern Biology." *Science* 56, no. 1456 (Nov 24 1922): 581–92.

Pennisi, Elizabeth. "Proposal to 'Wikify' GenBank Meets Stiff Resistance." *Science* 319, no. 5870 (Mar 21 2008): 1598–99.

Peres, Sara. "Saving the Gene Pool for the Future: Seed Banks as Archives." *Studies in History and Philosophy of Biological and Biomedical Sciences* 55 (2016): 96–104.

Perutz, Max F. "An X-Ray Study of Horse Methaemoglobin.2." *Proceedings of the Royal Society of London A—Mathematical and Physical Sciences* 195, no. 1043 (1949): 474–99.

Phillips, David C. "Protein Crystallography 1971: Coming of Age." *Cold Spring Harbor Symposia on Quantitative Biology* 36 (1972): 589–92.

Pickstone, John V. "Museological Science? The Place of the Analytical/Comparative in Nineteenth-Century Science, Technology and Medicine." *History of Science* 32 (1994): 111–38.

———. "Natural Histories, Analyses and Experimentation: Three Afterwords." *History of Science* 49, no. 164 (Sep 2011): 349–74.

———. *Ways of Knowing: A New History of Science, Technology and Medicine*. Manchester: Manchester University Press, 2000.

———. "Working Knowledges before and after circa 1800: Practices and Disciplines in the History of Science, Technology and Medicine." *Isis* 98 (2007): 489–516.

Pomian, Krzysztof, *Collectors and Curiosities: Paris and Venice, 1500–1800*. Cambridge, UK: Polity Press, 1990.

Ponting, Chris P., and Robert R. Russell. "The Natural History of Protein Domains." *Annual Review of Biophysics and Biomolecular Structure* 31 (2002): 45–71.

Porter, Theodore M. *Trust: The Pursuit of Objectivity in Science and Public Life*. Princeton: Princeton University Press, 1995.

Pribram, Ernst. *Der gegenwärtige Bestand der vorm. Krālschen Sammlung von Mikroorganismen*. Vienna, 1919.

Prosser, C. Ladd, ed. *Comparative Animal Physiology*. Philadelphia: W. B. Saunders, 1950.

"Protein Data Bank." *Nature New Biology* 233, no. 42 (Oct 20 1971): 223.

"Proteins: Yet More Sequences." *Nature* 224 (1969): 313.

Quinn, Stephen C. *Windows on Nature: The Great Habitat Dioramas of the American Museum of Natural History*. New York: Harry N. Abrams, 2006.

Rader, Karen A. *Making Mice: Standardizing Animals for American Biomedical Research, 1900–1955*. Princeton: Princeton University Press, 2004.

———. "'The Mouse People': Murine Genetics Work at the Bussey Institution, 1909–1936." *Journal of the History of Biology* 31, no. 3 (Fall 1998): 327–54.

———. "The Origins of Mouse Genetics: Beyond the Bussey Institution. II. Defining the Problem of Mouse Supply: The 1928 National Research Council Committee on Experimental Plants and Animals." *Mammalian Genome* 13, no. 1 (Jan 2002): 2–4.

Rader, Karen A., and Victoria E. M. Cain. *Life on Display: Revolutionizing U.S. Museums of Science and Natural History in the Twentieth Century*. Chicago: University of Chicago Press, 2014.

Radin, Joanna M. "Life on Ice: Frozen Blood and Biological Variation in a Genomic Age, 1950–2010." PhD diss., University of Pennsylvania, 2012.

Rainger, Ronald, Keith R. Benson, and Jane Maienschein, eds. *The American Development of Biology*. Philadelphia: University of Pennsylvania Press, 1988.

Raper, Kenneth B., Robert E. Buchanan, Paul R. Burkholder, Ralph E. Cleland, Robert D. Coghill, John F. Enders, Carl Lamanna, et al. "Culture Collections of Microorganisms." *Science* 116, no. 3007 (1952): 179–80.

Rasmussen, Nicolas. *Gene Jockeys: Life Science and the Rise of Biotech Enterprise*. Baltimore: Johns Hopkins University Press, 2014.

———. "The Mid-century Biophysics Bubble: Hiroshima and the Biological Revolution in America, Revisited." *History of Science* 35 (1997): 245–93.

Ratcliff, Marc. *The Quest for the Invisible: Microscopy in the Enlightenment*. Farnham, Surrey, UK: Ashgate, 2009.

"Recent Advances in the Collection of Serum from Marine Animals." *Serological Museum Bulletin* 5 (1950): 3–4.

Reichert, Edward Tyson, and Amos P. Brown. *The Differentiation and Specificity of Corresponding Proteins and Other Vital Substances in Relation to Biological Classification and Organic Evolution: The Crystallography of Hemoglobins*. Washington, DC: Carnegie Institution of Washington, 1909.

Rexer, Lyle, and Rachel Klein. *American Museum of Natural History: 125 Years of Expedition and Discovery*. New York: H. N. Abrams in association with the American Museum of Natural History, 1995.

Rheinberger, Hans-Jörg. *Toward a History of Epistemic Things: Synthesizing Proteins in the Test Tube*. Stanford: Stanford University Press, 1997.

Rhoades, Marcus M. "The Early Years of Maize Genetics." *Annual Review of Genetics* 18 (1984): 1–29.

———. "Rollins Adams Emerson, 1873–1947." *Biographical Memoirs of the National Academy of Science of the United States of America* 25 (1949): 311–23.

Rhodes, Richard. *Dark Sun: The Making of the Hydrogen Bomb*. New York: Simon & Schuster, 1995.

Richards, Frederic M. "The Matching of Physical Models to Three-Dimensional Electron-Density Maps: A Simple Optical Device." *Journal of Molecular Biology* 37, no. 1 (Oct 14 1968): 225–30.

Richardson, Jane S. "The Anatomy and Taxonomy of Protein Structure." *Advances in Protein Chemistry* 34 (1981): 167–339.

———. "β-Sheet Topology and the Relatedness of Proteins." *Nature* 268, no. 5620 (Aug 11 1977): 495–500.

Richardson, Sarah S., and Hallam Stevens. *Postgenomics: Perspectives on Biology after the Genome*. Durham: Duke University Press, 2015.

Riddle, Donald L., and P. Swanson. "Caenorhabditis Genetics Center." *Worm Breeder's Gazette* 5, no. 1 (1980): 4.

Rieppel, Lukas. "Dinosaurs: Assembling an Icon of Science." PhD diss., Harvard University, 2012.

Rindone, Wayne P., Harold M. Perry, Walter B. Goad, Howard S. Bilofsky, and Christine K. Carrico. "GenBank Tm: The Genetic Sequence Data-Bank." *DNA: A Journal of Molecular and Cellular Biology* 2, no. 2 (1983): 173–73.

Roberts, Leslie. "Bioinformatics: Private Pact Ends the DNA Data War." *Science* 299, no. 5606 (Jan 24 2003): 487–89.

———. "Genome Backlash Going Full Force." *Science* 248 (1990): 804.

———. "Genome Research: A Tussle over the Rules for DNA Data Sharing." *Science* 298, no. 5597 (Nov 15 2002): 1312–13.

———. "NIH, DOE Battle for Custody of DNA Sequence Data." *Science* 262, no. 5133 (Oct 22 1993): 504–5.

———. "Watson versus Japan." *Science* 246, no. 4930 (Nov 3 1989): 576–78.

Roberts, Richard J. "Benefits of Databases." *Nature* 342, no. 6246 (Nov 9 1989): 114.

———. "The Early Days of Bioinformatics Publishing." *Bioinformatics* 16, no. 1 (Jan 2000): 2–4.

———. "Pubmed Central: The GenBank of the Published Literature." *Proceedings of the National Academy of Sciences* 98, no. 2 (Jan 16 2001): 381–82.

Roberts, Richard J., and Dieter Söll. Preface. *Nucleic Acids Research* 12, no. 1 (1984).

———. Preface. *Nucleic Acids Research* 14, no. 1 (1986).

Roberts, Richard J., Harold E. Varmus, Michael Ashburner, Patrick O. Brown, Michael B. Eisen, Chaitan Khosla, Marc Kirschner, et al. "Building a 'GenBank' of the Published Literature." *Science* 291, no. 5512 (Mar 23 2001): 2318–19.

Roderick, Thomas. "Listing of Genetic Stock Centers and Newsletters." *Journal of Heredity* 66 (1975): 104–12.

Rogers, Lore A. "The American Type-Culture Collection." *Science* 62, no. 1603 (1925): 267.

———. "Type Cultures." *Science* 66, no. 1710 (1927): 329.

Roos, David S. "Computational Biology. Bioinformatics: Trying to Swim in a Sea of Data." *Science* 291, no. 5507 (Feb 16 2001): 1260–61.

Rosenberg, Charles E. *No Other Gods: On Science and American Social Thought*. Baltimore: Johns Hopkins University Press, 1997.

Rosenberg, Daniel. "Early Modern Information Overload." *Journal of the History of Ideas* 64, no. 1 (2003): 1–9.

Rosenthal, Nadia, and Michael Ashburner. "Taking Stock of Our Models: The Function and Future of Stock Centres." *Nature Reviews Genetics* 3, no. 9 (Sep 2002): 711–17.

Rossiter, Margaret W. *Women Scientists in America: Before Affirmative Action, 1940–1972*. Baltimore: Johns Hopkins University Press, 1995.

"Rutgers Unit Gets $5000." *New York Times*, Apr 7 1951, 6.

Salzberg, Steven L. "Genome Re-annotation: A Wiki Solution?" *Genome Biology* 8, no. 1 (2007): 102.

Sanger, Frederick. "Sequences, Sequences, and Sequences." *Annual Review of Biochemistry* 57 (1988): 1–28.

Sanger, Frederick, and Alan R. Coulson. "A Rapid Method for Determining Sequences in DNA by Primed Synthesis with DNA Polymerase." *Journal of Molecular Biology* 94, no. 3 (May 25 1975): 441–48.

Sanger, Frederick, Steve Nicklen, and Alan R. Coulson. "DNA Sequencing with Chain-Terminating Inhibitors." *Proceedings of the National Academy of Sciences* 74, no. 12 (Dec 1977): 5463–67.

Santesmases, Maria Jesús. "Severo Ochoa and the Biomedical Sciences in Spain under Franco (1959–1975)." *Isis* 91 (2000): 27–45.

Schekman, R. "The Nine Lives of Daniel E. Koshland, Jr. (1920–2007)." *Proceedings of the National Academy of Sciences* 104, no. 37 (Sep 11 2007): 14551–52.

Schickore, Jutta. "The 'Philosophical Grasp of the Appearances' and Experimental Microscopy: Johannes Müller's Microscopical Research, 1824–1832." *Studies in History and Philosophy of Biological and Biomedical Sciences* 34 (2003): 569–92.

Schiebinger, Londa L. *Plants and Empire: Colonial Bioprospecting in the Atlantic World.* Cambridge: Harvard University Press, 2004.

Schiebinger, Londa L., and Claudia Swan, eds. *Colonial Botany: Science, Commerce, and Politics in the Early Modern World.* Philadelphia: University of Pennsylvania Press, 2005.

Schneider, William H. "The History of Research on Blood Group Genetics: Initial Discovery and Diffusion." *History and Philosophy of the Life Sciences* 18 (1996): 277–303.

Schodde, Richard. "Obituary: Charles Sibley (1911–1998)," *Emu* 100, no. 1 (2000): 75–76.

"Science Exhibition." *Science* 106, no. 2763 (Dec 12 1947): 567–75.

Secord, Anne. "Science in the Pub: Artisan Botanists in Early Nineteenth-Century Lancashire." *History of Science* 32 (1994): 269–315.

"Sectional Meetings." *Lancet* 138, no. 3546 (1891): 371–401.

Sedgwick, William T., and Charles-Edward Winslow. "On the Relative Importance of Public Water Supplies and Other Factors in the Causation of Typhoid Fever." *Public Health Papers and Reports* 28 (1902): 288–95.

Sepkoski, David. "Towards 'a Natural History of Data': Evolving Practices and Epistemologies of Data in Paleontology, 1800–2000." *Journal of the History of Biology* 46, no. 3 (Aug 2013): 401–44.

Sequeira, E., J. McEntyre, and D. Lipman. "Pubmed Central Decentralized." *Nature* 410, no. 6830 (Apr 12 2001): 740.

"Sequences Add Up." *Nature* 297 (1982): 96.

"Serological Museum of Rutgers University." *Nature* 161, no. 4090 (Mar 20 1948): 428.

Shapin, Steven. "The House of Experiment in Seventeenth-Century England." *Isis* 79, no. 298 (Sep 1988): 373–404.

———. *The Scientific Life: A Moral History of a Late Modern Vocation.* Chicago: University of Chicago Press, 2008.

———. *The Scientific Revolution.* Chicago: University of Chicago Press, 1996.

———. *A Social History of Truth: Civility and Science in Seventeenth-Century England.* Chicago: University of Chicago Press, 1995.

Shapin, Steven, and Simon Schaffer. *Leviathan and the Air-Pump: Hobbes, Boyle, and the Experimental Life.* Princeton: Princeton University Press, 1985.

Shapiro, Marvin B., Carl R. Merril, Dan F. Bradley, and James E. Mosimann. "Reconstruction of Protein and Nucleic Acid Sequences: Alamine Transfer Ribonucleic Acid." *Science* 150, no. 698 (Nov 12 1965): 918–21.

Sheets-Pyenson, Susan. *Cathedrals of Science: The Development of Colonial Natural History Museums during the Late Nineteenth Century*. Kingston, Ont.: McGill-Queen's University Press, 1988.

Sherry, Stephen T., Minghong Ward, and Karl Sirotkin. "dbSNP-Database for Single Nucleotide Polymorphisms and Other Classes of Minor Genetic Variation." *Genome Research* 9, no. 8 (Aug 1999): 677–79.

Sibley, Charles G. "The 'Alpha Helix' Expedition to New Guinea." *Discovery* 4, no. 1 (1968): 45–52.

———. "The Comparative Morphology of Protein Molecules as Data for Classification." *Systematic Zoology* 11, no. 3 (1962): 108–18.

———. "The Electrophoretic Patterns of Avian Egg-White Proteins as Taxonomic Characters." *Ibis* 102 (1960): 215–84.

———. "Proteins and DNA in Systematic Biology." *Trends in Biochemical Sciences* 22, no. 9 (Sep 1997): 364–67.

———. "A Report on Program B of the Alpha Helix Expedition to New Guinea." *Discovery* 5, no. 1 (1969): 39–46.

Sibley, Charles G., and Jon E. Ahlquist. *Phylogeny and Classification of Birds: A Study in Molecular Evolution*. New Haven: Yale University Press, 1990.

———. "The Phylogeny of the Hominoid Primates, as Indicated by DNA-DNA Hybridization." *Journal of Molecular Evolution* 20, no. 1 (Jan 1 1984): 2–15.

Sibley, Charles G., and Herbert T. Hendrickson. "A Comparative Electrophoretic Study of Avian Plasma Proteins." *Condor* 72, no. 1 (1970): 43–49.

Sibley, Charles G., and Paul A. Johnsgard. "An Electrophoretic Study of Egg-White Proteins in Twenty-Three Breeds of the Domestic Fowl." *American Naturalist* 93, no. 869 (1959): 107–15.

———. "Variability in the Electrophoretic Patterns of Avian Serum Proteins." *Condor* 61, no. 2 (1959): 85–95.

Silberman, Steve. "Inside the High Tech Hunt for a Missing Silicon Valley Legend." *Wired*, Jul 24 2007.

Silverstein, Arthur M. *A History of Immunology*. London: Academic Press, 2009.

Simpson, George G. *Principles of Animal Taxonomy*. New York: Columbia University Press, 1961.

———. "The Principles of Classification and a Classification of Mammals." *Bulletin of the American Museum of Natural History* 85 (1945): 1–307.

Simpson, George G., and Anne Roe. *Quantitative Zoology; Numerical Concepts and Methods in the Study of Recent and Fossil Animals*. New York: McGraw-Hill, 1939.

Slack, Jonathan M. W. "Emerging Market Organisms." *Science* 323 (2009): 1674–75.

Slocum, Anthony H. "Sequences On-Line." *Nature* 299 (1982): 482.

Sly, Lindsay I., Teiji Iijima, and B. Kirsop, eds. *100 Years of Culture Collections*. Osaka: Institute for Fermentation, 1990.

Smith, Emil L. "Nucleotide Base Coding and Amino Acid Replacements in Proteins." *Proceedings of the National Academy of Sciences* 48 (Apr 15 1962): 677–84.

———. "Nucleotide Base Coding and Amino Acid Replacements in Proteins, II." *Proceedings of the National Academy of Sciences* 48 (May 15 1962): 859–64.

Smith, Emil L., and Emanuel Margoliash. "Evolution of Cytochrome C." *Federation Proceedings* 23 (Nov–Dec 1964): 1243–47.

Smith, Pamela H. "Laboratories." In *The Cambridge History of Science: Early Modern Science*, ed. Katherine Park and Lorraine Daston, 290–305. Cambridge: Cambridge University Press, 2006.

Smith, Temple F. "The History of the Genetic Sequence Databases." *Genomics* 6 (1990): 701–7.

Smith, Temple F., and Christian Burks. "Searching for Sequence Similarities." *Nature* 301 (1983): 194.

Smith, Temple F., and Michael S. Waterman. "Identification of Common Molecular Subsequences." *Journal of Molecular Biology* 147, no. 1 (1981): 195–97.

Somerville, Chris, and Maarten Koornneef. "A Fortunate Choice: The History of *Arabidopsis* as a Model Plant." *Nature Reviews Genetics* 3, no. 11 (Nov 2002): 883–89.

Sommer, Marianne. "History in the Gene: Negotiations between Molecular and Organismal Anthropology." *Journal of the History of Biology* 41, no. 3 (2008): 473–528.

———. *History Within: The Science, Culture, and Politics of Bones, Organisms, and Molecules*. Chicago: University of Chicago Press, 2016.

Spary, Emma C. *Utopia's Garden: French Natural History from Old Regime to Revolution*. Chicago: University of Chicago Press, 2000.

Spath, Susan Barbara. "C. B. Van Niel and the Culture of Microbiology, 1920–1965." PhD diss., University of California, Berkeley, 1999.

Speyer, Joseph F., Peter Lengyel, Carlos Basilio, and Severo Ochoa. "Synthetic Polynucleotides and the Amino Acid Code. II." *Proceedings of the National Academy of Sciences* 48 (Jan 15 1962): 63–68.

———. "Synthetic Polynucleotides and the Amino Acid Code. IV." *Proceedings of the National Academy of Sciences* 48 (Mar 15 1962): 441–48.

Stacy, Ralph W., and Bruce D. Waxman, eds. *Computers in Biomedical Research*. New York: Academic Press, 1965.

Sanford, Glenn M., William I. Lutterschmidt, and Victor H. Hutchison. "The Comparative Method Revisited." *BioScience* 52, no. 9 (2002): 830–36.

Sterling, Theodor D., and Seymour V. Pollack. *Computers and the Life Sciences*. New York: Columbia University Press, 1965.

Stevens, Hallam. "Coding Sequences: A History of Sequence Comparison Algorithms as a Scientific Instrument." *Perspectives on Science* 19, no. 3 (2011): 263–99.

———. "From Bomb to Bank: Walter Goad and the Introduction of Computers into Biology." In *Outsider Scientists: Routes to Innovation in Biology*, ed. Oren Harmon and Michael Dietrich, 128–44. Chicago: University of Chicago Press, 2013.

———. *Life out of Sequence: A Data-Driven History of Bioinformatics*. Chicago: University of Chicago Press, 2013.

———. "Life out of Sequence: An Ethnographic Account of Bioinformatics from the Arpanet to Postgenomics." PhD diss, Harvard University, 2010.

———. "Networking Biology: The Origins of Sequence-Sharing Practices in Genomics." *Technology and Culture* 56, no. 4 (Oct 2015): 839–67.

———. "The Politics of Sequence: Data Sharing and the Open Source Software Movement." *Information & Culture: A Journal of History* 50, no. 4 (2015): 465–503.

Stevens, Peter F. "Haüy and A.-P. De Candolle: Crystallography, Botanical Systematics, and Comparative Morphology, 1780–1840," *Journal of the History of Biology* 17, no. 1 (1984): 49–82.

Strasser, Bruno J. "Collecting, Comparing, and Computing Sequences: The Making of Margaret O. Dayhoff's Atlas of Protein Sequences and Structure, 1954–1965." *Journal of the History of Biology* 43, no. 4 (2010): 623–60.

———. "Collecting Nature: Practices, Styles, and Narratives." *Osiris* 27, no. 1 (2012): 303–40.

———. "The Experimenter's Museum: GenBank, Natural History, and the Moral Economies of Biomedicine." *Isis* 102, no. 1 (2011): 60–96.

———. *La fabrique d'une nouvelle science: La biologie moléculaire à l'âge atomique (1945–1964)*. Florence: Olschki, 2006.

———. "A World in One Dimension: Linus Pauling, Francis Crick and the Central Dogma of Molecular Biology." *History and Philosophy of the Life Science* 28 (2006): 491–512.

Strasser, Bruno J., Jérôme Baudry, Dana Mahr, Gabriela Sanchez, and Elise Tancoigne. "'Citizen Science'? Rethinking Science and Public Participation." *Science and Technology Studies* (2019, *in press*).

Strasser, Bruno J., and Soraya de Chadarevian. "The Comparative and the Exemplary: Revisiting the Early History of Molecular Biology." *History of Science* 49 (2011): 317–36.

Strasser, Bruno J., and Paul Edwards. *Open Access: Publishing, Commerce, and the Scientific Ethos*. Bern: Swiss Science and Innovation Council, 2016.

Strick, James E. "Creating a Cosmic Discipline: The Crystallization and Consolidation of Exobiology, 1957–1973." *Journal of the History of Biology* 37, no. 1 (2004): 131–80.

Stuessy, Tod F. *Plant Taxonomy: The Systematic Evaluation of Comparative Data*. New York: Columbia University Press, 2009.

Sturdy, Steve. "Looking for Trouble: Medical Science and Clinical Practice in the Historiography of Modern Medicine." *Social History of Medicine* 24, no. 3 (2011): 739–57.

Su, Andrew I., Benjamin M. Good, and Andre J. van Wijnen. "Gene Wiki Reviews: Marrying Crowdsourcing with Traditional Peer Review." *Gene* 531, no. 2 (Dec 1 2013): 125.

Suárez-Díaz, Edna. "The Long and Winding Road of Molecular Data in Phylogenetic Analysis." *Journal of the History of Biology* 47, no. 3 (Fall 2014): 443–78.

———. "Molecular Evolution: Concepts and the Origin of Disciplines." *Studies in History and Philosophy of Science Part C* 40, no. 1 (Mar 2009): 43–53.

Suárez-Díaz, Edna, and Victor H. Anaya-Muñoz. "History, Objectivity, and the Construction of Molecular Phylogenies." *Studies in History and Philosophy of Science Part C* 39, no. 4 (Dec 2008): 451–68.

Suber, Peter. *Open Access*. Cambridge: MIT Press, 2012.

Sulston, John, and Georgina Ferry. *The Common Thread: A Story of Science, Politics, Ethics and the Human Genome*. London: Bantam Press, 2002.

Summers, Neena L., and Martin Karplus. "Construction of Side-Chains in Homology Modelling. Application to the C-Terminal Lobe of Rhizopuspepsin." *Journal of Molecular Biology* 210, no. 4 (Dec 20 1989): 785–811.

Summers, William C. *Félix d'Hérelle and the Origins of Molecular Biology* (New Haven: Yale University Press, 1999).

———. "How Bacteriophage Came to Be Used by the Phage Group." *Journal of the History of Biology* 26, no. 2 (1993): 255–67.

Sunder Rajan, Kaushik. *Biocapital: The Constitution of Postgenomic Life*. Durham: Duke University Press, 2006.

Sunderland, M. E. "Regeneration: Thomas Hunt Morgan's Window into Development." *Journal of the History of Biology* 43, no. 2 (May 2010): 325–61.

Sussman, Joel L. "Protein Data Bank Deposits." *Science* 282, no. 5396 (Dec 11 1998): 1993.

Swanson, Kara. "Biotech in Court: A Legal Lesson on the Unity of Science." *Social Studies of Science* 37, no. 3 (2007): 357–84.

Swanson, Kara W. *Banking on the Body: The Market in Blood, Milk, and Sperm in Modern America*. Cambridge: Harvard University Press, 2014.

Teitelman, Robert. *Gene Dreams: Wall Street, Academia, and the Rise of Biotechnology*. New York: Basic Books, 1989.

Terrall, Mary. *Catching Nature in the Act: Réaumur and the Practice of Natural History in the Eighteenth Century*. Chicago: University of Chicago Press, 2014.

"This Museum Banks on Blood." *Home News (New Jersey)*, Mar 2 1971.

Thompson, Edward P. *The Making of the English Working Class*. New York: Penguin Books, 1979 [1963].

Throckmorton, Lynn H. "Reviewed Work: A Comparative Study of the Egg-White Proteins of Passerine Birds by Sibley, Charles G." *Quarterly Review of Biology* 47 (1973): 93–94.

Tkacz, Nathaniel. *Wikipedia and the Politics of Openness*. Chicago: University of Chicago Press, 2015.

Toffler, Alvin. *The Third Wave*. New York: Morrow, 1980.

Tomes, Nancy. *The Gospel of Germs: Men, Women, and the Microbe in American Life*. Cambridge: Harvard University Press, 1998.

Tozzi, Christopher J. *For Fun and Profit: A History of the Free and Open Source Software Revolution*. Cambridge: MIT Press, 2017.

Tuppy, Hans. "Aminosaure-Sequenzen in Proteinen." *Naturwissenschaften* 46, no. 2 (1959): 35–43.

Turner, Fred. *From Counterculture to Cyberculture: Stewart Brand, the Whole Earth Network, and the Rise of Digital Utopianism*. Chicago: University of Chicago Press, 2006.

Turrill, William Bertram. "The Subjective Element in Plant Taxonomy." *Bulletin du Jardin botanique de l'État à Bruxelles* 27, no. 1 (1957): 1–8.

Tyfield, David. "Transition to Science 2.0: 'Remoralizing' the Economy of Science." *Spontaneous Generations: A Journal for the History and Philosophy of Science* 7, no. 1 (2013): 29–48.

Uhlir, Paul F. *For Attribution: Developing Data Attribution and Citation Practices and Standards: Summary of an International Workshop*. Washington, DC: National Academies Press, 2012.

Van Noorden, Richard. "Funders Punish Open-Access Dodgers." *Nature* 508, no. 7495 (Apr 10 2014): 161.

Varmus, Harold. *The Art and Politics of Science.* New York: W. W. Norton, 2009.

Venter, J. Craig. *A Life Decoded: My Genome, My Life.* New York: Viking, 2007.

Vernon, Keith. "Desperately Seeking Status: Evolutionary Systematics and the Taxonomists' Search for Respectability, 1940–60." *British Journal for the History of Science* 26, no. 89 (Jun 1993): 207–27.

———. "The Founding of Numerical Taxonomy," *British Journal for the History of Science* 21, no. 69 (Jun 1988): 143–59.

Vettel, Eric J. *Biotech: The Countercultural Origins of an Industry.* Philadelphia: University of Pennsylvania Press, 2006.

von Buddenbrock, Wolfgang. *Grundriss der vergleichenden Physiologie.* Berlin: Borntraeger, 1924–28.

Waldrop, M. Mitchell. "Big Data: Wikiomics." *Nature* 455, no. 7209 (Sep 4 2008): 22–25.

———. "On-Line Archives Let Biologists Interrogate the Genome." *Science* 269, no. 5229 (Sep 8 1995): 1356–58.

Walker, Richard T. "A Method for the Rapid and Accurate Deposition of Nucleic Acid Sequence Data in an Acceptably Annotated Form." In *Biomolecular Data: A Resource in Transition*, ed. Rita Colwell, 45–51. Oxford: Oxford University Press, 1989.

Wang, Kai. "Gene-Function Wiki Would Let Biologists Pool Worldwide Resources." *Nature* 439, no. 7076 (Feb 2 2006): 534.

Wang, Yufeng F., and Timothy G. Lilburn. "Biological Resource Centers and Systems Biology." *Bioscience* 59, no. 2 (Feb 2009): 113–25.

Warner, John Harley. "Ideals of Science and Their Discontents in Late Nineteenth-Century American Medicine." *Isis* 82 (1991): 454–78.

Waterman, Michael S. *Skiing the Sun*, 2007, available at https://cseweb.ucsd.edu/classes/sp16/cse182-a/notes/newmex.pdf (accessed Oct 7 2018).

Waterman, Michael S., and Temple F. Smith. "On the Similarity of Dendrograms." *Journal of Theoretical Biology* 73, no. 4 (Aug 21 1978): 789–800.

Waterman, Michael S., Temple F. Smith, Mona Singh, and William A. Beyer. "Additive Evolutionary Trees." *Journal of Theoretical Biology* 64, no. 2 (Jan 21 1977): 199–213.

Waterton, Claire, Rebecca Ellis, and Brian Wynne. *Barcoding Nature: Shifting Cultures of Taxonomy in an Age of Biodiversity Loss.* Milton Park, Oxon, UK: Routledge, 2013.

Watson, James D. *The Double Helix.* London: Weidenfeld and Nicolson, 1968.

———. *The Double Helix: A Personal Account of the Discovery of the Structure of DNA: Text, Commentary, Reviews, Original Papers.* New York: Touchstone, 2001 [1968].

———. *Genes, Girls and Gamow.* Oxford: Oxford University Press, 2001.

Weatherall, Mark W., and Harmke Kamminga. "The Making of a Biochemist. II: The Construction of Frederick Gowland Hopkins' Reputation." *Medical History* 40, no. 4 (Oct 1996): 415–36.

Weinberg, Robert A. "The Case against Gene Sequencing." *Scientist* 1, no. 25 (Nov 16 1987): 11.

———. "Reflections on the Current State of Data and Reagent Exchange among Biomedical Researchers." In *Responsible Science: Ensuring the Integrity of the Research Process*, ed. Committee on Science Engineering and Public Policy (US), 66–78. Washington, DC: National Academy Press, 1993.

Weinberger, David. *Too Big to Know: Rethinking Knowledge Now That the Facts Aren't the Facts, Experts Are Everywhere, and the Smartest Person in the Room Is the Room*. New York: Basic Books, 2012.

Weiss, Freeman A. "The American Type Culture Collection." *AIBS Bulletin* 4, no. 1 (1954): 15–16.

Weitz, Bernard. "The Collection of Serum from Game in East Africa." *Serological Museum Bulletin* 4 (1950): 1–3.

———. "A Mobile Laboratory for the Collection of Sterile Serum from Game in East Africa." *Serological Museum Bulletin* 10 (1953): 4–5.

Westwick, Peter J. *The National Labs: Science in an American System, 1947–1974*. Cambridge: Harvard University Press, 2003.

Wheeler, William Morton. "The Dry-Rot of Our Academic Biology." *Science* 57 (1923): 61–71.

Whitfield, N., and T. Schlich. "Skills through History." *Medical History* 59, no. 3 (Jul 2015): 349–60.

Wilhelmi, Raymond W. "Serological Relationships between the Mollusca and Other Invertebrates." *Biological Bulletin* 87, no. 1 (1944): 96–105.

Williams, James G. "Governance of Special Information-Centers: Knowledge Availability Systems Center at University of Pittsburgh." *Library Trends* 26, no. 2 (1977): 241–54.

Williams, Nigel. "Europe Opens Institute to Deal with Gene Data Deluge." *Science* 269, no. 5224 (Aug 4 1995): 630.

Wilson, Allan C., Steven S. Carlson, and Thomas J. White. "Biochemical Evolution." *Annual Review of Biochemistry* 46 (1977): 573–639.

Wilson, Edmund B. "Aims and Methods of Study in Natural History." *Science* 13, no. 314 (1901): 14–23.

Wilson, Edward Osborne. *Naturalist*. Washington, DC: Island Books, 1994.

Winslow, Charles-Edward A. *Bacterial Collection and Bureau for the Distribution of Bacterial Cultures*. New York: American Museum of Natural History, 1913.

———. "Bacterial Collection and Bureau for the Distribution of Bacterial Cultures at the American Museum of Natural History, New York." *Science* 38, no. 976 (1913): 374–75.

———. "The First Forty Years of the Society of American Bacteriologists." *Science* 91, no. 2354 (Feb 9 1940): 125–29.

———. "The Importance of Preserving the Original Types of Newly Described Species of Bacteria." *Journal of Bacteriology* 6, no. 1 (Jan 1921): 133–34.

Winslow, Charles-Edward A., and Anne F. Rogers. "A Revision of the Coccaceae." *Science* 21, no. 539 (Apr 28 1905): 669–72.

Winsor, Mary P. *Reading the Shape of Nature: Comparative Zoology at the Agassiz Museum*. Chicago: University of Chicago Press, 1991.

Wise, M. Norton. *The Values of Precision*. Princeton: Princeton University Press, 1995.

Witkin, Evelyn M. "Chances and Choices: Cold Spring Harbor, 1944–1955." *Annual Review of Microbiology* 56 (2002): 1–15.

Wlodawer, Alexander. "Deposition of Macromolecular Coordinates Resulting from Crystallographic and NMR Studies." *Nature Structural Biology* 4, no. 3 (Mar 1997): 173–74.

———. "Deposition of Structural Data Redux." *Acta Crystallographica Section D—Biological Crystallography* 63, no. 3 (Mar 2007): 421–23.

Wlodawer, Alexander, David Davies, Gregory Petsko, Michael Rossmann, Arthur Olson, and Joel L. Sussman. "Immediate Release of Crystallographic Data: A Proposal." *Science* 279, no. 5349 (Jan 16 1998): 306–7.

Wodak, Shoshana J., and Joel Janin. "Location of Structural Domains in Protein." *Biochemistry* 20, no. 23 (Nov 10 1981): 6544–52.

Wolfe, Audra J. "Germs in Space: Joshua Lederberg, Exobiology, and the Public Imagination, 1958–1964." *Isis* 93 (2002): 183–205.

Wolfe, Harold R. "Standardization of the Precipitin Technique and Its Application to Studies of Relationships in Mammals, Birds and Reptiles." *Biological Bulletin* 76, no. 1 (1939): 108–20.

Wolpert, Lewis. *The Unnatural Nature of Science*. London: Faber and Faber, 1993.

"Woman Scientist Brings Rare Blood." *New York Times*, May 6 1950, 31.

Wonders, Karen. *Habitat Dioramas: Illusions of Wilderness in Museums of Natural History*. Uppsala: Acta Universitatis Upsaliensis, 1993.

Wright, Alex. *Glut: Mastering Information through the Ages*. Washington, DC: Joseph Henry Press, 2007.

Wright, Christopher A. *Biochemical and Immunological Taxonomy of Animals*. London: Academic Press, 1974.

Wright, Susan. *Molecular Politics: Developing American and British Regulatory Policy for Genetic Engineering, 1972–1982*. Chicago: University of Chicago Press, 1994.

Yates, JoAnne. *Control through Communication: The Rise of System in American Management*. Baltimore: Johns Hopkins University Press, 1989.

Yčas, Martynas. "The Protein Text." In *Symposium on Information Theory in Biology*, ed. Hubert P. Yockey, 70–102. New York: Pergamon Press, 1958.

———. "Replacement of Amino Acids in Proteins." *Journal of Theoretical Biology* 1, no. 2 (1961): 244–57.

Yoon, Carol Kaesuk. *Naming Nature: The Clash between Instinct and Science*. New York: W. W. Norton, 2009.

Zallen, Doris T. "The 'Light' Organism for the Job." *Journal of the History of Biology* 26, no. 3 (1993): 269–79.

Zelen, Marvin. "Review: Sharing Research Data." *Journal of the American Statistical Association* 83, no. 398 (1987): 685–86.

Zuckerkandl, Emile, and Linus Pauling. "Molecular Disease, Evolution, and Genic Heterogeneity." In *Horizons in Biochemistry*, ed. M. Kasha and B. Pullman, 189–224. New York: Academic Press, 1962.

———. "Molecules as Documents of Evolutionary History." *Journal of Theoretical Biology* 8 (1965): 357–66.

Notes

Introduction

1. The expression "data deluge" is widely used, at least since the early 1990s. See Peter Aldhous, "Managing the Genome Data Deluge," *Science* 262, no. 5133 (Oct 22 1993): 502–3, and more recently Gordon Bell, Tony Hey, and Alex Szalay, "Computer Science: Beyond the Data Deluge," *Science* 323, no. 5919 (Mar 6 2009): 1297–98. "Data tsunami": see Anthony J. G. Hey, Stuart Tansley, and Kristin Tolle, *The Fourth Paradigm: Data-Intensive Scientific Discovery* (Redmond, WA: Microsoft Research, 2009), 117, 131. "Floods": see Bell, Hey, and Szalay, "Computer Science," 1297. "Swim in a sea of data": see David S. Roos, "Computational Biology. Bioinformatics: Trying to Swim in a Sea of Data," *Science* 291, no. 5507 (Feb 16 2001): 1260–61. "Big Data": see *Nature* 455, no. 7209 (2008), cover.

2. The initial study comparing both encyclopedias was conducted by the journal Nature: Jim Giles, "Internet Encyclopaedias Go Head to Head," *Nature* 438, no. 7070 (Dec 15 2005): 900–901. For a discussion about the reliability of Wikipedia, see Nathaniel Tkacz, *Wikipedia and the Politics of Openness* (Chicago: University of Chicago Press, 2015).

3. Frank Zappa, "Packard Goose," *Joe's Garage: Act III*, 1978.

4. Bruno Latour, *Pandora's Hope: Essays on the Reality of Science Studies* (Cambridge: Harvard University Press, 1999), chap. 2.

5. *The Economist* 394, no. 8671 (2010).

6. Felice Frankel and Rosalind Reid, "Big Data: Distilling Meaning from Data," *Nature* 455, no. 7209 (Sep 4 2008): 30.

7. On the 9/11 terrorist attacks, see National Commission on Terrorist Attacks upon the United States, *The 9/11 Commission*

Report : Final Report of the National Commission on Terrorist Attacks upon the United States (New York: Norton, 2004), and on the financial crisis, Tomio Geron, *Forbes/ Tech*, Apr 26 2011, http://www.forbes.com/sites/tomiogeron/2011/04/26/can -the-u-n-use-big-data-to-respond-to-global-disasters/.

8. *Nature*, 455, no. 1 (Sep 4, 2008).

9. *Wired* (Jul 16 2008), cover.

10. See the special issue on "data-driven sciences," Sabina Leonelli, "Introduction: Making Sense of Data-Driven Research in the Biological and Biomedical Sciences," *Studies in History and Philosophy of Science Part C* 43, no. 1 (Mar 2012): 1–3.

11. Aldhous, "Managing the Genome Data Deluge."

12. Leslie Roberts, "Genome Backlash Going Full Force," *Science* 248 (1990): 804.

13. On this distinction, Rob Kitchin, *The Data Revolution : Big Data, Open Data, Data Infrastructures and Their Consequences* (Los Angeles: SAGE Publications, 2014), chap. 8.

14. Patrick O. Brown and David Botstein, "Exploring the New World of the Genome with DNA Microarrays," *Nature Genetics* 21, no. 1 suppl. (Jan 1999): 33–37.

15. Maureen A. O'Malley et al., "Philosophies of Funding," *Cell* 138, no. 4 (Aug 21 2009): 611–15.

16. Hey, Tansley, and Tolle, *Fourth Paradigm*.

17. Ibid., xvii–xxxi.

18. Steve Silberman, "Inside the High Tech Hunt for a Missing Silicon Valley Legend," *Wired*, Jul 24 2007.

19. Hey, Tansley, and Tolle, *Fourth Paradigm*, xv; Bell, Hey, and Szalay, "Computer Science."

20. J. F. Allen, "Bioinformatics and Discovery: Induction Beckons Again," *Bioessays* 23, no. 1 (Jan 2001): 104–7.

21. J. F. Allen, "*In Silico Veritas*. Data-Mining and Automated Discovery: The Truth Is in There," *EMBO Reports* 2, no. 7 (Jul 2001): 542–44.

22. The best critical discussion of "data driven" science and elaboration of an alternative framework, "data-centric biology," can be found in Sabina Leonelli, *Data-centric Biology: A Philosophical Study* (Chicago: University of Chicago Press, 2016).

23. Alvin Toffler, *The Third Wave* (New York: Morrow, 1980).

24. Eliot Marshall, "Data Sharing: DNA Sequencer Protests Being Scooped with His Own Data," *Science* 295, no. 5558 (Feb 15 2002): 1206–7.

25. Bruno J. Strasser et al., "'Citizen Science'? Rethinking Science and Public Participation," *Science and Technology Studies* (2019, in press).

26. Joseph A. November, *"Digitizing Life: The Introduction of Computers to Biology and Medicine"* (PhD diss., Princeton University, 2006); November, *Biomedical Computing: Digitizing Life in the United States* (Baltimore: Johns Hopkins University Press, 2012).

27. Miguel Garcia-Sancho, *Biology, Computing, and the History of Molecular Sequencing* (New York: Palgrave Macmillan, 2012).

28. Hallam Stevens, "Life out of Sequence: An Ethnographic Account of Bioinformatics from the Arpanet to Postgenomics" (PhD diss, Harvard University, 2010), and *Life out of Sequence: A Data-Driven History of Bioinformatics* (Chicago: University of Chicago Press, 2013), 225, 168, 203.

29. Leonelli, *Data-centric Biology*, 32, 191, 170–71.

30. See for example Michael A. Nielsen, *Reinventing Discovery: The New Era of Networked Science* (Princeton: Princeton University Press, 2012); Viktor Mayer-Schönberger and Kenneth Cukier, *Big Data: A Revolution That Will Transform How We Live, Work, and Think* (Boston: Houghton Mifflin Harcourt, 2013).

31. A preliminary version of this argument was first presented at the 2005 Ischia Summer School, "Gathering Things, Collecting Data, Producing Knowledge: The Use of Collections in Biological and Medical Knowledge Production from Early Modern Natural History to Genome Databases," organized by Janet Browne, Bernardino Fantini, and Hans-Jörg Rheinberger. The comments of the participants were most helpful in reframing the argument presented here.

32. There are many case studies, but a few scholars have taken "collecting" as a more general and defining practice. See however Anke te Heesen and Emma C. Spary, eds., *Sammeln als Wissen: Das Sammeln und seine wissenschaftsgeschichtliche Bedeutung* (Göttingen: Wallstein, 2001); Robert E. Kohler, "Finders, Keepers: Collecting Sciences and Collecting Practice," *History of Science* 45, no. 4 (2007): 428–54; and the discussion in Bruno J. Strasser, "Collecting Nature: Practices, Styles, and Narratives," *Osiris* 27, no. 1 (2012): 303–40.

33. Robert Darnton, "An Early Information Society: News and the Media in Eighteenth-Century Paris," *American Historical Review* 105, no. 1 (2000): 1–35.

34. For the example of Linnaeus in the eighteenth century, see Isabelle Charmantier and Staffan Müller-Wille, "Natural History and Information Overload: The Case of Linnaeus," *Studies in History and Philosophy of Biological and Biomedical Sciences* 43, no. 1 (2012): 4–15.

35. Ann Blair, *Too Much to Know: Managing Scholarly Information before the Modern Age* (New Haven: Yale University Press, 2010). However, as Nick Levine argues, there was also something specific about the current concerns with information overload that emerged when the self began to be conceptualized in the 1960s as an information-processing machine of limited capacity. Levine, "The Nature of the Glut: Information Overload in Postwar America," *History of the Human Sciences* 30, no. 1 (Feb 2017): 32–49.

36. JoAnne Yates, *Control through Communication: The Rise of System in American Management* (Baltimore: Johns Hopkins University Press, 1989); Delphine Gardey, *Ecrire, calculer, classer: Comment une révolution de papier a transformé les sociétés contemporaines, 1800–1940* (Paris: Découverte, 2008); Alex Wright, *Glut: Mastering Information through the Ages* (Washington, DC: Joseph Henry Press, 2007); Markus Krajewski, *Paper Machines: About Cards and Catalogs, 1548–1929* (Cambridge: MIT Press, 2011).

37. On the "dual arrangement" in Germany, see Lynn K. Nyhart, *Modern Nature: The Rise of the Biological Perspective in Germany* (Chicago: University of Chicago Press, 2009), chap. 6; in the colonial context, Susan Sheets-Pyenson, *Cathedrals of Science: The Development of Colonial Natural History Museums during the Late Nineteenth Century* (Kingston, Ont.: McGill-Queen's University Press, 1988); and the United States, Karen A. Rader and Victoria E. M. Cain, *Life on Display: Revolutionizing U.S. Museums of Science and Natural History in the Twentieth Century* (Chicago: University of Chicago Press, 2014).

38. In his classic *Biology in the Nineteenth Century*, the historian of biology William R. Coleman summarized his argument in the follow terms: "In its name—experiment—was

set in motion a campaign to revolutionize the goals and methods of biology." In his no less classic *Life Sciences in the Twentieth Century*, Garland E. Allen put it like this: "It was the twentieth century that saw the fanning out of the experimental method in all areas of biology." Coleman, *Biology in the Nineteenth Century: Problems of Form, Function and Transformation* (Cambridge: Cambridge University Press, 1971), 2; 1 Garland E. Allen, *Life Science in the Twentieth Century* (Cambridge: Cambridge University Press, 1978) xvi.

39. For a discussion of the standard narrative, its strengths and limitations, see Strasser, "Collecting Nature." On postgenomics, Sarah S. Richardson and Hallam Stevens, *Postgenomics: Perspectives on Biology after the Genome* (Durham: Duke University Press, 2015).

40. Lynn K. Nyhart, "Natural History and the 'New' Biology," in *Cultures of Natural History*, ed. Nicholas Jardine, James A. Secord, and Emma C. Spary (London: Cambridge University Press, 1996), 433. Similarly, Jane Maienschein has stressed the continuities in methods and problems among American biologists in the two decades around 1900 and the fact that they tried to overcome their old division. In 1901, the experimental cell biologist Edmund B. Wilson was not alone when he attempted to ease the tensions resulting from "a lack of mutual understanding . . . between the field naturalist and the laboratory workers." Maienschein, "Shifting Assumptions in American Biology: Embryology, 1890–1910," *Journal of the History of Biology* 14, no. 1 (Spring 1981): 89–113, and Maienschein, *Transforming Traditions in American Biology, 1880–1915* (Baltimore: Johns Hopkins University Press, 1991). Wilson's quote is from Edmund B. Wilson, "Aims and Methods of Study in Natural History," *Science* 13, no. 314 (1901): 19.

41. Nyhart, "Natural History."

42. Keith R. Benson, "From Museum Research to Laboratory Research: The Transformation of Natural History into Academic Biology," in *The American Development of Biology*, ed. Ronald Rainger, Keith R. Benson, and Jane Maienschein (Philadelphia: University of Pennsylvania Press, 1988), 49–83.

43. Sally Gregory Kohlstedt, "Museums on Campus: A Tradition of Inquiry and Teaching," in Rainger, Benson, and Maienschein, *American Development of Biology*, 15–47.

44. Alison Kraft and Samuel J. M. M. Alberti, "'Equal though Different': Laboratories, Museums and the Institutional Development of Biology in Late-Victorian Northern England," *Studies in History and Philosophy of Biological and Biomedical Sciences* 34, no. 2 (2003): 203; see also the struggles for space between museums and laboratories in Cambridge described by Sophie Forgan, "The Architecture of Display: Museums, Universities and Objects in 19th-Century Britain," *History of Science* 32, no. 96 (Jun 1994): 139–62.

45. On experimental taxonomy, see Joel B. Hagen, "Experimentalists and Naturalists in 20th-Century Botany: Experimental Taxonomy, 1920–1950," *Journal of the History of Biology* 17, no. 2 (1984): 249–70, and on French "experimental transformism," Laurent Loison, *Qu'est-ce que le néolamarckisme? Les biologistes français et la question de l'évolution des espèces* (Paris: Vuibert, 2010).

46. On the boundaries between field and lab, see Robert E. Kohler, *Landscapes and Labscapes: Exploring the Lab-Field Border in Biology* (Chicago: University of Chicago Press, 2002). On naturalists as "stamp collectors," Kristin Johnson, "Natural History as Stamp Collecting: A Brief History," *Archives of Natural History* 34, no. 2 (2007): 244–58, and on their actual practices, see Johnson, *Ordering Life: Karl Jordan and the Naturalist Tradition*

(Baltimore: Johns Hopkins University Press, 2012). Some have criticized Kohler's too neat division between lab and field (Mitman) and questioned whether his sample of historical actors was representative (Kingsland) or if the broader issue was about academic biology and natural professionals rather than about lab and field (Pauly). Sharon Kingsland, "Robert E. Kohler, *Landscapes and Labscapes: Exploring the Lab-Field Border in Biology*," *Isis* 95 (Sep 2004): 509–10; Gregg Mitman, "Robert E. Kohler, *Landscapes and Labscapes: Exploring the Lab-Field Border in Biology*," *Journal of the History of Biology* 36 (Sep 1 2003): 599–629; Philip J. Pauly, "*Landscapes and Labscapes: Exploring the Lab-Field Border in Biology*, by Robert E Kohler," *Quarterly Review of Biology* 78 (Jun 2003): 214–15. For experimental morphology, see for example Thomas H. Morgan's praise for Charles B. Davenport's *Experimental Morphology* (1899), T. H. Morgan, "Review of Experimental Morphology," *Science* 9 (1899): 648–50.

47. In Vienna in the 1790s, for example, the physician Franz Joseph Gall was actively collecting skulls of humans and animals, as well as plaster casts and wax replicas of brains, particularly of exceptional individuals, such as renowned artists, philosophers, and criminals. By the time of his death, in 1828, he had amassed hundreds of skulls and brains. It was by meticulously comparing the items in his collection that he elaborated the theory that came to be know as "phrenology." On practices in nineteenth- and twentieth-century medical museums, see Samuel J. M. M. Alberti, *Morbid Curiosities: Medical Museums in Nineteenth-Century Britain* (Oxford: Oxford University Press, 2011), and Samuel J. M. M. Alberti, and Elizabeth Hallam, eds., *Medical Museums: Past, Present, Future* (London: Royal College of Surgeons of England, 2013); on comparative anatomy before Cuvier, see Andrew Cunningham, *The Anatomist Anatomis'd: An Experimental Discipline in Enlightenment Europe* (Farnham, Surrey, UK: Ashgate, 2010); on the practice of collecting human skulls in the nineteenth century, see Ann Fabian, *The Skull Collectors: Race, Science, and America's Unburied Dead* (Chicago: University of Chicago Press, 2010), and Michael Hagner, *Geniale Gehirne: Zur Geschichte der Elitegehirnforschung* (Göttingen: Wallstein, 2004), and in the colonial context Elizabeth Hallam, *Anatomy Museum: Death and the Body Displayed* (London: Reaktion Books, 2016), and Warwick Anderson, *The Collectors of Lost Souls: Turning Kuru Scientists into Whitemen* (Baltimore: Johns Hopkins University Press, 2008); for a broad collection of essays about the role of anatomical collections of the last three centuries, see Rina Knoeff and Robert Zwijnenberg, *The Fate of Anatomical Collections* (Farnham, Surrey, UK: Ashgate, 2015).

48. As early as 1970, the molecular biologist and late historian of biology François Jacob told the same story, focusing on the rise of experimentalism with Claude Bernard and the decline of comparative sciences; see François Jacob, *La logique du vivant: Une histoire de l'hérédité* (Paris: Gallimard, 1970). Other fault lines have also been discussed for France, for example between Claude Bernard's style of experimentation, aiming at discovering fundamental mechanisms, and Pasteurian experimentation. I thank Michel Morange for pointing this out. The picture is indeed more complex, as French neolamarckian naturalists embraced experimentation and developed an original program of "experimental transformism" in an attempt to counter Darwinism. On how French neolamarckism prompted naturalists to adopt experimental methods, see Loison, *Qu'est-ce que le néolamarckisme?*; on the debate over Claude Bernard's experimentalism, "Controverses sur la méthode dans

les sciences du vivant: Physiologie, zoologie, botanique (1865–1931)," in *Claude Bernard: La méthode de la physiologie*, ed. François Deschenaux, Jean-Jacques Kupiec, and Michel Morange (Paris: Editions Rue d'Ulm, 2013), 63–82, and for a broader account of experimentation in France and the decline of the Muséum, see Harry W. Paul, *From Knowledge to Power: The Rise of the Science Empire in France, 1860–1939* (Cambridge: Cambridge University Press, 1985), 57.

49. "The Laboratory in Modern Science," *Science* 3 (1884): 173.

50. Susan E. Lederer, *Frankenstein: Penetrating the Secrets of Nature* (New Brunswick, NJ: Rutgers University Press, 2002); Christopher Frayling, *Mad, Bad and Dangerous?: The Scientist and the Cinema* (London: Reaktion, 2005); Charles E. Rosenberg, *No Other Gods: On Science and American Social Thought* (Baltimore: Johns Hopkins University Press, 1997), chap. 7.

51. J. F. A. Adams, "Is Botany a Suitable Study for Young Men," *Science* 9, no. 209 (1887): 117. On the gendering of biology at the turn of the century, see Philip J. Pauly, "Summer Resort and Scientific Discipline: Woods Hole and the Structure of American Biology," in Rainger, Benson, and Maienschein, *American Development of Biology*, 121–50; on women in surveys, Robert E. Kohler, *All Creatures: Naturalists, Collectors, and Biodiversity, 1850–1950* (Princeton: Princeton University Press, 2006), 215–20; and more generally on women in science, Margaret W. Rossiter, *Women Scientists in America: Before Affirmative Action, 1940–1972* (Baltimore: Johns Hopkins University Press, 1995), chap. 3.

52. R. Pearl, "Trends of Modern Biology," *Science* 56, no. 1456 (Nov 24 1922): 583, 585.

53. C. Hart Merriam, "Biology in Our Colleges: A Plea for a Broader and More Liberal Biology," *Science* 21 (1893): 352–353.

54. James G. Needham, "Methods of Securing Better Co-operation between Government and Laboratory Zoologists in the Solution of Problems of General or National Importance," *Science* 49 (1919): 457.

55. William Morton Wheeler, "The Dry-Rot of Our Academic Biology," *Science* 57 (1923): 67–68.

56. As early as 1981, Frederick B. Churchill gave a systematic overview of the arguments involved, including the confusion between methods and objects of study implicit in these analytical categories. Churchill, "In Search of the New Biology: An Epilogue," *Journal of the History of Biology* 14 (1981): 177–91, and the other contributions to this issue.

57. Mary Terrall, *Catching Nature in the Act: Réaumur and the Practice of Natural History in the Eighteenth Century* (Chicago: University of Chicago Press, 2014).

58. Georges Cuvier, *Le règne animal distribué d'après son organisation, pour servir de base à l'histoire naturelle des animaux et d'introduction à l'anatomie comparée* (Paris: Deterville, 1817), translation mine. I thank Lynn Nyhart for drawing my attention to this illuminating passage.

59. Merriam, "Biology in Our Colleges," 352–53.

60. On the multiple meanings of experimentation, including in "observational sciences," see for example Jutta Schickore, "The 'Philosophical Grasp of the Appearances' and Experimental Microscopy: Johannes Müller's Microscopical Research, 1824–1832," *Studies in History and Philosophy of Biological and Biomedical Sciences* 34 (2003): 569–92; Nyhart, "Natural History."

61. On "experimentation" as opposed to "observation," see Lorraine Daston, "The Empire of Observation, 1600–1800," in *Histories of Scientific Observation*, ed. Daston and Elizabeth Lunbeck (Chicago: University of Chicago Press, 2011), 81–113. On experimentation as practice and rhetoric, see Nick Hopwood, "Embryology," in *The Cambridge History of Science: The Modern Biological and Earth Sciences*, ed. Peter J. Bowler and John V. Pickstone (Cambridge: Cambridge University Press, 2009), 304, and on the ways in which "experimentation" was used practically and discursively by American biologists around 1900, Maienschein, *Transforming Traditions in American Biology*.

62. Catherine M. Jackson, "The Laboratory," in *A Companion to the History of Science*, ed. Bernard V. Lightman (West Sussex, UK: John Wiley & Sons, 2016), 296–309; Graeme Gooday, "Placing or Replacing the Laboratory in the History of Science?," *Isis* 99, no. 4 (Dec 2008): 783–95.

63. Kohler, *Landscapes and Labscapes*; on the opposition to the study of dead specimens, see Raf de Bon, "Between the Laboratory and the Deep Blue Sea: Space Issues in the Marine Stations of Naples and Wimereux," *Social Studies of Science* 39, no. 2 (2009): 199–227; Raf de Bont, *Stations in the Field: A History of Place-Based Animal Research, 1870–1930* (Chicago: University of Chicago Press, 2015); on American marine stations resulting from "the twin legacy of natural history in the Agassiz tradition and research laboratories on the Johns Hopkins model," Keith R. Benson, "Laboratories on the New England Shore: The 'Somewhat Different Direction' of American Marine Biology," *New England Quarterly* 61, no. 1 (1988): 55; on a broad view of the networks of American "liquid laboratories," Samantha K. Muka, "Working at Water's Edge: Life Sciences at American Marine Stations, 1880–1930," (PhD diss., University of Pennsylvania, 2014); and for a primary source on European stations, Charles Atwood Kofoid, *The Biological Stations of Europe*. (Washington, DC: Government Printing Office, 1910). On the different picture, when drawn from an institutional perspective, see the perceptive comments in Churchill, "In Search of the New Biology."

64. On Morgan's work on regeneration, see M. E. Sunderland, "Regeneration: Thomas Hunt Morgan's Window into Development," *Journal of the History of Biology* 43, no. 2 (May 2010): 325–61; Thomas Hunt Morgan, *Regeneration* (New York: Macmillan, 1901). As Frederick B. Churchill has shown, at August Weismann's Institute of Zoology in Germany, researchers also worked experimentally on a similarly wide range of organisms in the late nineteenth century. Churchill, "Life before Model Systems: General Zoology at August Weismann's Institute," *American Zoologist* 37, no. 3 (Jun 1997): 260–68. And the same holds true for Germany more generally; see Lynn K. Nyhart, *Biology Takes Form: Animal Morphology and the German Universities, 1800–1900* (Chicago: University of Chicago Press, 1995).

65. Although they valued biological diversity and comparison, taxonomists also adopted the exemplary way of knowing, for example when they used a single specimen, a type specimen, to stand for all the individuals of a species. See Lorraine Daston. "Type Specimens and Scientific Memory," *Critical Inquiry* 31 (2004): 153–82. I thank Lynn Nyhart for pointing this out to me.

66. For a more detailed discussion of Pickstone's narrow definition of "experimentation" (as distinct from "analytical" in his early work) and how it is adapted here, see Bruno J.

Strasser and Soraya de Chadarevian, "The Comparative and the Exemplary: Revisiting the Early History of Molecular Biology," *History of Science* 49 (2011): 317–36. Pickstone's categories are most worked out in John V. Pickstone, "Museological Science? The Place of the Analytical/Comparative in Nineteenth-Century Science, Technology and Medicine," ibid., 32 (1994): 111–38; Pickstone, *Ways of Knowing: A New History of Science, Technology and Medicine* (Manchester: Manchester University Press, 2000); Pickstone, "Working Knowledges before and after circa 1800: Practices and Disciplines in the History of Science, Technology and Medicine," *Isis* 98 (2007): 489–516, and discussed in John V. Pickstone, "Natural Histories, Analyses and Experimentation: Three Afterwords," *History of Science* 49, no. 164 (Sep 2011): 349–74. On experimentation in the eighteenth century, see Terrall, *Catching Nature in the Act,* and Marc Ratcliff, *The Quest for the Invisible: Microscopy in the Enlightenment* (Farnham, Surrey, UK: Ashgate, 2009).

67. A summary of these debates can be found in Churchill, "In Search of the New Biology," 181–82.

68. Nick Hopwood, *Haeckel's Embryos: Images, Evolution, and Fraud* (Chicago: University of Chicago Press, 2015), especially chap. 2 and 281–82; Louis Agassiz, *Methods of Study in Natural History* (Boston: Fields, Osgood, 1869), 21. Elsewhere, Hopwood gives a rich illustration of the ways in which descriptive (and comparative) embryology coexisted with experimental embryology, and how proponents of the latter created the former to demarcate themselves around 1900. Hopwood, "Embryology."

69. For other suggestions that these two categories cannot simply be opposed, see Nyhart, "Natural History"; Hagen, "Experimentalists and Naturalists"; Kohler, *Landscapes and Labscapes.*

70. Benson, "From Museum Research to Laboratory Research," 77.

71. As John Pickstone argues, comparative traditions have also played a role in the physical sciences, especially in astronomy in the eighteenth and nineteenth centuries, but overall the focus on exemplary experimental systems came to define empirical physics in the twentieth century; Pickstone, *Ways of Knowing.*

72. Emma C. Spary, *Utopia's Garden: French Natural History from Old Regime to Revolution* (Chicago: University of Chicago Press, 2000).

73. For a general introduction to patenting in the life sciences, see Graham Dutfield, *Intellectual Property Rights and the Life Science Industries: A Twentieth Century History* (Hackensack, NJ: World Scientific, 2009), and on the importance of patents for academic biology and the rise of the biotech industry, Nicolas Rasmussen, *Gene Jockeys: Life Science and the Rise of Biotech Enterprise* (Baltimore: Johns Hopkins University Press, 2014). How this historical paradox is to be explained is unclear. Were the scientists promoting open access as a form of "resistance" against the excesses of the commercialization of public research, following the Bayh-Dole act, or, less plausibly, are both processes the result of the (neo)liberal attempt to maximize the availability of public research for private corporations?

74. Fred Turner, *From Counterculture to Cyberculture: Stewart Brand, the Whole Earth Network, and the Rise of Digital Utopianism* (Chicago: University of Chicago Press, 2006).

75. Ernst Mayr, *The Growth of Biological Thought: Diversity, Evolution, and Inheritance* (Cambridge, MA: Belknap Press, 1982).

76. Robert E. Kohler has looked at the converse: how experimental methods transformed field practice. Kohler, *Landscapes and Labscapes*.

77. Ibid., 18.

78. Steven Shapin, "The House of Experiment in Seventeenth-Century England," *Isis* 79, no. 298 (Sep 1988): 373–404; Pamela H. Smith, "Laboratories," in *The Cambridge History of Science: Early Modern Science*, ed. Katherine Park and Lorraine Daston (Cambridge: Cambridge University Press, 2006), 290–305.

79. On the regulation of access to the laboratory, see Steven Shapin and Simon Schaffer, *Leviathan and the Air-Pump: Hobbes, Boyle, and the Experimental Life* (Princeton: Princeton University Press, 1985), and on the link between credibility and social status, Steven Shapin, *A Social History of Truth : Civility and Science in Seventeenth-Century England* (Chicago: University of Chicago Press, 1995).

80. On experimentation as production of phenomena, see Ian Hacking, *Representing and Intervening* (Cambridge: Cambridge University Press, 1983), and on the genealogy of this notion from Gaston Bachelard's "phenomenotechnique," Teresa Castelao-Lawless, "Phenomenotechnique in Historical Perspective: Its Origins and Implications for Philosophy of Science," *Philosophy of Science* 62, no. 1 (Mar 1995): 44–59.

81. Robert E. Kohler, "Place and Practice in Field Biology," *History of Science* 40, no. 128, pt. 2 (Jun 2002): 189–210, and Thomas F. Gieryn, "Three Truth-Spots," *Journal of the History of the Behavioral Sciences* 38, no. 2 (Spr 2002): 113–32.

82. Shapin, "House of Experiment."

83. Bruno Latour, "Give Me a Laboratory and I Will Raise the World," in *Science Observed* (New York: Routledge, 1983), 258–75.

84. Jacques Monod and François Jacob, "General Conclusions: Teleonomic Mechanisms in Cellular Metabolism, Growth, and Differentiation," *Cold Spring Harbor Symposia on Quantitative Biology* 21 (1961): 389–401. On the origins of this saying, see Herbert C. Friedmann, "From 'Butyribacterium' to 'E. Coli': An Essay on Unity in Biochemistry," *Perspectives in Biology and Medicine* 47, no. 1 (Win 2004): 47–66.

85. For the distinction between research organisms and model organisms, see Rachel A. Ankeny and Sabina Leonelli, "What's So Special about Model Organisms?," *Studies in History and Philosophy of Science Part C* 42, no. 2 (Jun 2011): 313–23, and for a history of biology seen from the perspective of its model organisms, see Jim Endersby, *A Guinea Pig's History of Biology* (Cambridge: Harvard University Press, 2007).

86. On objectivity, see Lorraine Daston and Peter Galison, *Objectivity* (New York: Zone Books, 2007); on precision, M. Norton Wise, *The Values of Precision* (Princeton: Princeton University Press, 1995); on quantification, Theodore M. Porter, *Trust: The Pursuit of Objectivity in Science and Public Life* (Princeton: Princeton University Press, 1995); on observation, Lorraine Daston and Elizabeth Lunbeck, eds., *Histories of Scientific Observation* (Chicago: University of Chicago Press, 2011).

87. David L. Hull, "Openness and Secrecy in Science: Their Origins and Limitations," *Science, Technology, and Human Values* 10, no. 2 (1985): 4–13; on scientific authorship more generally, Mario Biagioli and Peter Galison, eds., *Scientific Authorship: Credit and Intellectual Property in Science* (New York: Routledge, 2003).

88. Robert E. Kohler, "Lab History Reflections," *Isis* 99, no. 4 (Dec 2008): 761–68; Jackson, "The Laboratory"; Gooday, "Placing or Replacing the Laboratory"; Ursula Klein, "The Laboratory Challenge: Some Revisions of the Standard View of Early Modern Experimentation," *Isis* 99, no. 4 (Dec 2008): 769–82; Tomas F. Gieryn, "Laboratory Design for Post-Fordist Science," ibid., 796–802.

89. A few resources on collecting practices include David E. Allen, *The Naturalist in Britain: A Social History* (London: A. Lane, 1976); Douglas Cole, *Captured Heritage: The Scramble for Northwest Coast Artifacts* (Norman: University of Oklahoma Press, 1995); Simon J. Knell, *The Culture of English Geology, 1815–1851: A Science Revealed through Its Collecting* (Aldershot, UK: Ashgate, 2000); Anderson, *Collectors of Lost Souls*; Harold J. Cook, *Matters of Exchange: Commerce, Medicine, and Science in the Dutch Golden Age* (New Haven: Yale University Press, 2007); Londa L Schiebinger, *Plants and Empire: Colonial Bioprospecting in the Atlantic World* (Cambridge: Harvard University Press, 2004); Londa L. Schiebinger and Claudia Swan, eds., *Colonial Botany: Science, Commerce, and Politics in the Early Modern World* (Philadelphia: University of Pennsylvania Press, 2005); Lucile Brockway, *Science and Colonial Expansion: The Role of the British Royal Botanic Gardens* (New Haven: Yale University Press, 2002); Sheets-Pyenson, *Cathedrals of Science*.

90. Charles C. Gillispie. "Scientific Aspects of the French Egyptian Expedition, 1798–1801," *Proceedings of the American Philosophical Society* 133, no. 4 (1989): 447–74.

91. Jim Endersby, *Imperial Nature: Joseph Hooker and the Practices of Victorian Science* (Chicago: University of Chicago Press, 2008).

92. Fa-ti Fan, *British Naturalists in Qing China: Science, Empire, and Cultural Encounter* (Cambridge: Harvard University Press, 2003); Mark V. Barrow, "The Specimen Dealer: Entrepreneurial Natural History in America's Gilded Age," *Journal of the History of Biology* 33 (2000): 493–534; Elizabeth Hanson, *Animal Attractions: Nature on Display in American Zoos* (Princeton: Princeton University Press, 2002), chap. 3; Cook, *Matters of Exchange*. And on the regulation of standard animal production in the UK, see R. G. W. Kirk, "'Wanted—Standard Guinea Pigs': Standardisation and the Experimental Animal Market in Britain ca. 1919–1947," *Studies in the History and Philosophy of Biological and Biomedical Sciences* 39 (Mar 2008): 280–91, and Kirk, "A Brave New Animal for a Brave New World: The British Laboratory Animals Bureau and the Constitution of International Standards of Laboratory Animal Production and Use, circa 1947–1968," *Isis* 101, no. 1 (Mar 2010): 62–94.

93. Lyle Rexer and Rachel Klein, *American Museum of Natural History: 125 Years of Expedition and Discovery* (New York: H. N. Abrams in association with the American Museum of Natural History, 1995).

94. Paula Findlen, "The Economy of Scientific Exchange in Early Modern Italy," in *Patronage and Institutions: Science, Technology, and Medicine at the European Court, 1500–1750*, ed. Bruce T. Moran (Rochester, NY: Boydell Press, 1991), 5–24; Findlen, *Possessing Nature: Museums, Collecting, and Scientific Culture in Early Modern Italy* (Berkeley: University of California Press, 1994); Philipp Blom, *To Have and to Hold: An Intimate History of Collectors and Collecting* (Woodstock, NY: Overlook Press, 2003); Endersby, *Imperial Nature*; Allen, *Naturalist in Britain*; Mark V. Barrow, *A Passion for Birds: American Ornithology after Audubon* (Princeton: Princeton University Press, 1998).

95. Spary, *Utopia's Garden*, 77.

96. On dioramas, see Donna Haraway, "Teddy Bear Patriarchy: Taxidermy in the Garden of Eden, New York City, 1908–1936," *Social Text* 11 (1984–85): 20–64; Stephen C. Quinn, *Windows on Nature: The Great Habitat Dioramas of the American Museum of Natural History* (New York: Harry N. Abrams, 2006); Karen Wonders, *Habitat Dioramas: Illusions of Wilderness in Museums of Natural History* (Uppsala: Acta Universitatis Upsaliensis, 1993); Rader and Cain, *Life on Display*. For an essay on this theme, see Strasser, "Collecting Nature." On the use of collections for the production of knowledge, see Endersby, *Imperial Nature*, and Johnson, *Ordering Life*.

97. Lorraine Daston and Peter Galison, "The Image of Objectivity," *Representations* 40 (1992): 81–128.

98. Hopwood, *Haeckel's Embryos*, 12; see also Hopwood, "Producing Development: The Anatomy of Human Embryos and the Norms of Wilhelm His," *Bulletin of the History of Medicine* 74, no. 1 (Spr 2000): 29–79.

99. For a more detailed discussion, see Strasser, "Collecting Nature"; on epistemic things, Hans-Jörg Rheinberger, *Toward a History of Epistemic Things: Synthesizing Proteins in the Test Tube* (Stanford: Stanford University Press, 1997); on the experimentalists studying their own creations, Ian Hacking, "The Self-Vindication of the Laboratory Sciences," in *Science as Practice and Culture*, ed. Andrew Pickering (Chicago: University of Chicago Press, 1992), 29–64; on taxidermy, Paul L. Farber, "Development of Taxidermy and History of Ornithology," *Isis* 68, no. 244 (1977): 550–66; and on making kuru brain collections, Anderson, *Collectors of Lost Souls*.

100. Georges Louis Leclerc Buffon, *Histoire naturelle, générale et particulière, avec la description du Cabinet du Roy* (Paris, 1749), 4.

101. On second nature, see Strasser, "Collecting Nature"; a similar view in Kitchin, *Data Revolution*, and Lorraine Daston, *Science in the Archives: Pasts, Presents, Futures* (Chicago: University of Chicago Press, 2017), 1.

102. The notion of moral economy has been popularized by social historian E. P. Thompson as an alternative to economic and mob psychology explanations of peasant food riots in eighteenth-century England. See Thompson, *The Making of the English Working Class* (New York: Penguin Books, 1979 [1963]). He argued that the riots were driven, not just by unfocused anger, but by a sentiment of injustice and betrayal of a system of moral norms defining "just price" and exchange and the distribution of resources. The notion of moral economy has been imported into science studies and used in a variety of ways, most fruitfully (pun intended) in Robert E. Kohler, *Lords of the Fly: Drosophila Genetics and the Experimental Life* (Chicago: University of Chicago Press, 1994). Here, close to Thompson's and Robert E. Kohler's usage, I will define it as the system of values that underlies the exchange of scientific knowledge, with particular regard to how knowledge is tied to issues of property, privacy, and priority. It is essential to remember that moral economies, unlike Mertonian norms, are locally and historically situated and can thus differ between scientific communities, in our case, between experimentalists and collectors of experimental data.

103. On "Ethics in Taxonomy," see Ernst Mayr, E. Gorton Linsley, and Robert Leslie Usinger, *Methods and Principles of Systematic Zoology* (New York,: McGraw-Hill, 1953), 71.

These "collaborations" between naturalists and field collectors were not, however, exempt from tensions; see Endersby, *Imperial Nature.*

104. On curiosity, see Krzysztof Pomian, *Collectors and Curiosities: Paris and Venice, 1500–1800* (Cambridge, UK: Polity Press, 1990); Palmira Fontes da Costa, "The Culture of Curiosity at the Royal Society in the First Half of the Eighteenth Century," *Notes and Records of the Royal Society* 56, no. 2 (May 2002): 147–66; on wonders, Lorraine Daston and Katharine Park, eds., *Wonders and the Order of Nature, 1150–1750* (Cambridge, MA: Zone Books, 1998), and on cabinets of curiosity, Findlen, *Possessing Nature.*

105. Kohler, "Lab History Reflections," and Graeme Gooday, "Placing or Replacing the Laboratory in the History of Science?," *Isis* 99, no. 4 (Dec 2008): 783–95.

Chapter One

1. On the history and epistemology of model organisms, see Richard M. Burian, "How the Choice of Experimental Organism Matters: Epistemological Reflections on an Aspect of Biological Practice," *Journal of the History of Biology* 26, no. 2 (1993): 351–67; Gerald L. Geison and Manfred D Laubichler, "The Varied Lives of Organisms: Variation in the Historiography of the Biological Sciences," *Studies in History and Philosophy of Biological and Biomedical Sciences* 32, no. 1 (2001): 1–29; Gabriel Gachelin, *Les organismes modèles dans la recherche médicale* (Paris: Presses universitaires de France, 2006); Angela N. H. Creager, Elizabeth Lunbeck, and M. Norton Wise, *Science without Laws: Model Systems, Cases, Exemplary Narratives* (Durham: Duke University Press, 2007); Endersby, *Guinea Pig's History*; and especially the extensive work by Sabina Leonelli and Rachel A. Ankeny: Ankeny and Leonelli, "What's So Special about Model Organisms?"; Leonelli, "When Humans Are the Exception: Cross-Species Databases at the Interface of Biological and Clinical Research," *Social Studies of Science* 42, no. 2 (Apr 2012): 214–36; Leonelli and Ankeny, "Re-thinking Organisms: The Impact of Databases on Model Organism Biology," *Studies in History and Philosophy of Science Part C* 43, no. 1 (Mar 2012): 29–36; Leonelli and Ankeny, "What Makes a Model Organism?," *Endeavour* 37, no. 4 (Dec 2013): 209–12; Ankeny et al., "Making Organisms Model Human Behavior: Situated Models in North American Alcohol Research, since 1950," *Science in Context* 27, no. 3 (Sep 2014): 485–509. On the related issue of how biologists use model organisms, see J. A. Bolker, "Exemplary and Surrogate Models: Two Modes of Representation in Biology," *Perspectives in Biology and Medicine* 52, no. 4 (Autumn 2009): 485–99. For model organisms not discussed in this chapter, the following studies are particularly illuminating: Angela N. H. Creager, *The Life of a Virus: Tobacco Mosaic Virus as an Experimental Model, 1930–1965* (Chicago: University of Chicago Press, 2002); Gregg Mitman and Anne Fausto-Sterling, "Whatever Happened to Planaria? C. M. Child and the Physiology of Inheritance," in *The Right Tool for the Job*, ed. A. E. Clarke and J. H. Fujimura (Princeton: Princeton University Press, 1992), 172–97; on Chlamydomonas, Doris T. Zallen, "The 'Light' Organism for the Job," *Journal of the History of Biology* 26, no. 3 (1993): 269–79.

2. Endersby, *Guinea Pig's History.*

3. Long after writing this chapter, I came across Christopher M. Kelty's excellent article which makes a similar point about how newsletters "closed" communities and required "openness." See Christopher M. Kelty's "This Is Not an Article: Model Organism Newsletters and the Question of 'Open Science,'" *Biosocieties* 7, no. 2 (Jun 2012): 140–68.

4. Yates, *Control through Communication*, chap. 2; Gardey, *Ecrire, calculer, classer*, 134–38.

5. On the supply of organisms for experimental research from the Stazione Zoologica in Naples, see Christiane Groeben, "The Stazione Zoologica: A Clearing House for Marine Organisms," in *Oceanographic History: The Pacific and Beyond*, ed. Keith Rodney Benson and Philip F. Rehbock (Seattle: University of Washington Press, 2002), 537–47.

6. Staffan Müller-Wille and Hans-Jörg Rheinberger, *A Cultural History of Heredity* (Chicago: University of Chicago Press, 2012), 148ff.

7. Kohler, *Lords of the Fly*.

8. Garland E. Allen, *Thomas Hunt Morgan: The Man and His Science* (Princeton: Princeton University Press, 1978).

9. On genetic mapping, see Jean-Paul Gaudillière and Hans-Jörg Rheinberger, *From Molecular Genetics to Genomics: The Mapping Cultures of Twentieth-Century Genetics* (London: Routledge, 2004).

10. The career of Carl Jordan (and his collection) provides a good example. Johnson, *Ordering Life*.

11. Clarke and Fujimura, *Right Tool*.

12. Kohler, *Lords of the Fly*.

13. Lily E. Kay, *The Molecular Vision of Life: Caltech, the Rockefeller Foundation and the Rise of the New Biology* (New York: Oxford University Press, 1993), conclusion.

14. Surprisingly, the ATCC has not yet received any detailed attention from historians. For insider histories, see Freeman A. Weiss, "The American Type Culture Collection," *AIBS Bulletin* 4, no. 1 (1954): 15–16; Robert E. Buchanan, "History and Development of the Amercian Type Culture Collection," *Quarterly Review of Biology* 41, no. 2 (1966): 101–4; William A. Clark and Dorothy H. Geary, "The Story of the American Type Culture Collection: Its History and Development (1899–1973)," *Advances in Applied Microbiology* 17 (1974): 295–309; and various unpublished historical outlines held in the American Society of Microbiology Archives, Section 12—Other Orgs, IB folders 1–10 ATCC 1935–1973, University of Maryland Baltimore County (ASM Archives hereafter). For an account of the late history of the ATCC, see Hannah Landecker, "Sending Cells Around: How to Exchange Biological Matter," in *The Moment of Conversion: Exchange Networks in Modern Biomedical Science*, Department of Anthropology, History, and Social Medicine, UC San Francisco, Apr 4–5 2003.

15. *Annual Report* (New York: American Museum of Natural History, 1911).

16. "Charles-Edward Amory Winslow, February 4, 1887–January 8, 1957," *American Journal of Public Health and the Nation's Health* 47, no. 2 (Feb 1957): 153–67; John F. Fulton. "C.-E.A. Winslow, Leader in Public Health," *Science* 125, no. 3260 (Jun 21 1957): 1236; Ira V. Hiscock, "Charles-Edward Amory Winslow; February 4, 1877–January 8, 1957," *Journal of Bacteriology* 73, no. 3 (Mar 1957): 295–96.

17. William T. Sedgwick and Charles-Edward Winslow, "On the Relative Importance of Public Water Supplies and Other Factors in the Causation of Typhoid Fever," *Public Health Papers and Reports* 28 (1902): 288–95.

18. Charles-Edward A. Winslow and Anne F. Rogers, "A Revision of the Coccaceae," *Science* 21, no. 539 (Apr 28 1905): 669–72.

19. Charles-Edward A. Winslow, *Bacterial Collection and Bureau for the Distribution of Bacterial Cultures* (New York: American Museum of Natural History, 1913). An abridged

version of this text was published in *Science*, "Bacterial Collection and Bureau for the Distribution of Bacterial Cultures at the American Museum of Natural History, New York," *Science* 38, no. 976 (1913): 374–75. The *Annual Report* of the AMNH (1911) mentions that the department began its work in September 1910, *Annual Report* (1911), 52.

20. Winslow, "Bacterial Collection," 375. *Annual Report* (New York: American Museum of Natural History, 1912).

21. Winslow, "Bacterial Collection," 375. A list of every single donation of bacterial cultures can be found in the *Annual Reports* of the AMNH between 1912 and 1923.

22. Charles-Edward A. Winslow and Anne F. Rogers, "A Revision of the Coccaceae," *Science* 21, no. 539 (Apr 28 1905): 669–72

23. Winslow, *Bacterial Collection*, 1.

24. Ibid.

25. Nancy Tomes, *The Gospel of Germs: Men, Women, and the Microbe in American Life* (Cambridge: Harvard University Press, 1998).

26. Michael Kennedy, *Philanthropy and Science in New York City: The American Museum of Natural History, 1868–1968* (1968), microform.

27. Rader and Cain, *Life on Display*, 27.

28. *Annual Report* (1911), 63–65.

29. *Annual Report* (1912), 63–64.

30. Kennedy, *Philanthropy and Science*, 173.

31. *Annual Report* (1912), 64.

32. Miloslav Kocur, "History of the Král Collection," in *100 Years of Culture Collections*, ed. Lindsay I. Sly, Teiji Iijima, and Barbara E. Kirsop (Osaka: Institute for Fermentation, 1990), 4–12.

33. "Sectional Meetings," *Lancet* 138, no. 3546 (1891): 371–401. For another report of Král's displays, see "The International Exhibition at Rome," *Lancet* 143, no. 3686 (1894): 1030–31.

34. Kocur, "History of the Král Collection," 6.

35. The introduction of the 1900 edition of his catalogue is reproduced in Lindsay I. Sly, Teiji Iijima, and B. Kirsop, eds., *100 Years of Culture Collections* (Osaka: Institute for Fermentation, 1990).

36. Ernst Pribram, *Der gegenwärtige Bestand der vorm. Králschen Sammlung von Mikroorganismen* (Vienna, 1919). The 1919 edition of Král's collection contains strains given by Winslow, most likely in exchange for other strains.

37. Winslow, *Bacterial Collection.*

38. Ibid., 3.

39. "The First Forty Years of the Society of American Bacteriologists," *Science* 91, no. 2354 (Feb 9 1940): 125–29.

40. See the *Annual Report* of the AMNH between 1912 and 1923.

41. Winslow, "Bacterial Collection," 375.

42. See for example, Charles-Edward A. Winslow, "The Importance of Preserving the Original Types of Newly Described Species of Bacteria," *Journal of Bacteriology* 6, no. 1 (Jan 1921): 133–34.

43. Daston, "Type Specimens."

44. Winslow, "Importance of Preserving the Original Types."

45. *Annual Report* (New York: American Museum of Natural History, 1923).

46. Clark and Geary, "Story of the American Type Culture Collection," 297.

47. Lore A. Rogers. "The American Type-Culture Collection," *Science* 62, no. 1603 (1925): 267.

48. Clark and Geary, "Story of the American Type Culture Collection," 304.

49. Summary notes of meeting, April 30, May 1, 1965, R. W. Barratt, May 6 1965, 6, Genetics Society of America, box 25, APS Archives.

50. Diana Morgan, "American Type Culture Collection Seeks to Expand Research Effort," *Scientist* 4, no. 16 (Aug 20 1990): 1–7. On the importance of freezing techniques in biomedicine more generally, see Joanna M. Radin, "Life on Ice: Frozen Blood and Biological Variation in a Genomic Age, 1950–2010" (PhD diss., University of Pennsylvania, 2012).

51. Especially Felix M. Warburg and Walter B. James. See *Annual Report* (New York: American Museum of Natural History, 1915).

52. Clark and Geary, "Story of the American Type Culture Collection."

53. "Summary of financial history of ATCC," [n.d., probably 1953], ASM Archives.

54. Lore A. Rogers, "Type Cultures," *Science* 66, no. 1710 (1927): 329. The same year, the ATCC began running ads in *Science* to promote its services. See *Science* 66 no. 1707 (1927): xiii; 76 (1932): 8.

55. "The American Type Culture Collection," *Science* 79, no. 2050 (1934): 336.

56. See Weiss, "American Type Culture Collection"; Clark and Geary, "Story of the American Type Culture Collection"; Lore A. Rogers, "The Chemical Foundation, Inc." [n.d., 1928], ASM Archives.

57. George H. Weaver to H. G. Dunham, Apr 11 1935, ASM Archives.

58. Joseph E. Smadel to W. McD. Hammon, Nov 29 1949, ASM Archives

59. Kenneth B. Raper et al., "Culture Collections of Microorganisms," *Science* 116, no. 3007 (1952): 179.

60. Ibid., 180.

61. Clark and Geary, "Story of the American Type Culture Collection," 304.

62. Karen A. Rader, *Making Mice: Standardizing Animals for American Biomedical Research, 1900–1955* (Princeton: Princeton University Press, 2004), chap. 1.

63. Leslie C. Dunn, "William Ernest Castle," *Biographical Memoirs of the National Academy of Science of the United States of America* 38 (1965): 33–80.

64. Ibid., 52–53.

65. Karen A. Rader. "'The Mouse People': Murine Genetics Work at the Bussey Institution, 1909–1936," *Journal of the History of Biology* 31, no. 3 (Fal 1998): 327–54; Rader, *Making Mice*, 29.

66. Rader, *Making Mice*, 54–55.

67. Ibid.; "'Mouse People,'" 350–51.

68. Rader, *Making Mice*, 111.

69. Ibid.

70. Ibid., 123.

71. Ibid., 257.

72. Bonnie T. Clause, "The Wistar Rat as a Right Choice: Establishing Mammalian Standards and the Ideal of a Standardized Mammal," *Journal of the History of Biology* 26, no. 2 (Sum 1993): 329–49.

73. On rat sales at the Wistar Institute, see J. Russell Lindsey, "Historical Foundations," in *The Laboratory Rat*, ed. Henry J. Baker, J. Russell Lindsey, and Steven H. Weisbroth (San Diego: Academic Press, 1979), 1–36.

74. On the hybrid corn revolution, see Jack Ralph Kloppenburg, *First the Seed: The Political Economy of Plant Biotechnology, 1492–2000* (Cambridge: Cambridge University Press, 1988); Deborah Kay Fitzgerald, *The Business of Breeding: Hybrid Corn in Illinois, 1890–1940* (Ithaca: Cornell University Press, 1990), and on its relation to the rise of Mendelism in America, see Diane B. Paul, "Mendel in America: Theory and Practice, 1900–1919," in Rainger, Benson, and Maienschein, *American Development of Biology*, 281–310. On East, see Donald F. Jones, "Edward Murray East, 1879–1938," *Biographical Memoirs of the National Academy of Science of the United States of America* 23 (1944): 215–42.

75. Marcus M. Rhoades, "Rollins Adams Emerson, 1873–1947," *Biographical Memoirs of the National Academy of Science of the United States of America* 25 (1949): 311–23.

76. For a comparison of Emerson's and Morgan's programs, see Barbara A. Kimmelman, "Organisms and Interests: R. A. Emerson's Claims for the Unique Contributions of Agricultural Genetics," in Clarke and Fujimura, *Right Tool*, 198–232, and Lee B. Kass and Christophe Bonneuil, "Mapping and Seeing: Barbara McClintock and the Linking of Genetics and Cytology in Maize Genetics, 1928–35," in *Classical Genetic Research and Its Legacy: The Mapping Cultures of Twentieth-Century Genetics*, ed. Hans-Jörg Rheinberger and Jean-Paul Gaudillière (London: Routledge, 2004), 91–118.

77. On Emerson's "democratic" cooperative style and his commitment to agricultural improvement, see Kimmelman, "Organisms and Interests," 220–22; and for a more detailed description of his cooperative style, Lee B. Kass, Christophe Bonneuil, and Edward H. Coe Jr., "Cornfests, Cornfabs and Cooperation: The Origins and Beginnings of the Maize Genetics Cooperation News Letter," *Genetics* 169, no. 4 (Apr 2005): 1787–97.

78. Kass, Bonneuil, and Coe, "Cornfests, Cornfabs and Cooperation"; Emerson to Jones, Nov 8 1918, cited in ibid., 1788.

79. Emerson, "Materials supporting a request for funds to further the cooperation in maize genetics studies, Exhibit A," Rockefeller Archives Center (RAC hereafter), series 200.D, box 136, folder 1684.

80. Royal Alexander Brink, according to Warren Weaver, Diary, Aug 8 1933, RAC, series 200.D, box 136, folder 1679.

81. Kass, Bonneuil, and Coe, "Cornfests, Cornfabs and Cooperation."

82. Barbara McClintock. "Chromosome Morphology in Zea Mays," *Science* 69, no. 1798 (Jun 14 1929): 629.

83. Kass and Bonneuil, "Mapping and Seeing," 108.

84. Marcus M. Rhoades, "The Early Years of Maize Genetics," *Annual Review of Genetics* 18 (1984): 1–29; Kass, Bonneuil, and Coe, "Cornfests, Cornfabs and Cooperation"; *Maize Genetics Coöperation Newsletter* (MNL hereafter) 7, Sep 13 1934, p. 1, available at https://www.maizegdb.org/mnl.

85. Kass, Bonneuil, and Coe, "Cornfests, Cornfabs and Cooperation," 1793.

86. Rollins A. Emerson, "The Present Status of Maize Genetics," *Proceedings of the Sixth International Congress of Genetics* 1 (1932): 141–52.

87. Warren Weaver, "Cornell University Maize Stock," Mar 16 1934, RAC, series 200.D, box 136, folder 1679.

88. Marcus M. Rhoades, "To corn geneticists," Oct5 1932, available at www.agron.missouri.edu (accessed Sep 11 2009).

89. MNL 7, Sep 13 1934, p. 1.

90. Royal Alexander Brink, according to Warren Weaver, Diary, Aug 8 1933, RAC, series 200.D, box 136, folder 1679.

91. MNL 5, Jan 25 1934.

92. MNL 7, Sep 13 1934, p. 1.

93. George W. Beadle to Milislav Demerec, Apr 13 1929, box 2, folder George W. Beadle, Milislav Demerec Papers, American Philosophical Society (APS Archives hereafter).

94. MNL 14, Mar 5 1940.

95. Kay, *Molecular Vision of Life.*

96. "Grant in AID–New York, RA-NS–4107," Feb 10 1941, RAC, series 200.D, box 136, folder 1679.

97. MNL 1, Oct 5 1932.

98. MNL 5, Jan 25 1934.

99. MNL 4, Dec 18 1933, p. 1.

100. Kelty, "This Is Not an Article," 142.

101. Kass, Bonneuil, and Coe, "Cornfests, Cornfabs and Cooperation," 1793; Rollins A. Emerson, George W. Beadle, and Allan C. Fraser. "A Summary of Linkage Studies in Maize," *Cornell University Agricultural Experiment Station Memoirs* 180 (1935).

102. Edward H. Coe Jr., "East, Emerson, and the Birth of Maize Genetics," in *The Maize Handbook, vol. II: Genetics and Genomics*, ed. J. L. Bennetzen and S. Hake (New York: Springer-Verlag, 2009), 3–15 ; "The Origins of Maize Genetics," *Nature Reviews Genetics* 2, no. 11 (Nov 2001): 898–905; Kass, Bonneuil, and Coe, "Cornfests, Cornfabs and Cooperation."

103. MNL 8, Nov 25 1934, p. 1.

104. MNL 18, Jan 31 1944.

105. MNL 18, Jan 31 1942.

106. Emerson to Weaver, Jan 3 1941, RAC, series 200.D, box 136, folder 1681.

107. Kohler, *Lords of the Fly*.

108. On the importance of the newsletter, especially the *Drosophila Information Service*, for building scientific collectives, see Kelty, "This Is Not an Article."

109. *Drosophila Information Service* (DIS hereafter) 1, Mar 1934, p. 2.

110. DIS 1, Mar 1934, p. 2; Kohler, *Lords of the Fly*, 167.

111. DIS 1, Mar 1934, p. 2.

112. DIS 1, Mar 1934.

113. DIS 1, Mar 1934, cover. For a detailed discussion of what a "nonpublication" might mean, see Kelty, "This Is Not an Article."

114. DIS 2, Aug 1934, pp. 66–67.

115. DIS 2, Aug 1934, pp. 3–4.

116. DIS 2, Aug 1934, p. 4; Emerson, "Present Status," 141–42.

117. DIS 1, Mar 1934, p. 3.

118. DIS 4, Sep 1935, p. 3.

119. DIS 11, Jan 1939, p. 6.

120. DIS 7, Mar 1937, p. 3.

121. DIS Jan 1939, p. 5.

122. Demerec to I. J. Wachtel, Sep 6 1935, box 8, APS Archives.

123. L. C. Dunn to Demerec, Dec 26 1933, box 8, APS Archives.

124. Demerec to L. J. Stadler, Mar 19 1934, box 7, APS Archives.

125. A modest grant of $9,000 for 5 years. W. Weaver to E. W. Lindstrom, Mar 28 1934, box 8, APS Archives.

126. Demerec to L. C. Dunn, J. W. Gowen, M. H. Harnley, R. Pearl, H. H. Plough, K. Stern, Apr 2 1934, box 8, APS Archives.

127. Demerec to I. J. Wachtel, Sep 11 1935, box 8, APS Archives.

128. Demerec to V. Bush, Apr 8 1939, box 8, APS Archives.

129. Demerec to V. Bush, Apr 8 1939, box 8, APS Archives.

130. Milislav Demerec, "Department of Genetics," *Carnegie Institution of Washington Year Book* 47 (1948): 139–43.

131. Confidential monthly report to trustees, May 1937, p. 13, RAC, series 200.D, box 137.

132. Vannevar Bush to Warren Weaver, Mar 25 1939, RAC, box 137, folder 137, Drosophila stock center, 1933–37.

133. Frank B. Hanson, Diary, Aug 21–29 1939, RAC 200.D, box 137; Morgan also expressed his feeling directly to Demerec; see Kohler, *Lords of the Fly*, 160, and, more generally, on the rivalries between Morgan and Demerec, 157–62.

134. Frank B. Hanson, Diary, Jul–Aug 1939, RAC 200.D, box 137.

135. Ibid.

136. Ibid.

137. William C Summers, *Félix d'Hérelle and the Origins of Molecular Biology* (New Haven: Yale University Press, 1999).

138. Max Delbrück and Emory L. Ellis, "How Bacteriophage Came to Be Used by the Phage Group," *Journal of the History of Biology* 26, no. 2 (1993): 255–67.

139. Creager, *Life of a Virus*, chap. 6.

140. Ernst P. Fischer and Carol Lipson, *Thinking about Science: Max Delbrück and the Origins of Molecular Biology* (New York: W. W. Norton, 1988).

141. On the rise of phage genetics, see Michel Morange, *A History of Molecular Biology* (Cambridge: Harvard University Press, 2000); Frederic L. Holmes, *Reconceiving the Gene: Seymour Benzer's Adventures in Phage Genetics* (New Haven: Yale University Press, 2006); Fischer and Lipson, *Thinking about Science*; Kay, *Molecular Vision of Life*.

142. Fischer and Lipson, *Thinking about Science*, 154; Kay, *Molecular Vision of Life*, chap. 8.

143. Kay, *Molecular Vision of Life*, 245.

144. Evelyn M. Witkin, "Chances and Choices: Cold Spring Harbor, 1944–1955," *Annual Review of Microbiology* 56 (2002): 1–15.

145. John Cairns, Gunther S. Stent, and James D. Watson, eds., *Phage and the Origins of Molecular Biology* (New York: Cold Spring Harbor Press, 1966).

146. Jean-Paul Gaudillière, "Paris-New York Roundtrip: Transatlantic Crossings and the Reconstruction of the Biological Sciences in Post-War France," *Studies in the History and Philosophy of Biological and Biomedical Sciences* 33C (2002): 389–417.

147. William A. Clark to J. Lederberg, Nov 16 1955, Profiles in Science, profiles.nlm.nih.gov. There is no reply, and a year later a researchers asked Lederberg for the strain because it was not available at the ATCC.

148. Bruno J. Strasser, *La fabrique d'une nouvelle science: La biologie moléculaire à l'âge atomique (1945–1964)* (Florence: Olschki, 2006), chap. 2.

149. B. Bachman to J. Lederberg, Jul 25 1988, Profiles in Science, profiles.nlm.nih.gov.

150. Barbara J Bachmann, "Pedigrees of Some Mutant Strains of Escherichia Coli K-12," *Bacteriological Reviews* 36, no. 4 (Dec 1972): 525–57.

151. B. Bachman to J. Lederberg, Sep 15 1971, Profiles in Science, profiles.nlm.nih.gov.

152. Bachmann, "Pedigrees," 526.

153. Ibid.

154. Barbara J. Bachmann and K. Brooks Low, "Linkage Map of Escherichia Coli K-12, Edition 6," *Microbiological Reviews* 44, no. 1 (Mar 1980): 1–56.

155. Barbara J. Bachmann, "This Weeks's Citation Classic," *Current Contents, Life Sciences* 8 (Feb 22 1982): 22.

156. J. Lederberg to L. E. Rosenberg, Dec 13 1974, Profiles in Science, profiles.nlm.nih.gov.

157. Soraya de Chadarevian, "Of Worms and Programmes: *Caenorhabditis Elegans* and the Study of Development," *Studies in History and Philosophy of Biological and Biomedical Sciences* 29, no. 1 (1998): 81–105.

158. Donald L. Riddle and P. Swanson, "Caenorhabditis Genetics Center," *Worm Breeder's Gazette* 5, no. 1 (1980): 4.

159. On *Arabidopsis*, see Sabina Leonelli, "*Arabidopsis*, the Botanical *Drosophila*: From Mouse Cress to Model Organism," *Endeavour* 31, no. 1 (Mar 2007): 34–38; Chris Somerville and Maarten Koornneef, "A Fortunate Choice: The History of *Arabidopsis* as a Model Plant," *Nature Reviews Genetics* 3, no. 11 (Nov 2002): 883–89; Endersby, *Guinea Pig's History*, chap. 10; on zebrafish, ibid., chap. 11.

160. Karen A. Rader, "The Origins of Mouse Genetics: Beyond the Bussey Institution, II. Defining the Problem of Mouse Supply: The 1928 National Research Council Committee on Experimental Plants and Animals," *Mammalian Genome* 13, no. 1 (Jan 2002): 2–4; Melinda Gormley, "Geneticist L. C. Dunn: Politics, Activism, and Community" (PhD diss., Oregon State University, 2006), 107–10.

161. Vannevar Bush to Warren Weaver, Mar 25 1939, RAC, box 137, folder 137, Drosophila stock center, 1933–1937.

162. Weaver to Bush, Mar 27 1939, RAC, box 137, folder 137, Drosophila stock center, 1933–1937.

163. Walter Landauer, "Shall We Lose or Keep Our Plant and Animal Stocks," *Science* 101, no. 2629 (May 18 1945): 497–99.

164. Report of the AIBS Committee on Maintenance of Genetic Stocks, 1958, Genetics Society of America, box 17, APS Archives.

165. C. P. Oliver (GSA president) and R. P. Wagner (GSA chairman), "Proposal to the NSF from the GSA," Dec 22 1958, APS Genetics Society of America, box 18, 1959 Committee on Maintenance of Genetic Stocks.

166. Newsletter, Aug 1 1965, attached to R. W. Barratt to R. P. Wagner, Aug 7 1965, Genetics Society of America, box 25, APS Archives.

167. Thomas Roderick. "Listing of Genetic Stock Centers and Newsletters," *Journal of Heredity* 66 (1975): 104–12.

168. Toby A. Appel, *Shaping Biology: The National Science Foundation and American Biological Research, 1945–1975* (Baltimore: Johns Hopkins University Press, 2000).

169. National Science Foundation, *Twentieth Annual Report of the National Science Foundation* (Washington, DC: National Science Foundation, 1970).

170. Weiss, "American Type Culture Collection," 16.

171. Ibid.

172. Carl Lamanna, "Microbiology, Museums, and the American Type Culture Collection," *Quarterly Review of Biology* 41, no. 2 (1966): 95–97.

173. On exhibitions of *Volkskrankheiten* in Imperial Germany, see Christine Brecht and Sybilla Nikolow, "Displaying the Invisible: *Volkskrankheiten* on Exhibition in Imperial Germany," *Studies in History and Philosophy of Biological and Biomedical Sciences* 31, no. 4 (2000): 511–30.

174. Lamanna, "Microbiology, Museums, and the American Type Culture Collection," 95.

175. Alan T. Waterman (NSF) to Karl Sax (GSA), Mar 6 1959, Genetics Society of America, box 18, APS Archives.

176. Colin M. MacLeod, "Some Thoughts on Microbial Taxonomy," *Quarterly Review of Biology* 41, no. 2 (1966): 98–100.

177. AMNH, *Report for Fiscal Years 2001 through 2003* (New York: AMNH, 2005), 8.

178. A similar rhetoric was at work for seed banks; see Sara Peres, "Saving the Gene Pool for the Future: Seed Banks as Archives," *Studies in History and Philosophy of Biological and Biomedical Sciences* 55 (2016): 96–104, and for the broader history of the concerns with the loss of genetic diversity, Helen A. Curry, "From Working Collections to the World Germplasm Project: Agricultural Modernization and Genetic Conservation at the Rockefeller Foundation," *History and Philosophy of the Life Science* 39, no. 2 (Jun 2017): 5. AMNH, *Report for Fiscal Years 2001 through 2003* (New York: AMNH, 2005), 9.

179. Except for the Wistar rats distributed since 1906; see Clause, "Wistar Rat."

180. Microbial culture collections are listed by the World Federation for Culture Collections, http://www.wfcc.info/about/ (accessed Aug 4 2017). On the more recent status of stock collections, see Nadia Rosenthal and Michael Ashburner, "Taking Stock of Our Models: The Function and Future of Stock Centres," *Nature Reviews Genetics* 3, no. 9 (Sep 2002): 711–17; Yufeng F. Wang and Timothy G. Lilburn, "Biological Resource Centers and Systems Biology," *Bioscience* 59, no. 2 (Feb 2009): 113–25; Michael A. Mares, "Natural Science Collections: America's Irreplaceable Resource," ibid., no. 7 (Jul–Aug 2009): 544–45.

181. Frederic L Holmes, *Claude Bernard and Animal Chemistry: The Emergence of a Scientist* (Cambridge: Harvard University Press, 1974).

182. Christoph Gradmann and Elborg Forster, *Laboratory Disease: Robert Koch's Medical Bacteriology* (Baltimore: Johns Hopkins University Press, 2009).

183. Hopwood, *Haeckel's Embryos*; Muka, "Working at Water's Edge."

184. The case of the Naples Stazione Zoologica does not fit this pattern, since it was, as Christianne Groeben put it, a "clearing house for model organisms." Groeben, "Stazione Zoologica."

185. Michael R. Dietrich and Brandi H. Tambasco, "Beyond the Boss and the Boys: Women and the Division of Labor in Drosophila Genetics in the United States, 1934–1970," *Journal of the History of Biology* 40, no. 3 (2007): 509–28.

186. Again, this transition toward centralized and public collection station was similar to what took place at the Naples Stazione Zoologica, leading to the creation of the Central Embryological Collection in Utrecht. Hopwood, *Haeckel's Embryos*, 185.

187. Allen, *Thomas Hunt Morgan.*

188. Demerec to L. J. Cole, Mar 22 1934, box 7, APS Archives.

189. For a list of newsletters, see Kelty, "This Is Not an Article," table 1.

190. Philip J. Pauly, "Summer Resort and Scientific Discipline: Woods Hole and the Structure of American Biology," in Rainger, Benson, and Maienschein, *American Development of Biology*, 121–50.

191. Kohler, *Lords of the Fly*, 162.

Chapter Two

1. On objectivity, see Daston and Galison, *Objectivity*; on precision, Wise, *Values of Precision*; on quantification, Porter, *Trust.*

2. Joel B. Hagen, "Naturalists, Molecular Biology, and the Challenge of Molecular Evolution," *Journal of the History of Biology* 32 (1999): 321–41; Hagen, "Experimentalists and Naturalists."

3. George S. Graham-Smith, "George Henry Falkiner Nuttall (5 July 1862–16 December 1937)," *Journal of Hygiene* 38, no. 2 (1938): 129–40.

4. George S. Graham-Smith and David Keilin, "George Henry Falkiner Nuttall, 1862–1937," *Obituary Notice of Fellows of the Royal Society* 2, no. 7 (1939): 493–99. For his later work in parasitology, see Francis E. G. Cox, "George Henry Falkiner Nuttall and the Origins of Parasitology and *Parasitology*," *Parasitology* 136, no. 12 (Mar 30 2009): 1389–94.

5. On the history of serum therapy, see Derek S. Linton, *Emil von Behring: Infectious Disease, Immunology, Serum Therapy* (Philadelphia: American Philosophical Society, 2005); Arthur M. Silverstein, *A History of Immunology* (London: Academic Press, 2009); Gradmann and Forster, *Laboratory Disease.*

6. Karl Landsteiner had just published his paper on the inheritance of human blood groups in 1901; Louis K. Diamond, "The Story of Our Blood Groups," in *Blood, Pure and Eloquent*, ed. Maxwell M. Wintrobe (New York: McGraw-Hill, 1980), 691–717.

7. George H. F. Nuttall and Edgar M. Dinkelspiel, "Experiments upon the New Specific Test for Blood (Preliminary Note)," *British Medical Journal* 1 (Jan–Jun 1901): 1141; George H. F. Nuttall, "On the Formation of Specific Anti-bodies in the Blood, Following upon

Treatment with the Sera of Different Animals," *American Naturalist* 35, no. 419 (1901): 927–32; George H. F. Nuttall and Edgar M. Dinkelspiel, "On the Formation of Specific Antibodies in the Blood Following upon Treatment with the Sera of Different Animals, Together with Their Use in Legal Medicine," *Journal of Hygiene* 1, no. 3 (1901): 367–87.

8. George H. F. Nuttall, "A Further Note on the Biological Test for Blood and Its Importance in Zoological Classification," *British Medical Journal* 2 (1901): 669; Nuttall, "The New Biological Test for Blood in Relation to Zoological Classification.," *Proceedings of the Royal Society of London B—Biological Sciences* 69, no. 453 (Dec 1901): 150–53; Nuttall, "Progress Report upon the Biological Test for Blood as Applied to over 500 Bloods from Various Sources, Together with a Preliminary Note upon a Method for Measuring the Degree of Reaction," *British Medical Journal* 1 (1902): 825–27.

9. Nuttall, "Progress Report," 827; Nuttall, *Blood Immunity and Blood Relationship: A Demonstration of Certain Blood-Relationships amongst Animals by Means of the Precipitin Test for Blood* (Cambridge: Cambridge University Press, 1904).

10. Nuttall, *Blood Immunity*, 411–13.

11. Ibid., 213.

12. Ibid., 2.

13. Ibid., 362.

14. For an overview of these critiques, see Alan A. Boyden, "Systematic Serology: A Critical Appreciation," *Physiological Zoölogy* 15, no. 2 (1942): 109–45. The lack of standardization of the antisera, and the difficulty of visually determining if a precipitation had actually occurred, were some of the reasons for the inconsistent results.

15. See for example Tod F. Stuessy, *Plant Taxonomy: The Systematic Evaluation of Comparative Data* (New York: Columbia University Press, 2009); Walter S. Judd, *Plant Systematics: A Phylogenetic Approach* (Sunderland, MA: Sinauer Associates, 2008); Roderic D. M. Page and Edward C. Holmes, *Molecular Evolution: A Phylogenetic Approach* (Oxford: Blackwell Science, 1998).

16. Silverstein, *History of Immunology*, chap. 12.

17. "Reichert, Edward Tyson," in *Marquis Who Was Who in America 1607–1984*, www.credoreference.com/ (last accessed Mar 1 2009).

18. Edward Tyson Reichert and Amos P. Brown, *The Differentiation and Specificity of Corresponding Proteins and Other Vital Substances in Relation to Biological Classification and Organic Evolution: The Crystallography of Hemoglobins* (Washington, DC: Carnegie Institution, 1909).

19. Ibid., xvi–xvii.

20. Ibid., xvii.

21. Jacques Loeb, "Is Species Specificity a Mendelian Character?," *Science* 45 (Jan–Jun 1917): 191–93; Leo Loeb, "Scientific Books," ibid., 33 (1911): 147–50.

22. Michael Heidelberger and Karl Landsteiner, "On the Antigenic Properties of Hemoglobin," *Journal of Experimental Medicine* 38, no. 5 (Oct 31 1923): 561–71.

23. There is no mention of Reichert's work in George G. Simpson, "The Principles of Classification and a Classification of Mammals," *Bulletin of the American Museum of Natural History* 85 (1945): 1–307, or Mayr, Linsley, and Usinger, *Methods and Principles*, 106–7, for example.

24. "Blood Will Really Tell," *Washington Post*, Sep 28 1913, MS3.

25. Anna H. Ewing, "Blood Will Tell," *Popular Science* 93, no. 6 (1918): 72–73; "Blood Crystals Aid Detectives," *Washington Post*, Sep 1 1912, M1.

26. "Blood Will Really Tell."

27. However, the chemist David Keilin used Nuttall's blood samples forty years later to test the stability of hemoglobin. See David Keilin and Yin-Lai Wang, "Stability of Haemoglobin and of Certain Endoerythrocytic Enzymes in Vitro," *Biochemical Journal* 41, no. 4 (1947): 491–500.

28. William H. Schneider, "The History of Research on Blood Group Genetics: Initial Discovery and Diffusion," *History and Philosophy of the Life Sciences* 18 (1996): 277–303; Jonathan Marks, "The Legacy of Serological Studies in American Physical Anthropology," ibid., 345–62. For an overview, see William C. Boyd, "Systematics, Evolution, and Anthropology in the Light of Immunology," *Quarterly Review of Biology* 24, no. 2 (1949): 102–8, and for a critical analysis of postwar blood group research, Marianne Sommer, *History Within: The Science, Culture, and Politics of Bones, Organisms, and Molecules* (Chicago: University of Chicago Press, 2016), chap. 11.

29. Joel B. Hagen, "Experimental Taxonomy, 1920–1950: The Impact of Cytology, Ecology, and Genetics on the Ideas of Biological Classification" (PhD diss., Oregon State University, 1984).

30. Ibid., 260.

31. In the first half of the twentieth century, the biological sciences were still strictly divided between botany, zoology, microbiology, and anthropology, with often little communication between these areas. I thank Joel B. Hagen for pointing out this distinction to me, and on this point see "The Statistical Frame of Mind in Systematic Biology from *Quantitative Zoology* to *Biometry*," *Journal of the History of Biology* 36, no. 2 (Sum 2003): 353–84.

32. For Boyden's biographical data, see James G. Crowther, *Famous American Men of Science* (Harmondsworth, Middlesex, UK, 1944), 189, and for his personal file, Alan Arthur Boyden, "Faculty Data," Jan 1 1942, "Faculty Biographical Information," Sep 1 1946, and "Faculty Biographical Information," Aug 20 1957, Rutgers Archives, RG 04/A14, Special Collections and University Archives, Rutgers University Libraries.

33. Alan A. Boyden. "The Precipitin Reaction in the Study of Animal Relationships," *Biological Bulletin* 50, no. 2 (Feb 1926): 73–107.

34. Michael F. Guyer, "Blood Reactions of Man and Animals," *Scientific Monthly* 21, no. 10 (1925): 145–46. On Guyer's later work, see Patrick Bungener and Marino Buscaglia, "Early Connection between Cytology and Mendelism: Michael F. Guyer's Contribution," *History and Philosophy of the Life Sciences* 25, no. 1 (2003): 27–50.

35. Michael F. Guyer, "Immune Sera and Certain Biological Problems," *American Naturalist* 55, no. 637 (1921): 97–115.

36. Karl Landsteiner, *The Specificity of Serological Reactions* (Springfield, IL: C. C. Thomas, 1936).

37. Karl Landsteiner and C. Philip Miller Jr., "Serological Observations on the Relationship of the Bloods of Man and the Anthropoid Apes," *Science* 61, no. 1584 (May 8 1925): 492–93.

38. Karl Landsteiner, "Cell Antigens and Individual Specificity," *Journal of Immunology* 15, no. 6 (Nov 1928): 589–600.

39. For a review, see for example Landsteiner and Miller, "Serological Observations," chap. 2.

40. Boyden, "Precipitin Reaction"; Boyden and Joseph G. Baier Jr., "A Rapid Quantitative Precipitin Technic," *Journal of Immunology* 17, no. 1 (Jul 1929): 29–37; Boyden, "Precipitin Tests as a Basis for a Quantitative Phylogeny," *Proceeding of the Society for Experimental Biology* 29 (1932): 955–57; Boyden, "Precipitins and Phylogeny in Animals," *American Naturalist* 68 (1934): 516–36.

41. For a description of the technique, see Alan A. Boyden and Ralph J. DeFalco, "Report on the Use of the Photoreflectometer in Serological Comparisons," *Physiological Zoölogy* 16, no. 3 (1943): 229–41.

42. Basically, Boyden took the integral of the titration curve; see Alan A. Boyden, "Systematic Serology: A Critical Appreciation," *Physiological Zoölogy* 15, no. 2 (1942): 109–45; and Boyden, "Serology and Animal Systematics," *American Naturalist* 77 (1943): 234–55.

43. See for example the summary in Boyden, "Systematic Serology."

44. Boyden, "Precipitin Reaction," 103.

45. Boyden, "Precipitins and Phylogeny," 518.

46. Boyden, "Systematic Serology," 118.

47. Boyden, "Precipitin Reaction," 102.

48. Boyden, "Precipitin Tests," 957.

49. Boyden, "Serology and Animal Systematics," 253.

50. Boyden, "Systematic Serology," 115.

51. Hagen, "Statistical Frame of Mind."

52. George G. Simpson and Anne Roe, *Quantitative Zoology; Numerical Concepts and Methods in the Study of Recent and Fossil Animals* (New York: McGraw-Hill, 1939); on quantification in paleontology, see Léo F. Laporte, *George Gaylord Simpson: Paleontologist and Evolutionist* (New York: Columbia University Press, 2000), chap. 3.

53. Hagen, "Statistical Frame of Mind," 358.

54. George G. Simpson, *Principles of Animal Taxonomy* (New York: Columbia University Press, 1961).

55. George Gaylord Simpson to Alan A. Boyden, Jan 23 1947, George Gaylord Simpson Papers, American Philosophical Society, Philadelphia, MS Coll. 31, Series I (APS Archives hereafter). In addition, of course, Simpson, having the fossil record at his disposal, was on safer ground to make claims about homologies than Boyden and any other systematists working only on present forms. Ernst Mayr, the leading American systematist, also squarely rejected Boyden's interpretation of homology as mere similarity, reaffirming that "the term homology has no meaning whatsoever if not a phylogenetic one." Ernst Mayr to Alan A. Boyden, Feb 15 1944, Harvard University Archives, Boston, MA, HUG (HUG hereafter) (FP) 14.7, box 2, folder 92.

56. George Gaylord Simpson to Alan A. Boyden, 30 Oct 1944, APS Archives.

57. Mayr, Linsley, and Usinger, *Methods and Principles*, 106–7.

58. N. Whitfield and T. Schlich, "Skills through History," *Medical History* 59, no. 3 (Jul 2015): 349–60. The relationship between scientists and clinicians was far more complex,

as many combined ways of knowing from both medical science and clinical practice, and the argument of a "hybrid culture" developed here could apply to medicine as well. On this point, see Steve Sturdy, "Looking for Trouble: Medical Science and Clinical Practice in the Historiography of Modern Medicine," *Social History of Medicine* 24, no. 3 (2011): 739–57. I thank Nick Hopwood for pointing this out.

59. William Bertram Turrill, "The Subjective Element in Plant Taxonomy," *Bulletin du Jardin botanique de l'État à Bruxelles* 27, no. 1 (1957): 8.

60. Boyden, "Systematic Serology," 116.

61. Alan A. Boyden to Robert C. Clothier, Jun 2 1944, Rutgers Archives, RG 04/A14, box 85.

62. Boyden, "Systematic Serology," 120.

63. Keith Vernon, "Desperately Seeking Status: Evolutionary Systematics and the Taxonomists' Search for Respectability, 1940–60," *British Journal for the History of Science* 26, no. 89 (Jun 1993): 207–27.

64. Even though they shared a common goal, there were sharp intellectual divisions among systematists, especially between traditional taxonomists and "new systematists"; see David L. Hull, *Science as a Process* (Chicago: University of Chicago Press, 1988), and Joseph Allen Cain, "Launching the Society of Systematic Zoology in 1947," in *Milestones in Systematics*, ed. David M. Williams and Peter L. Forey (Boca Raton: CRC Press, 2004), 19–48. The later development of numerical taxonomy is not considered here but can be seen to be in continuity with this movement. See Keith Vernon, "The Founding of Numerical Taxonomy," *British Journal for the History of Science* 21, no. 69 (Jun 1988): 143–59; Joel B. Hagen. "The Introduction of Computers into Systematic Research in the United States during the 1960s," *Studies in the History and Philosophy of Biological and Biomedical Sciences* 32, no. 2 (2001): 291–314.

65. Boyden, "Serology and Animal Systematics," 241.

66. In the United States, until 1945, most of the systematic serologists working on animal taxonomy were colleagues of Boyden at Rutgers, most importantly Charles A. Leone, but also Ralph J. DeFalco and Douglas G. Gemeroy. For some exceptions, see the work of Raymond W. Wilhelmi at Woods Hole, who insisted like Boyden on the objectivity of the serological data. Wilhelmi, "Serological Relationships between the Mollusca and Other Invertebrates," *Biological Bulletin* 87, no. 1 (1944): 96–105. See also Russell W. Cumley, at the University of Texas: Cumley. "Comparison of Serologic and Taxonomic Relationships of Drosophila Species," *Journal of the New York Entomological Society* 48, no. 3 (1940): 265–74. In the 1930s, serology seems to have been more popular, especially in Germany. On the German context, in the field of plant taxonomy, see for example Carl Mez, "Die Bedeutung der Sero-Diagnostik für die stammesgeschichtliche Forschung," *Botanisches Archiv* 16 (1926): 1–23, and for a general review of serological taxonomy and phylogeny, see Albert Erhardt, "Die Verwandtschaftsbestimmungen mittels der Immunitätsreaktionen in der Zoologie und ihr Wert für phylogenetische Untersuchungen," *Ergebnisse und Fortschritte der Zoologie* 7 (1931): 279–377. In the United States, see K. Starr Chester, "A Critique of Plant Serology, Part II. Application of Serology to the Classification of Plants and the Identification of Plant Products," *Quarterly Review of Biology* 12, no. 2 (1937): 165–90; "A Critique of Plant Serology, Part I. The Nature and Utilization of Phytoserological Procedures," *Quarterly Review of Biology* 12, no. 1 (1937): 19–46; "A Critique of Plant Serology, Part III. Phytoserology

in Medicine and General Biology. Bibliography," *Quarterly Review of Biology* 12, no. 3 (1937): 294–321.

67. Boyden's work is not mentioned in Julian Huxley, ed., *The New Systematics* (Oxford: Clarendon Press, 1940), or in Ernst Mayr, *Systematics and the Origin of Species from the Viewpoint of a Zoologist* (New York: Columbia University Press, 1942), or Simpson, "Principles of Classification." On the other hand, it is mentioned in passing in Mayr, Linsley, and Usinger, *Methods and Principles,* and in Simpson, *Principles of Animal Taxonomy,* and, on the other side of the systematists' community, in Richard E. Blackwelder, *Taxonomy: A Text and Reference Book* (New York: Wiley, 1967). On the debates between Blackwelder and Simpson, see Cain, "Launching the Society of Systematic Zoology."

68. E. Mayr to W. F. Loomis, Mar 19 1951, HUG(FP) 14.7, box 8, folder 388.

69. "Serological Museum of Rutgers University," *Nature* 161, no. 4090 (Mar 20 1948): 428; "New and Unique Kind of Museum," *Science* 107, no. 2774 (Feb 27 1948): 217. The connection with the growing popularity of blood banks, if any, is unclear. On blood banks, see Susan E. Lederer, *Flesh and Blood: Organ Transplantation and Blood Transfusion in Twentieth-Century America* (Oxford: Oxford University Press, 2008), chap. 3 and Kara W. Swanson, *Banking on the Body: The Market in Blood, Milk, and Sperm in Modern America* (Cambridge: Harvard University Press, 2014).

70. For late nineteenth-century examples of continuities between laboratories and museums in Britain, see Alison Kraft and Samuel J. M. M. Alberti, "'Equal Though Different': Laboratories, Museums and the Institutional Development of Biology in Late-Victorian Northern England," *Studies in History and Philosophy of Biological and Biomedical Sciences* 34, no. 2 (2003): 203–36; Forgan, "Architecture of Display."

71. On this episode, see Gregg Mitman and Richard W. Burkhardt Jr., "Struggling for Identity: The Study of Animal Behavior in America, 1930–1950," in *The American Expansion of Biology,* ed. Keith R. Benson, Ronald Rainger, and Jane Maienschein (New Brunswick: Rutgers University Press, 1987), 164–94 ; Joseph Allen Cain, "Common Problems and Cooperative Solutions: Organizational Activity in Evolutionary Studies, 1936–1947," *Isis* 84, no. 1 (Mar 1993): 1–25.

72. *Annual Report* (New York: American Museum of Natural History, 1929). Noble and Boyden later collaborated on a serological study of amphibians, Noble's specialty; Alan A. Boyden and G. Kingsley Noble, "The Relationships of Some Common Amphibia as Determined by Serological Study," *American Museum Novitates* 606 (1933): 1–24.

73. "New and Unique Kind of Museum."

74. Ibid.

75. "Blood Proteins to Be Kept in New, Special Museum," *Science News Letter,* Jun 12 1948, 377.

76. Alan A. Boyden, "Introductory Remarks," in *Serological and Biochemical Comparisons of Proteins,* ed. William H. Cole (New Brunswick: Rutgers University Press, 1958), 1–2.

77. George H. Holsten (Rutgers News Service), "Press Release," 6–7 Apr 1951, RG 04/A14, box 101, Rutgers Archives; and "Rutgers Unit Gets $5000," *New York Times,* Apr 7 1951, 6.

78. *Serological Museum Bulletin* 1 (1948).

79. *Serological Museum Bulletin* 5 (1950): 1.

80. "Triple Play," undated newspaper clipping, Folder Faculty Bio, Rutgers Archives.

81. Ibid.

82. Boyden, "Introductory Remarks," 1; Findlen, *Possessing Nature*, 48.

83. See, for example, Endersby, *Imperial Nature*; and Richard W. Burkhardt Jr., "Naturalists' Practices and Nature's Empire: Paris and the Platypus, 1815–1833," *Pacific Science* 55 (2001): 327–41.

84. "New and Unique Kind of Museum."

85. Alan A. Boyden, "Fifty Years of Systematic Serology," *Systematic Zoology* 2, no. 1 (1953): 19–30.

86. Alan A. Boyden to George Gaylord Simpson, 13 Nov 1947, APS Archives.

87. Charles A. Leone, "Comparative Serology of Some Branchyuran Crustacea and Studies in Hemocyanin Correspondence," *Biological Bulletin* 97, no. 3 (Dec 1949): 273–86.

88. "New and Unique Kind of Museum."

89. Alan A. Boyden, "Zoological Collecting Expeditions and the Salvage of Animal Bloods for Comparative Serology," *Science* 118, no. 3054 (Jul 10 1953): 57–58.

90. On the disciplining of individual collectors, see Endersby, *Imperial Nature*, chap. 8.

91. Bernard Weitz, "The Collection of Serum from Game in East Africa," *Serological Museum Bulletin* 4 (1950): 1–3.

92. "Recent Advances in the Collection of Serum from Marine Animals," *Serological Museum Bulletin* 5 (1950): 3. See also Alan A. Boyden, "Collecting Serological Samples," *Atoll Research Bulletin* 17 (1953): 96–99.

93. Bernard Weitz, "A Mobile Laboratory for the Collection of Sterile Serum from Game in East Africa," *Serological Museum Bulletin* 10 (1953): 4–5.

94. "Woman Scientist Brings Rare Blood," *New York Times*, May 6 1950, 31.

95. See for example Benjamin Adelman, "Blood Museum Traces Animal Cousins," *Science Digest* 30 (1951): 6–8.

96. Rosenberg, *No Other Gods*, chap. 7.

97. Boyden, "Zoological Collecting Expeditions," 58.

98. Boyden, "Serology and Animal Systematics," 242.

99. Boyden, "Zoological Collecting Expeditions," 58.

100. Alan A. Boyden, "The 'Flying Lemur' Comes to the Museum," *Serological Museum Bulletin* 4 (1950): 4–5.

101. Alan A. Boyden, "Major Objectives," *Serological Museum Bulletin* 4 (1950): 1.

102. Elias Cohen and Gunnar B. Stickler, "Absence of Albuminlike Serum Proteins in Turtles," *Science* 127 (1958): 1392.

103. Alan A. Boyden, "Our Contacts Grow," *Serological Museum Bulletin* 3, (1949): 1.

104. William Kingston, "Streptomycin, Schatz V. Waksman, and the Balance of Credit for Discovery," *Journal of the History of Medicine and Allied Sciences* 59, no. 3 (2004): 441–62.

105. On the fly group, see Kohler, *Lords of the Fly*; on the phage group and more generally on the cooperative individualism fostered by the Rockefeller Foundation at Caltech, see Kay, *Molecular Vision of Life*.

106. Alan A. Boyden to Robert C. Clothier, 21 Jul 1943, and Alan A. Boyden, "A Plan for the Training of American Youth in Responsible Citizenship," attached to Alan A. Boyden to Robert C. Clothier, 2 Oct 1943, Rutgers Archives, RG 04/A14, box 85.

107. Alan A. Boyden to Robert C. Clothier, 23 Oct 1950, Rutgers Archives, RG 04/A14, box 84.

108. "This Museum Banks on Blood," *Home News (New Jersey)*, Mar 2 1971.

109. Joel B. Hagen, "Waiting for Sequences: Morris Goodman, Immunodiffusion Experiments, and the Origins of Molecular Anthropology," *Journal of the History of Biology* 43, no. 4 (Dec 2010): 697–725.

110. Morris Goodman and G. William Moore, "Immunodiffusion Systematics of the Primates, I. The Catarrhini," *Systematic Biology* 20, no. 1 (1971): 19–62.

111. See for example, Harold R. Wolfe, "Standardization of the Precipitin Technique and Its Application to Studies of Relationships in Mammals, Birds and Reptiles," *Biological Bulletin* 76, no. 1 (1939): 108–20.

112. Hagen, "Waiting for Sequences."

113. Ibid.

114. Ibid.

115. For a good overview of systematic serology in the 1960s, see Charles A. Leone, *Taxonomic Biochemistry and Serology* (New York: Ronald Press, 1964), and, with a focus on botany, Ralph E. Alston and Billie Lee Turner, *Biochemical Systematics* (Englewood Cliffs, NJ: Prentice-Hall, 1963), chap. 5. The work of Herbert C. Dessauer continued this tradition most directly. On its progressive marginalization among biochemical systematics in the 1970s and 1980s, see Christopher A. Wright, *Biochemical and Immunological Taxonomy of Animals* (London: Academic Press, 1974); Andrew Ferguson, *Biochemical Systematics and Evolution* (New York: Wiley, 1980).

116. On the place of serology in physical anthropology, see Marks, "Legacy of Serological Studies."

117. For biographical information about Sibley, see Alan H. Brush, "Charles Gald Sibley," *Biographical Memoirs of the National Academy of Sciences* 83 (2003): 216–39; Kendall W. Corbin and Alan H. Brush, "In Memoriam: Charles Gald Sibley, 1917–1998," *Auk* 116, no. 3 (1999): 806–14; Jon E. Ahlquist, "Charles G. Sibley: A Commentary on 30 Years of Collaboration," *Auk* 116, no. 3 (Jul 1999): 856–60; Robert B. Payne, "Charles Gald Sibley (1917–1998): Obituary," *Ibis* 140, no. 4 (Oct 1998): 697–99. On Sibley's career at the interface between molecular biology and natural history, see Hagen, "Naturalists, Molecular Biology, and the Challenge of Molecular Evolution," 335–39.

118. Payne, "Charles Gald Sibley."

119. Corbin and Brush, "In Memoriam," 809.

120. Karl Landsteiner, Lewis G. Longsworth, and James van der Scheer, "Electrophoresis Experiments with Egg Albumins and Hemoglobins," *Science* 88, no. 2273 (Jul 22 1938): 83–85.

121. Harold F. Deutsch and Martha B. Goodloe, "An Electrophoretic Survey of Various Animal Plasmas," *Journal of Biological Chemistry* 161 (1945): 1–20.

122. Herbert C. Dessauer and Wade Fox, "Characteristic Electrophoretic Patterns of Plasma Proteins of Orders of Amphibia and Reptilia," *Science* 124, no. 3214 (Aug 3 1956): 225–26.

123. Charles G. Sibley and Paul A. Johnsgard, "Variability in the Electrophoretic Patterns of Avian Serum Proteins," *Condor* 61, no. 2 (1959): 85–95.

124. Paul A. Johnsgard, "In Memoriam: Charles G. Sibley," *Nebraska Bird Review* 66, no. 2 (1998): 68–69.

125. Charles G. Sibley and Paul A. Johnsgard, "An Electrophoretic Study of Egg-White Proteins in Twenty-Three Breeds of the Domestic Fowl," *American Naturalist* 93, no. 869 (1959): 107–15.

126. Charles G. Sibley, "Proteins and DNA in Systematic Biology," *Trends in Biochemical Sciences* 22, no. 9 (Sep 1997): 364–67. Sibley mentions 1957, but this is unlikely to be true, as he was then still working with serum proteins, not egg-white proteins. On bird collecting, see Barrow, *Passion for Birds*.

127. Sibley, "Proteins and DNA," 365.

128. Corbin and Brush, "In Memoriam," 813.

129. Charles G. Sibley, "The Electrophoretic Patterns of Avian Egg-White Proteins as Taxonomic Characters," *Ibis* 102 (1960): 215–84.

130. Allen, *Naturalist in Britain*; Brockway, *Science and Colonial Expansion*.

131. Robert E. Kohler, "Labscapes: Naturalizing the Lab," *History of Science* 40, no. 130 (Dec 2002): 473–501.

132. Sibley to Mayr, Apr 4 1963, HUG(FP) 74.7, box 9, folder 826.

133. Sibley to Mayr, Jul 22 1963, HUG(FP) 74.7, box 10, folder 842.

134. Sibley to Mayr, Apr 4 1963, HUG(FP) 74.7, box 9, folder 826.

135. Sibley to Mayr, Dec 17 1963, HUG(FP) 74.7, box 10, folder 842.

136. Sibley to Mayr, Apr 4 1963, HUG(FP) 74.7, box 9, folder 826.

137. Charles G. Sibley and Herbert T. Hendrickson, "A Comparative Electrophoretic Study of Avian Plasma Proteins," *Condor* 72, no. 1 (1970): 43–49.

138. Charles G. Sibley, "The 'Alpha Helix' Expedition to New Guinea," *Discovery* 4, no. 1 (1968): 45. For a report of the expedition, see "A Report on Program B of the Alpha Helix Expedition to New Guinea," *Discovery* 5, no. 1 (1969): 39–46. On the *Alpha Helix*, see Deborah Day, "Alpha Helix Program Administrative History" (2009), UC San Diego: Scripps Institution of Oceanography Archives, retrieved from http://escholarship.org/uc/item/7nd5x74v.

139. Sibley, "'Alpha Helix' Expedition."

140. Joanna M. Radin, "Life on Ice: Frozen Blood and Biological Variation in a Genomic Age, 1950–2010" (PhD diss., University of Pennsylvania, 2012), 172.

141. Corbin and Brush, "In Memoriam," 811.

142. Sibley to Mayr, Nov 11 1969, HUG(FP) 74.7, box 18, folder 1063.

143. Sibley to Mayr, Feb 14 1979, HUG(FP) 74.7, box 28, folder 1314.

144. Radin, "Life on Ice." Protein sequencing, and even more so DNA sequencing, required better preservation of biological tissues.

145. Sibley, "Electrophoretic Patterns," 252.

146. Sibley, "The Comparative Morphology of Protein Molecules as Data for Classification," *Systematic Zoology* 11, no. 3 (1962): 108–18.

147. Sibley to Mayr, Jul 6 1978, HUG(FP) 74.7, box 26, folder 1299.

148. Sibley to Mayr, Jul 22 1963, HUG(FP) 74.7, box 10, folder 842.

149. Sibley to Mayr, Jul 22 1963, HUG(FP) 74.7, box 10, folder 842.

150. Ahlquist, "Charles G. Sibley," 857.

151. "For the Birds," *New York Times*, Jul 17 1974, 36.

152. Fred Ferretti, "Fining of Bird Scholar Stirs Colleagues," *New York Times*, Jul 13 1973, 55.

153. On this episode, see Ahlquist, "Charles G. Sibley," 857.

154. C. G. Sibley, untitled and undated memo in response to the press coverage about the Lacey Act violation. Peabody Archives, Yale University.

155. Sibley to Robert C Murphy, Jul 27 1955, Peabody Archives, Yale University.

156. Ahlquist, "Charles G. Sibley." The most detailed account of this controversy can be found in Clive Gammon, "The Case of the Absent Eggs," *Sports illustrated* 40, no. 25 (Jun 24 1974): 26–33.

157. Sibley, "Comparative Morphology," 115.

158. Alan Feduccia, "Reviewed Work: A Comparative Study of the Egg-White Proteins of Passerine Birds by Sibley, Charles G.; A Comparative Study of the Egg-White Proteins of Non-passerine Birds by Sibley, Charles G.; Jon E. Ahlquist," *Auk* 90, no. 4 (1973): 919–21.

159. Sibley to Mayr, Sep 19 1960, HUG(FP) 74.7, box 7, folder 752.

160. Lynn H. Throckmorton, "Reviewed Work: A Comparative Study of the Egg-White Proteins of Passerine Birds by Sibley, Charles G.," *Quarterly Review of Biology* 47 (1973): 93–94.

161. Sibley, "Comparative Morphology," 116.

162. "Egg White Clue to Evolution," *Hartford Courant*, Jan 11 1968, 43C.

163. Sibley and Johnsgard, "Variability," 86.

164. Sibley to Mayr, Jul 23 1981, HUG(FP) 74.7, box 29, folder 1342.

165. Sibley to Mayr, Sept 10 1981, HUG(FP) 74.7, box 29, folder 1342.

166. Sibley to Mayr, Jul 23 1981, HUG(FP) 74.7, box 29, folder 1342.

167. Sibley to Mayr, Sep 10 1981, HUG(FP) 74.7, box 29, folder 1342.

168. Sibley to Mayr, Sep 10 1981, HUG(FP) 74.7, box 29, folder 1342.

169. Sibley, "Comparative Morphology," 116.

170. Sibley, "Proteins and DNA," 365; Brian J. McCarthy and Ellis T. Bolton, "An Approach to the Measurement of Genetic Relatedness among Organisms," *Proceedings of the National Academy of Sciences* 50 (Jul 1963): 156–64; Bill H. Hoyer, Brian J. McCarthy, and Ellis T. Bolton, "A Molecular Approach in the Systematics of Higher Organisms: DNA Interactions Provide a Basis for Detecting Common Polynucleotide Sequences among Diverse Organisms," *Science* 144 (May 22 1964): 959–67.

171. Ahlquist, "Charles G. Sibley," 857.

172. Sibley to Mayr, Jul 6, 1978, HUG(FP) 74.7, box 26, folder 1299.

173. Sibley to Mayr, Nov 28 1978, HUG(FP) 74.7, box 26, folder 1299.

174. Sibley to Mayr, May 23, 1978, HUG(FP) 74.7, box 26, folder 1299.

175. Sibley to Mayr, Apr 23 1979, HUG(FP) 74.7, box 28, folder 1314.

176. Ibid.

177. Charles G. Sibley and Jon E. Ahlquist, *Phylogeny and Classification of Birds: A Study in Molecular Evolution* (New Haven: Yale University Press, 1990).

178. Ibid., 502.

179. Ibid., xx.

180. Ibid., xvii.

181. Ibid., xviii.

182. David P. Mindell, "Review: DNA-DNA Hybridization and Avian Phylogeny," *Systematic Biology* 41, no. 1 (1992): 126–34.

183. Sibley and Ahlquist, *Phylogeny*, 247.

184. Frank B. Gill and Frederick H. Sheldon, "The Birds Reclassified," *Science* 252, no. 5008 (May 17 1991): 1003–5.

185. Roger Lewin, "DNA Clock Conflict Continues," *Science* 241, no. 4874 (Sep 30 1988): 1756–59.

186. Sibley, "Electrophoretic Patterns," 227.

187. Charles G. Sibley and Jon E. Ahlquist, "The Phylogeny of the Hominoid Primates, as Indicated by DNA-DNA Hybridization," *Journal of Molecular Evolution* 20, no. 1 (Jan 1 1984): 2–15.

188. Lewin, "DNA Clock Conflict Continues"; Roger Lewin, "Conflict over DNA Clock Results," *Science* 241, no. 4873 (Sep 23 1988): 1598–1600. Scientific fraud (or scientific misconduct) had become a much-debated issue since the 1980s, and scientific journals played an important role in turning these cases into public controversies about different aspects of scientific methods, integrity, and institutions. See for example Daniel J. Kevles, *The Baltimore Case: A Trial of Politics, Science, and Character* (New York: W. W. Norton, 1998), and, Hopwood, *Haeckel's Embryos*, chap. 17.

189. Sibley, "Electrophoretic Patterns."

190. Peter Houde, "Review: *Phylogeny and Classification of Birds: A Study in Molecular Evolution*, by Charles G. Sibley and Jon E. Ahlquist," *Quarterly Review of Biology* 67, no. 1 (1992): 62–63.

191. Erika L Milam, "The Equally Wonderful Field: Ernst Mayr and Organismic Biology," *Historical Studies in the Natural Sciences* (2010): 279–317.

192. Mayr to Sibley, Apr 22 1963, HUG(FP) 74.7, box 9, folder 826.

193. Mayr to Sibley, Oct 21 1976, HUG(FP) 74.7, box 24, folder 1259.

194. Mayr to Sibley, May 16 1978 and Jun 28 1978, HUG(FP) 74.7, box 26, folder 1299.

195. Mayr to Sibley, Oct 21 1981, HUG(FP) 74.7, box 29, folder 1342.

196. Mayr to Sibley, Apr 17 1979, HUG(FP) 74.7, box 28, folder 1314.

197. Mayr to Sibley, Oct 21 1981, HUG(FP) 74.7, box 29, folder 1342.

198. Ernst Mayr, "A New Classification of the Living Birds of the World," *Auk* 106 (1989): 508–12.

199. On this point, see Hagen, "Naturalists, Molecular Biology, and the Challenge of Molecular Evolution."

200. Lloyd E. Rozeboom to Mayr, Aug 17, 1964, HUG(FP) 74.7, box 11, folder 857.

201. Mayr to Lloyd E. Rozeboom, Sep 8 1964, HUG(FP) 74.7, box 11, folder 857.

202. In reaction to the perceived invasion of their territory by experimentalists, naturalists called for a peace treaty and the recognition of the "unity" of biology around evolutionary theory; see for example Theodosius Dobzhansky, "Taxonomy, Molecular Biology, and the Peck Order," *Evolution* 15, no. 2 (1961): 263–64. Experimentalists, on the other hand, understood "unity" as the ultimate outcome of their reductionist agenda and

insisted that every biological process was ultimately to be explained at the molecular level. On this debate, see Michael R. Dietrich, "Paradox and Persuasion: Negotiating the Place of Molecular Evolution within Evolutionary Biology," *Journal of the History of Biology* 31 (1998): 85–111, and Milam, "Equally Wonderful Field."

203. For a structurally similar argument concerning clinicians and basic scientists in the late nineteenth century, see John Harley Warner, "Ideals of Science and Their Discontents in Late Nineteenth-Century American Medicine," *Isis* 82 (1991): 454–78.

204. The episodes described by Robert E. Kohler and by Joel B. Hagen are thus part of a much broader transformation of natural history in the twentieth century, which remains to be studied; see Kohler, *Landscapes and Labscapes*; Hagen, "Experimental Taxonomy."

205. Even physiologists, the earlier symbol of the experimental life sciences at the time of Claude Bernard, developed a broad comparative approach in the twentieth century, Wolfgang von Buddenbrock, *Grundriss der vergleichenden Physiologie* (Berlin: Borntraeger, 1924–28); C. Ladd Prosser, ed., *Comparative Animal Physiology* (Philadelphia: W. B. Saunders, 1950). A more precise argument would require further research, for a starter, see Glenn M. Sanford, William I. Lutterschmidt, and Victor H. Hutchison, "The Comparative Method Revisited," *BioScience* 52, no. 9 (2002): 830–36.

206. Bruno J. Strasser. "Collecting, Comparing, and Computing Sequences: The Making of Margaret O. Dayhoff's Atlas of Protein Sequences and Structure, 1954–1965," *Journal of the History of Biology* 43, no. 4 (2010): 623–60.

207. National Research Council (US), *Systematic Biology: Proceedings of an International Conference* (Washington, DC: National Academy of Sciences, 1969).

208. Sibley to Mayr, Jan 9 1970, HUG(FP) 74.7, box 18, folder 1063.

209. For the same reason, other experimental taxonomists were turning to the field experiment, for example, Hagen, "Experimentalists and Naturalists," 266.

210. Marianne Sommer, "History in the Gene: Negotiations between Molecular and Organismal Anthropology," *Journal of the History of Biology* 41, no. 3 (2008): 473–528; Edna Suárez-Díaz and Victor H. Anaya-Muñoz, "History, Objectivity, and the Construction of Molecular Phylogenies," *Studies in History and Philosophy of Science Part C* 39, no. 4 (Dec 2008): 451–68.

211. Similarly, the historiography of molecular biology has sometimes obscured the broader trend of molecularization in the twentieth century. For a longer view, see Soraya de Chadarevian and Harmke Kamminga, eds., *Molecularizing Biology and Medicine: New Practices and Alliances, 1910s–1970s* (Amsterdam: Harwood Academic Publishers, 1998).

212. Daston and Galison, *Objectivity*. However, Daston and Galison argue that by the end of the twentieth century, another regime of objectivity, "trained judgment," would become prominent, based on a new idea of the scientific self as an "expert" relying on personal judgment acquired by formal training, which was thus different from earlier forms of subjectivity, where judgment was believed to be universal, reflecting the principles of human reason. Ibid., chap. 6, pp. 370–71.

213. Simpson, *Principles of Animal Taxonomy*, 152.

214. Sommer, "History in the Gene."

215. See, for example, on mice, Rader, *Making Mice*; on TMV, Creager, *Life of a Virus*; on *Drosophila*, Kohler, *Lords of the Fly*; on *C. elegans*, de Chadarevian, "Of Worms and

Programmes"; on *Arabidopsis*, Leonelli, "Arabidopsis"; and more generally Clarke and Fujimura, *Right Tool*; and Creager, Lunbeck, and Wise, *Science without Laws*, part I.

216. On the collection of biological material for laboratory studies, see Herbert C. Dessauer and Mark S. Hafner, *Collections of Frozen Tissues: Value, Management, Field and Laboratory Procedures, and Directory of Existing Collections* (Lawrence, KS: Association of Systematics Collections, 1984).

217. Townes in National Research Council (US), *Systematic Biology*, 393.

218. Rogers in ibid., 392.

219. Beschel, in ibid., 401.

220. Sokal in ibid., 389.

221. Sibley in ibid., 405.

222. On these debates, the best source is Hull, *Science as a Process*, and for a popular summary, Carol Kaesuk Yoon, *Naming Nature: The Clash between Instinct and Science* (New York: W. W. Norton, 2009).

223. Allan C. Wilson, Steven S. Carlson, and Thomas J. White, "Biochemical Evolution," *Annual Review of Biochemistry* 46 (1977): 573–639.

224. David M. Hillis, Craig Moritz, and Barbara K. Mable, *Molecular Systematics* (Sunderland, MA: Sinauer Associates, 1996), chap. 1.

225. The literature of DNA barcoding is abundant; for an typical critique from taxonomists, see Laurence Packer et al., "DNA Barcoding and the Mediocrity of Morphology," *Molecular Ecology Resources* 9 suppl. s1 (May 2009): 42–50; for a historical perspective, Andrew Hamilton and Quentin D. Wheeler, "Taxonomy and Why History of Science Matters for Science: A Case Study," *Isis* 99, no. 2 (Jun 2008): 331–40; and for a science and technology studies (STS) perspective, Claire Waterton, Rebecca Ellis, and Brian Wynne, *Barcoding Nature: Shifting Cultures of Taxonomy in an Age of Biodiversity Loss* (Milton Park, Abingdon, Oxon: Routledge, 2013).

226. On the changing requirement of biological collecting, see Dessauer and Hafner, *Collections of Frozen Tissues*.

227. Richard Schodde, "Obituary: Charles Sibley (1911–1998)," *Emu* 100, no. 1 (2000): 75–76.

Chapter Three

1. Dennis Benson (NCBI) to the author, June 30, 2018.

2. Daston and Galison, *Objectivity*, 22–26.

3. Derek J. de Solla Price, *Little Science, Big Science* (New York: Columbia University Press, 1963). Other databases include the chemist Olga Kennard's Crystallographic Database, a collection of structures of small molecules established in Cambridge, UK (1965), discussed in chapter 4, and the medical geneticist Victor A. McKusick's *Mendelian Inheritance in Man*, a collection of phenotypes of inheritable human diseases (1966). See Victor A. McKusick, *Mendelian Inheritance in Man: Catalogs of Autosomal Dominant, Autosomal Recessive, and X-Linked Phenotypes* (Baltimore: Johns Hopkins University Press, 1966).

4. On identifying the origins of bioinformatics solely in the analysis of DNA sequences, see Stevens, *Life out of Sequence*; on a proteins as "epistemic objects," Rheinberger, *Toward a History of Epistemic Things*; on the role of protein sequences in the disciplinary

formation of molecular biology, Soraya de Chadarevian and Jean-Paul Gaudillière, "The Tools of the Discipline: Biochemists and Molecular Biologists," *Journal of the History of Biology* 29, no. 3 (1996): 327–30; Soraya de Chadarevian, "Protein Sequencing and the Making of Molecular Genetics," *Trends in Biochemical Sciences* 24 (1999): 203–6; and on sequencing as a "form of work," Garcia-Sancho, *Biology, Computing, and the History of Molecular Sequencing.*

5. Hopwood, *Haeckel's Embryos*, 281–82.

6. Edna Suárez-Díaz, "The Long and Winding Road of Molecular Data in Phylogenetic Analysis," *Journal of the History of Biology* 47, no. 3 (Fall 2014): 443–78; Dietrich, "Paradox and Persuasion"; Hagen, "Naturalists, Molecular Biology, and the Challenge of Molecular Evolution."

7. Emile Zuckerkandl and Linus Pauling, "Molecular Disease, Evolution, and Genic Heterogeneity," in *Horizons in Biochemistry*, ed. M. Kasha and B. Pullman (New York: Academic Press, 1962), 189–224.

8. Emile Zuckerkandl and Linus Pauling, "Molecules as Documents of Evolutionary History," *Journal of Theoretical Biology* 8 (1965): 357–66.

9. For different views about this episode, see Dietrich, "Paradox and Persuasion," and Hagen, "Naturalists, Molecular Biology, and the Challenge of Molecular Evolution."

10. Ernest Baldwin, *An Introduction to Comparative Biochemistry* (Cambridge: Cambridge University Press, 1937).

11. Baldwin was a student of Hopkins. On Hopkins, see Robert E. Kohler, *From Medical Chemistry to Biochemistry: The Making of a Biomedical Discipline* (Cambridge: Cambridge University Press, 1982); Harmke Kamminga and Mark W. Weatherall, "The Making of a Biochemist, I: Frederick Gowland Hopkins' Construction of Dynamic Biochemistry," *Medical History* 40, no. 3 (Jul 1996): 269–92; Mark W. Weatherall and Harmke Kamminga, "The Making of a Biochemist, II: The Construction of Frederick Gowland Hopkins' Reputation," ibid., no. 4 (Oct 1996): 415–36.

12. Baldwin, *Introduction to Comparative Biochemistry*, xiv.

13. Marcel Florkin, *L'évolution biochimique* (Paris: Masson, 1944); Florkin, *Biochemical Evolution* (New York: Academic Press, 1949).

14. Florkin, *L'évolution biochimique*, 11; translation is mine.

15. Ibid, 194–96; translation is mine. The biochemist Erwin Chargaff's studies on the regularities of nucleic acid composition were also derived from the examination of material from several species, including man, ox, rat, frog, sea urchins, yeast, and bacteria; D. Elson and E. Chargaff, "Evidence of Common Regularities in the Composition of Pentose Nucleic Acids," *Biochimica et Biophysica Acta* 17, no. 3 (Jul 1955): 367–76. In the United States, the comparative biochemistry tradition was also promoted by microbiologists, such as Cornelius B. van Niel, a student of Albert Jan Kluyver, from Delft, who had coined the expression "comparative biochemistry"; see Susan Barbara Spath, "C. B. Van Niel and the Culture of Microbiology, 1920–1965" (PhD diss., University of California, Berkeley, 1999).

16. Frederick Sanger, "Sequences, Sequences, and Sequences," *Annual Review of Biochemistry* 57 (1988): 1–28; George S. Graham-Smith and Frederick Sanger, "The Biological or Precipitin Test for Blood Considered Mainly from Its Medico-Legal Aspect," *Journal of Hygiene* 3, no. 2 (Apr 1903): 258–91.

17. J. Ieuan Harris, Michael A. Naughton, and Frederick Sanger, "Species Differences in Insulin," *Archives of Biochemistry and Biophysics* 65, no. 1 (Nov 1956): 427–38; H. Brown, Frederick Sanger, and Ruth Kitai, "The Structure of Pig and Sheep Insulins," *Biochemical Journal* 60, nos. 1–4 (1955): 556–65.

18. Harris, Naughton, and Sanger, "Species Differences," 437.

19. Sven Paléus and Hans Tuppy, "A Hemopeptide from a Tryptic Hydrolysate of *Rhodospirillum-rubrum* Cytochrome C," *Acta Chemica Scandinavica* 13, no. 4 (1959): 641–46.

20. Birger Blombäck apparently learned the technique from visiting scientist Ikuo Yamashina, who had just spent some time in Edman's laboratory. In addition, Margareta and Birger Blombäck worked in Edman's laboratory after Edman had moved to Melbourne in 1961. Birger Blombäck, "Travels with Fibrinogen," *Journal of Thrombosis and Haemostasis* 4, no. 8 (Aug 2006): 1653–60.

21. Birger Blombäck, Margareta Blombäck, and Nils Jakob Grondahl, "Studies on Fibrinopeptides from Mammals," *Acta Chemica Scandinavica* 19 (1965): 1789–91.

22. Soraya de Chadarevian, "Following Molecules: Haemoglobin between the Clinic and the Laboratory," in *Molecularizing Biology and Medicine: New Practices and Alliances, 1910s–1970s*, ed. de Chadarevian and Harmke Kamminga (Amsterdam: Harwood Academic Publishers, 1998), 171–201.

23. Christian B. Anfinsen et al., "A Comparative Study of the Structures of Bovine and Ovine Pancreatic Ribonucleases," *Journal of Biological Chemistry* 234, no. 5 (May 1959): 1118–23.

24. Christian B. Anfinsen, *The Molecular Basis of Evolution* (New York: Wiley, 1959).

25. Harris, Naughton, and Sanger, "Species Differences," 137.

26. Hans Tuppy, "Aminosaure-Sequenzen in Proteinen," *Naturwissenschaften* 46, no. 2 (1959): 35–43.

27. Anfinsen, *Molecular Basis*, 143.

28. Ibid., chaps. 7 and 11.

29. Following the work of William H. Stein and Stanford Moore, the instrument maker Beckman brought the automatic amino acid analyzer on the market, a Spinco Model 120; Stanford Moore, Darrel H. Spackman, and William H. Stein, "Automatic Recording Apparatus for Use in the Chromatography of Amino Acids," *Federation Proceedings* 17, no. 4 (Dec 1958): 1107–15; Pehr Edman and Geoffrey Begg, "A Protein Sequenator," *FEBS Journal* 1, no. 1 (Mar 1967): 80–91.

30. Bruno J. Strasser, "A World in One Dimension: Linus Pauling, Francis Crick and the Central Dogma of Molecular Biology," *History and Philosophy of the Life Science* 28 (2006): 491–512.

31. Anfinsen, *Molecular Basis*, 143, emphasis in original.

32. Zuckerkandl and Pauling, "Molecules as Documents."

33. George Gamow, "Possible Relation between Desoxyribonucleic Acid and Protein Structure," *Nature* 173 (1954): 318; George Gamow and Nicholas Metropolis, "Numerology of Polypeptide Chains," *Science* 120, no. 3124 (1954): 779–80.

34. Lily E. Kay, *Who Wrote the Book of Life: A History of the Genetic Code* (Stanford: Sanford University Press, 2000).

35. On the history of protein sequencing, see de Chadarevian and Gaudillière, "Tools of the Discipline"; de Chadarevian, "Protein Sequencing"; Garcia-Sancho, *Biology, Computing, and the History of Molecular Sequencing.*

36. If in the DNA sequence "abcd" includes two successive overlapping codons "abc" and "bcd," coding for amino acids X and Y, then X will frequently be followed by Y in protein sequences (a frequency of 9/20 for overlapping codes and of 1/20 in nonoverlapping codes). Sydney Brenner inferred from all the published sequences that in view of the random distribution of the amino acids, codes overlapping by 2 nucleotides out of 3 were impossible. Brenner, "On the Impossibility of All Overlapping Triplet Codes in Information Transfer from Nucleic Acid to Proteins," *Proceedings of the National Academy of Sciences* 43, no. 8 (Aug 15 1957): 687–94.

37. George Gamow, Alexander Rich, and Martynas Yčas, "The Problem of Information Transfer from the Nucleic Acids to Proteins," *Advances in Biological and Medical Physics* 4 (1956): 23–68.

38. Martynas Yčas, "Replacement of Amino Acids in Proteins," *Journal of Theoretical Biology* 1, no. 2 (1961): 244–57; Yčas, "The Protein Text," in *Symposium on Information Theory in Biology*, ed. Hubert P. Yockey (New York: Pergamon Press, 1958), 70–102.

39. Creager, *Life of a Virus*, 303–11; Kay, *Who Wrote the Book of Life*, 179–92; Christina Brandt, *Metapher und Experiment: Von der Virusforschung zum genetischen Code* (Göttingen: Wallstein, 2004), chap. 6.

40. Lily Kay points out that this information was used to confirm the code, but she does not make the point that it was produced with the coding problem in mind. Kay, *Who Wrote the Book of Life*, 187–89; however, Angela Creager makes this point, *Life of a Virus*, 303–11.

41. For example, if UUU coded for the amino acid phenylalanine, as Nirenberg and Matthaei had established, and phenylalanine was replaced by another amino acid in a mutant, one could deduce that this amino acid was coded by one of only nine different codons (all including two Us), thus excluding forty-four other possible combinations.

42. Joseph F. Speyer et al., "Synthetic Polynucleotides and the Amino Acid Code, IV," *Proceedings of the National Academy of Sciences* 48 (Mar 15 1962): 441–48; Speyer et al., "Synthetic Polynucleotides and the Amino Acid Code, II," *Proceedings of the National Academy of Sciences* 48 (Jan 15 1962): 63–68; Peter Lengyel et al., "Synthetic Polynucleotides and the Amino Acid Code, III," ibid., 48 (Feb 1962): 282–24; Carlos Basilio et al. "Synthetic Polynucleotides and the Amino Acid Code, V," ibid., 48 (Apr 15 1962): 613–16; Peter Lengyel, Joseph F. Speyer, and Severo Ochoa, "Synthetic Polynucleotides and the Amino Acid Code," ibid., 47 (Dec 15 1961): 1936–42. In 1962, Smith used sequence alignments to gather information about amino acid replacements and confirmed the genetic code assignments made by Ochoa and others. Emil L. Smith, "Nucleotide Base Coding and Amino Acid Replacements in Proteins," ibid., 48 (Apr 15 1962): 677–84; Smith, "Nucleotide Base Coding and Amino Acid Replacements in Proteins, II," *ibid.*, 48 (May 15 1962): 859–64. On Ochoa, see Maria Jesús Santesmases, "Severo Ochoa and the Biomedical Sciences in Spain under Franco (1959–1975)," *Isis* 91 (2000): 27–45.

43. Smith, "Nucleotide Base Coding, II," 863.

44. Thomas H. Jukes, "Beta Lactoglobulins and the Amino Acid Code," *Biochemical and Biophysical Research Communications* 7, no. 4 (1962): 281–83; Jukes, "Possible Base

Sequences in the Amino Acid Code," *Biochemical and Biophysical Research Communications* 7, no. 6 (1962): 497–502; Jukes, "Relations between Mutations and Base Sequences in Amino Acid Code," *Proceedings of the National Academy of Sciences* 48, no. 10 (1962): 1809–15.

45. Richard V. Eck, "Non-randomness in Amino-Acid 'Alleles,'" *Nature* 191 (Sep 23 1961): 1284–85.

46. Richard V. Eck, "The Protein Cryptogram, I: Non-random Occurrence of Amino Acid 'Alleles,'" *Journal of Theoretical Biology* 2 (1962): 139–51.

47. Walter M. Fitch, "The Relation between Frequencies of Amino Acids and Ordered Trinucleotides," *Journal of Molecular Biology* 16, no. 1 (Mar 1966): 1–8; Fitch, "The Probable Sequence of Nucleotides in Some Codons," *Proceedings of the National Academy of Sciences* 52 (Aug 1964): 298–305.

48. David Keilin, "Preparation of Pure Cytochrome C from Heart Muscle and Some of Its Properties," *Proceedings of the Royal Society of London Series B—Biological Sciences* 122 (1937): 298–308.

49. "Science Exhibition," *Science* 106, no. 2763 (Dec 12 1947): 567–75.

50. Although it was widely available, insulin produced by Eli Lilly from pigs does not seem to have been used for protein sequence studies.

51. Vernon M. Ingram, "Sickle-Cell Anemia Hemoglobin: The Molecular Biology of the First 'Molecular Disease'—the Crucial Importance of Serendipity," *Genetics* 167, no. 1 (May 2004): 1–7.

52. de Chadarevian, "Following Molecules."

53. John F. Henahan, "Dr. Doolittle: Making Big Changes in Small Steps," *Chemical and Engineering News*, Feb 9 1970. 22–32.

54. Margareta Blombäck to the author, May 19 2010; Blombäck, "Thrombosis and Haemostasis Research: Stimulating, Hard Work and Fun," *Journal of Thrombosis and Haemostasis* 98, no. 1 (Jul 2007): 8–15.

55. Margareta Blombäck to her parents, Sep 19 1963, Margareta Blombäck personal papers.

56. Ibid.

57. John Buettner-Janusch and Robert L. Hill, "Molecules and Monkeys," *Science* 147 (Feb 19 1965): 836–42.

58. Russell F. Doolittle, "Characterization of Lamprey Fibrinopeptides," *Biochemical Journal* 94 (Mar 1965): 742–50; Roger Acher, Jacqueline Chauvet, and Marie-Thérèse Chauvet, "Phylogeny of the Neurohypophysial Hormones," *Nature* 216, no. 5119 (Dec 9 1967): 1037–38.

59. Soraya de Chadarevian, *Designs for Life: Molecular Biology after World War II* (Cambridge: Cambridge University Press, 2002).

60. Om P. Bahl and Emil L. Smith, "Amino Acid Sequence of Rattlesnake Heart Cytochrome c," *Journal of Biological Chemistry* 240, no. 9 (Sep 1965): 3585–93.

61. Interview with Judith Dayhoff, Bethesda, Jun 25 2008.

62. Margaret O. Dayhoff, "Biographical sketch Margaret Oakley Dayhoff," 1965, National Biomedical Research Foundation Archives, currently processed at the National Library of Medicine, Bethesda (NBRF Archives hereafter).

63. Margaret B. Oakley and George E. Kimball, "Punched Card Calculation of Resonance Energies," *Journal of Chemical Physics* 17, no. 8 (1949): 706–17.

64. Willis E. Lamb, "Banquet Speech" (December 10, 1955), available at http://www.nobelprize.org/nobel_prizes/physics/laureates/1955/lamb-speech.html (accessed Dec 4 2015).

65. M. O. Dayhoff to Eugene M. Volpert, May 13 1968, Judith Dayhoff personal archives; Margaret O. Dayhoff, Gertrude E. Perlmann, and Duncan A. MacInnes, "The Partial Specific Volumes, in Aqueous Solution, of Three Proteins," *Journal of the American Chemical Society* 74, no. 10 (1952): 2515–17.

66. "Edward Dayhoff; Contributor to Nobel-Winning Work on Atom," *Washington Post*, Jan 11 2007.

67. Robert S. Ledley to Harvey E. Saveley, Jun 29 1960, NBRF Archives. The NBRF eventually moved to Georgetown University Medical Center, Washington, DC.

68. M. O. Dayhoff to Naomi Mendelsohn, Jun 28 1966, NBRF Archives.

69. On the early career of Ledley, see November, *"Digitizing Life,"* 59–76, and *Biomedical Computing.*

70. Robert S. Ledley, *The Use of Computers in Biology and Medicine* (New York: McGraw-Hill, 1965). On the genesis of this volume, see November, *"Digitizing Life,"* chap. 2.

71. George Gamow to James Watson, Dec 6 1954, reproduced in James D. Watson, *Genes, Girls and Gamow* (Oxford: Oxford University Press, 2001), annex 12.

72. Robert S. Ledley. "Digital Computational Methods in Symbolic Logic, with Examples in Biochemistry," *Proceedings of the National Academy of Sciences* 41, no. 7 (1955): 498–511. The paper was communicated by George Gamow.

73. Robert S. Ledley, "Digital Computational Methods in Symbolic Logic, with Examples in Biochemistry," *Proceedings of the National Academy of Sciences* 41, no. 7 (1955), 511. Similarly, at Los Alamos, George Gamow was using the MANIAC computer to make Monte Carlo simulations to produce randomly ordered protein sequences and compare them with the available empirical data in his study of the genetic code. George Gamow and Martynas Yčas, "Statistical Correlation of Protein and Ribonucleic Acid Composition," *Proceedings of the National Academy of Sciences* 41, no. 12 (Dec 15 1955): 1011–19. On this episode, see Kay, *Who Wrote the Book of Life,* 141.

74. Ledley's paper was almost never cited, except by Ledley himself. ISI Web of Science.

75. As cited in Margaret O. Dayhoff and Robert S. Ledley, "Comprotein: A Computer Program to Aid Primary Protein Structure Determination," in *Proceedings of the Fall Joint Computer Conference* (Santa Monica: American Federation of Information Processing Societies, 1962), 262–74.

76. Ledley, *Use of Computers,* 373.

77. Robert S. Ledley, "Memorandum," Nov 16 1960, NBRF Archives.

78. "Summary Progress Report of GM-08710," Jan 15 1963, NBRF Archives.

79. Ibid.

80. Eck, "Protein Cryptogram."

81. "Summary Progress Report of Grant Sequences of Amino Acids in Proteins by Computer Aids," Jan 15 1963, NBRF Archives.

82. Robert S. Ledley and Lee B. Lusted, "Reasoning Foundations of Medical Diagnosis: Symbolic Logic, Probability, and Value Theory Aid Our Understanding of How Physicians Reason," *Science* 130, no. 3366 (Jul 3 1959): 9–21; Ledley and Lusted, "Probability, Logic and Medical Diagnosis," *Science* 130, no. 3380 (Oct 9 1959): 892–930; and the reactions in Robert S. Ledley, "Letters to the Editor," ibid., 131, no. 3399 (Feb 19 1960): 474–564.

83. Dayhoff and Ledley, "Comprotein," 267. In the same paper, Dayhoff and Ledley suggest using the same approach for DNA and RNA sequencing once the experimental data becomes available, p. 274. See also Sidney A. Bernhard, D. F. Bradley, and William L. Duda, "Automatic Determination of Amino Acid Sequences," *IBM Journal of Research and Development* 7, no. 3 (1963): 246–51, for another computer approach to the same problem.

84. The articles written by Eck, Dayhoff, and Ledley were hardly ever cited, except by themselves in the 1960s and 1970s. See, however, the discussions between Margaret O. Dayhoff and Marvin Shapiro, NBRF Archives, December 1962, and Marvin B. Shapiro et al., "Reconstruction of Protein and Nucleic Acid Sequences: Alamine Transfer Ribonucleic Acid," *Science* 150, no. 698 (Nov 12 1965): 918–21.

85. By this time, most campuses in the United States had central computing facilities, but there is no evidence that biochemists used them. "Computing in the University," *Datamation* 8 (1962): 27–30.

86. M. O. Dayhoff to Naomi Mendelsohn, Jun 28 1966, NBRF Archives.

87. See table 1 in John A. Hunt and Vernon M Ingram, "The Chemical Effects of Gene Mutations in Some Abnormal Human Haemoglobins," in *Symposium on Protein Structure*, ed. Albert Neuberger (New York: Wiley, 1958), 148–51.

88. Dayhoff, "NIH GM 8710 Reports, 1962–1965," NBRF Archives.

89. de Chadarevian, *Designs for Life*, chap. 4.

90. Margaret O. Dayhoff, "Computer Search for Active Site Configurations," *Journal of the American Chemical Society* 86, no. 11 (1964): 2295.

91. Dayhoff, "Progress Report on Sequences of Amino Acids in Proteins by Computer Aids," Sep 1965, NBRF Archives.

92. Richard V. Eck, "Progress Report on the RNA-Amino Acid Code," July 1965, p. iii, NBRF Archives.

93. Richard V. Eck, "Appendix to Progress Report," July 1965, NBRF Archives.

94. Wyndham D. Miles, *A History of the National Library of Medicine: The Nation's Treasury of Medical Knowledge* (Washington, DC: US Dept. of Health and Human Services, 1982), chap. 13.

95. William Aspray and Bernard O. Williams, "Arming American Scientists: NSF and the Provision of Scientific Computing Facilities for Universities, 1950–1973," *IEEE Annals of the History of Computing* 16, no. 4 (Win 1994): 60–74.

96. Robert S. Ledley to James B. Wilson, "Report on 'A Tabledex Computer Program,'" Mar 1 1961, NBRF Archives.

97. Robert S. Ledley, "Final Report on SAph 71251," January 1961, NBRF Archives.

98. Robert S. Ledley, "Functional criteria for biomedical digital electronic computer design," Mar 1957, NBRF Archives.

99. M. O. Dayhoff to John C. Kendrew, Jan 28 1966, NBRF Archives.

100. Robert S. Ledley, "Functional criteria for biomedical digital electronic computer design," Mar 1957, NBRF Archives. This manuscript formed the basis of the influential piece published two years later in *Science*, Robert S. Ledley, "Digital Electronic Computers in Biomedical Sciences," *Science* 130 (1959): 1225–34.

101. See, for example, Ralph W. Stacy and Bruce D. Waxman, eds., *Computers in Biomedical Research* (New York: Academic Press, 1965); Theodor D. Sterling and Seymour V. Pollack, *Computers and the Life Sciences* (New York: Columbia University Press, 1965); Medical Research Council (London), *Mathematics and Computer Science in Biology and Medicine: Proceedings of Conference Held by Medical Research Council in Association with the Health Departments, Oxford, July 1964* (London: Her Majesty's Stationery Office, 1965).

102. Richard V. Eck, "Cryptogrammic Detection of a Pattern in Amino Acid 'Alleles': Its Use in Tracing the Evolution of Proteins," *Proceedings of the 17th Annual Conference on Engineering in Medicine and Biology* 6 (1964): 115.

103. On the history of exobiology, see Audra J. Wolfe, "Germs in Space: Joshua Lederberg, Exobiology, and the Public Imagination, 1958–1964," *Isis* 93 (2002): 183–205, and Steven J. Dick and James Edgar Strick, *The Living Universe: NASA and the Development of Astrobiology* (New Brunswick, NJ: Rutgers University Press, 2004).

104. Margaret O. Dayhoff et al., "Venus: Atmospheric Evolution," *Science* 155, no. 3762 (Feb 3 1967): 556–58.

105. Margaret O. Dayhoff, Ellis R. Lippincott, and Richard V. Eck, "Thermodynamic Equilibria in Prebiological Atmospheres," *Science* 146, no. 1461 (Dec 11 1964): 1461–64.

106. Stanley L. Miller and Harold C. Urey, "Organic Compound Synthesis on the Primitive Earth," *Science* 130, no. 3370 (Jul 31 1959): 245–51.

107. "Final Report to the Office of the Life Sciences Programs, October 1, 1965, to September 1, 1965," NBRF Archives.

108. Richard V. Eck and Margaret O. Dayhoff, "Evolution of the Structure of Ferredoxin Based on Living Relics of Primitive Amino Acid Sequences," *Science* 152, no. 3720 (Apr 15 1966): 363–66.

109. M. O. Dayhoff to George Jacobs, Jan 12 1966, NBRF Archives. This was perhaps true, but Dayhoff was not the only one to find internal duplication in proteins, as at least two other groups published the same conclusion in 1966. Russell F. Doolittle, Seymour J. Singer, and Henry Metzger, "Evolution of Immunoglobulin Polypeptide Chains: Carboxy-Terminal of an IgM Heavy Chain," *Science* 154, no. 756 (Dec 23): 1561–62; Walter M. Fitch, "Evidence Suggesting a Partial, Internal Duplication in the Ancestral Gene for Heme-Containing Globins," *Journal of Molecular Biology* 16, no. 1 (Mar 1966): 17–27.

110. M. O. Dayhoff to Carl Berkley, Feb 27 1967, NBRF Archives.

111. M. O. Dayhoff to S. Tideman, Oct 18 1968, NBRF Archives.

112. McKusick, *Mendelian Inheritance in Man*. On McKusick's collection, see Nathaniel C. Comfort, *The Science of Human Perfection: How Genes Became the Heart of American Medicine* (New Haven: Yale University Press, 2012), chap. 6, and M. Susan Lindee, *Moments of Truth in Genetic Medicine* (Baltimore: Johns Hopkins University Press, 2005), chap. 3.

113. Hopwood, *Haeckel's Embryos*, chap. 5.

114. Richard E. Dickerson to Margaret O. Dayhoff, Jan 9 1969, Judith Dayhoff personal archives.

115. Margaret O. Dayhoff et al., *Atlas of Protein Sequence and Structure* (Silver Spring, MD: National Biomedical Research Foundation, 1965), unnumbered page.

116. Ibid, p. 2. On the issue of precision and objectivity, see Suárez-Díaz and Anaya-Muñoz, "History, Objectivity, and the Construction of Molecular Phylogenies."

117. Dietrich, "Paradox and Persuasion"; Edna Suárez-Díaz, "The Long and Winding Road of Molecular Data in Phylogenetic Analysis," *Journal of the History of Biology* 47, no. 3 (Fall 2014): 443–78.

118. M. O. Dayhoff to D. C. Phillips, Dec 20 1965, Judith Dayhoff personal archives.

119. Walther M. Fitch to Richard V. Eck, Jul 1 1966; Dayhoff to Joel W. Adelson, Dec 11 1968, Judith Dayhoff personal archives.

120. M. O. Dayhoff, "LM 01206, Comprehensive progress report," Aug 23 1973, NBRF Archives.

121. M. O. Dayhoff to Robert G. Denkewalter, Feb 8 1971, NBRF Archives.

122. Dayhoff to P. Edman, Feb 14 1977, NBRF Archives.

123. Frances C. Bernstein et al., "Protein Data Bank: Computer-Based Archival File for Macromolecular Structures," *European Journal of Biochemistry* 80, no. 2 (1977): 319–24.

124. B. S. Guttman to Dayhoff, Jun 10 1968, NBRF Archives.

125. "Correspondence index card set," 1965, NBRF Archives.

126. Melvin Calvin to Dayhoff, Feb 11 1966, NBRF Archives.

127. Joshua Lederberg to Dayhoff, Mar 12 1964, NBRF Archives.

128. Ernst Mayr to Dayhoff, May 20 1968, Judith Dayhoff personal archives.

129. Emanuel Margoliash to Richard V. Eck, Feb 2 1966, NBRF Archives.

130. M. Laskowski to Dayhoff, Oct 31 1969, NBRF Archives.

131. Oliver Smithies to Winona Barker, Oct 5 1970, NBRF Archives.

132. Dayhoff to Donald DeVincenzi, Jun 10 1980, NBRF Archives.

133. Richard Synge to "Compilers," Apr 7 1966, NBRF Archives.

134. John T. Edsall to Dayhoff, Nov 4 1969, NBRF Archives.

135. William H. Stein to Dayhoff, Dec 4 1969, NBRF Archives.

136. Emanuel Margoliash to Dayhoff, Dec 10 1968, Judith Dayhoff personal archives. Eventually, Edsall supported Dayhoff's application, and she became a member in 1970. Dayhoff to Robert A. Harte, Apr 16 1970, Judith Dayhoff personal archives.

137. MEDLARS was the computerized version of the Index Medicus that had become available in 1964. Dayhoff and Richard V. Eck to the NSF, "Application for publication support," 1966, p. 5, NBRF Archives.

138. M. O. Dayhoff to D. Haas, Apr 11 1969, NBRF Archives.

139. Anfinsen, *Molecular Basis*, 146.

140. Dayhoff would hold to this position all her life, often in face of fierce opposition from funding agencies that believed the two activities should be kept separate.

141. Thomas Uzzell to Birger Blombäck, Jun 20 1969, Birger Blombäck personal archives.

142. John A. W. Kirsch to Dayhoff, Apr 24 1971, NBRF Archives.

143. Thomas H. Jukes. "Some Recent Advances in Studies of the Transcription of the Genetic Message," *Advances in Biological and Medical Physics* 9 (1963): 1–41. On theory

as a dividing line between biochemists and molecular biologists, see Pnina Abir-Am, "The Politics of Macromolecules: Molecular Biologists, Biochemists, and Rhetoric," *Osiris* 7 (1992): 164–91.

144. Gerhard Braunitzer to Dayhoff, Apr 18, 1968, NBRF Archives.

145. John T. Edsall to Dayhoff, Dec 3 1968, NBRF Archives. The interpretive material was presented in a discursive format and had not been peer reviewed, two factors that may have predisposed other scientists to ignore it.

146. Nabil G. Seidah and Michel Chrétien to Dayhoff, Jun 14 1979, NBRF Archives.

147. Sanger, "Sequences," 1.

148. Dayhoff, "LM Grant Application," May 31 1973, p. 24, NBRF Archives.

149. M. O. Dayhoff to D. M. Moore, Sep 24 1981, NBRF Archives.

150. Jeanne L. Brand (NLM) to Dayhoff, May 21 1970, NBRF Archives.

151. M. O. Dayhoff to Jeanne L. Brand (NLM), Mar 1 1974, NBRF Archives.

152. Susan Wright, *Molecular Politics: Developing American and British Regulatory Policy for Genetic Engineering, 1972–1982* (Chicago: University of Chicago Press, 1994).

153. Victor A. McKusick to Dayhoff, May 9 1979, NBRF Archives.

154. "Proteins: Yet More Sequences," *Nature* 224 (1969): 313.

155. Philip H Abelson, "Amino Acid Sequence in Proteins," *Science* 160 (1968): 951.

156. Dayhoff, "Notes of phone conversation with Emanuel Margoliash," Aug 20 1970, Judith Dayhoff personal archives.

157. Dayhoff, "NIH Grant 1206 Report, Section C," July 1973, NBRF Archives.

158. Richard V. Eck and Margaret O. Dayhoff, *Atlas of Protein Sequence and Structure* (Silver Spring, MD: National Biomedical Research Foundation, 1966).

159. Warren Hagstrom, "Gift Giving as an Organizing Principle in Science," in *Science in Context: Readings in the Sociology of Science*, ed. Barry Barnes and David Edge (Cambridge: MIT Press, 1982.), 21–34.

160. Alain E. Bussard, "Data Proliferation: A Challenge for Science and for Codata," in *Biomolecular Data: A Resource in Transition*, ed. Rita Colwell (Oxford: Oxford University Press, 1989), 11–15.

161. See the reviews included in the critical edition, James D. Watson, *The Double Helix: A Personal Account of the Discovery of the Structure of DNA: Text, Commentary, Reviews, Original Papers* (New York: Touchstone, 2001 [1968]).

162. R. C. Lewontin, *Chicago Sunday*, Feb 25 1968, 1–2, reproduced in Watson, *Double Helix* (2001).

163. R. Holmquist to Dayhoff, Dec 23 1979, NBRF Archives.

164. W. Salser to W. Goad, Dec 31 1979, APS Archives. See also R. Holmquist to Dayhoff, Dec 23 1979, NBRF Archives; Russell F. Doolittle, "On the Trail of Protein Sequences," *Bioinformatics* 16, no. 1 (Jan 2000): 24–33, and interview with Temple F. Smith, Boston, Feb 16 2006.

165. Jennifer S. Light, "When Computers Were Women," *Technology and Culture* 40, no. 3 (1999): 455–83.

166. M. O. Dayhoff to S. Tideman, Oct 18 1968, NBRF Archives.

167. M. O. Dayhoff to Eugene M. Volpert, May 13 1968, Judith Dayhoff personal archives.

168. M. O. Dayhoff to R. F. Doolittle, Oct 18 1968, Judith Dayhoff personal archives.

169. George H. Kennedy to Robert S. Ledley, Jan 11 1962, Judith Dayhoff personal archives.

170. Robert S. Ledley to Carl R. Brewer, Mar 13 1962, Judith Dayhoff personal archives.

171. Interview with Robert S. Ledley, Georgetown, Jan 26 2006.

172. "Survey," n.d. [1971], Judith Dayhoff personal archives.

173. Winona Barker to Ernst Knobil, Oct 3 1980, NBRF Archives.

174. Evelyn Fox Keller, *Secrets of Life, Secrets of Death: Essays on Language, Gender, and Science* (New York: Routledge, 1992).

175. Eck and Dayhoff, *Atlas (1966)*.

176. Ibid, 163.

177. Joseph Felsenstein, *Inferring Phylogenies* (Sunderland, MA: Sinauer Associates, 2004), chap. 10.

178. For a useful comparison of early tree-building methodologies, see ibid., chap. 10.

179. Ibid., 165.

180. Ibid., 165. Two years later, the NBRF would acquire its own mainframe computer, an IBM 360 Model 44, instead of using the shared IBM 7090. Robert S. Ledley, "Application (renewal), NASA 21-003-002," 1968, NBRF Archives.

181. For a later description, see Margaret O. Dayhoff, "Computer Analysis of Protein Evolution," *Scientific American* 221 (1969): 86–95.

182. Walter M. Fitch and Emanuel Margoliash, "Construction of Phylogenetic Trees," *Science* 155, no. 760 (Jan 20 1967): 279–84, fig. 2.

183. Russell F. Doolittle and Birger Blombäck, in a phylogeny published in 1964 based on variations in fibrinopeptide sequences, used an even more crude method than Fitch and Margoliash, simply counting the number of amino acid changes, without any weighting. Russell F. Doolittle and Birger Blombäck. "Amino-Acid Sequence Investigations of Fibrinopeptides from Various Mammals: Evolutionary Implications," *Nature* 202 (1964): 147–52. Margoliash had done the same a year earlier. Emil L. Smith and Emanuel Margoliash, "Evolution of Cytochrome C," *Federation Proceedings* 23 (Nov–Dec 1964): 1243–47.

184. Peter J. McLaughlin and Margaret O. Dayhoff, "Eukaryote Evolution: A View Based on Cytochrome C Sequence Data," *Journal of Molecular Evolution* 2, nos. 2–3 (1973): 99–116.

185. M. O. Dayhoff to Hoimar V. Ditfurth, Oct 30 1970, NBRF Archives.

186. M. O. Dayhoff to George G. Jacob, Mar 21 1969, Judith Dayhoff personal archives.

187. Margaret O. Dayhoff et al., "Evolution of Sequences within Protein Superfamilies," *Naturwissenschaften* 62, no. 4 (1975): 154–61.

188. Winona C. Barker and Margaret O. Dayhoff, "Viral SRC Gene Products Are Related to the Catalytic Chain of Mammalian cAMP-Dependent Protein Kinase," *Proceedings of the National Academy of Sciences* 79, no. 9 (May 1982): 2836–39.

189. Winona C. Barker, Lynne K. Ketcham, and Margaret O. Dayhoff, "A Comprehensive Examination of Protein Sequences for Evidence of Internal Gene Duplication," *Journal of Molecular Evolution* 10, no. 4 (Feb 21 1978): 265–81.

190. Mary P. Winsor, *Reading the Shape of Nature: Comparative Zoology at the Agassiz Museum* (Chicago: University of Chicago Press, 1991); Endersby, *Imperial Nature*; Johnson, *Ordering Life*.

191. See, for example, Johnson, "Ernst Mayr, Karl Jordan, and the History of Systematics," *History of Science* 43 (2005): 1–35; Johnson, *Ordering Life*.

192. M. O. Dayhoff to Martin D. Kamen, n.d. [ca. 1968], NBRF Archives.

193. Margaret O. Dayhoff, "Computer Analysis of Protein Sequences," *Federation Proceedings* 33, no. 12 (Dec 1974): 2314–16.

Chapter Four

1. See for example Max Delbrück's description of phage research as "a fine playground." Delbrück, "Experiments with Bacterial Viruses (Bacteriophages)," *Harvey Lectures* 41 (1946): 161–87; and more generally the expansion on this theme in Cairns, Stent, and Watson, *Phage*, analyzed in Pnina Abir-Am, "Themes, Genres and Orders of Legitimation in the Consolidation of New Scientific Disciplines: Deconstructing the Historiography of Molecular Biology," *History of Science* 23 (1985): 73–117, and Angela N. H. Creager. "The Paradox of the Phage Group: Essay Review," *Journal of the History of Biology* 43, no. 1 (Feb 2010): 183–93.

2. As Francis Crick and John Kendrew put it in 1957, protein structure determination is "a very long and tedious business"; Francis H. C. Crick and John C. Kendrew, "X-Ray Analysis and Protein Structure," *Advances in Protein Chemistry* 12 (1957): 133–214. "The photographing, indexing, measuring, correcting and correlating of some 7000 reflexions was a task whose length and tediousness it will be better not to describe"; Max F. Perutz, "An X-Ray Study of Horse Methaemoglobin 2," *Proceedings of the Royal Society of London Series A—Mathematical and Physical Sciences* 195, no. 1043 (1949): 474–99. For other typical examples, see Walter C. Hamilton, "The Revolution in Crystallography," *Science* 169, no. 941 (Jul 10 1970): 133–41; Helen M. Berman, "X-Ray Crystallography of Biological Molecules," in *Spectroscopy in Biology and Chemistry*, ed. Sow-Hsin Chen and Sidney Yip (New York: Academic Press, 1974), 145–75.

3. On the self-fashioning of postwar physicists, see Paul Forman, "Social Niche and Self-Image of the American Physicist," in *The Restructuring of Physical Sciences in Europe and the Unites States, 1945–1960*, ed. Michelangelo De Maria, Mario Grilli, and Fabio Sebastiani (Singapore: World Scientific Publishing, 1989), 96–104, and for molecular biologists, Strasser, *La fabrique d'une nouvelle science*.

4. See recollections of phage researchers in Cairns, Stent, and Watson, *Phage*.

5. Georgina Ferry, *Max Perutz and the Secret of Life* (New York: Cold Spring Harbor Laboratory Press, 2007); de Chadarevian, *Designs for Life*, chaps. 4–5.

6. The expression comes from Watson's 1966 autobiographical sketch, in Cairns, Stent, and Watson, *Phage*.

7. James D. Watson, *The Double Helix* (London: Weidenfeld and Nicholson, 1968).

8. Robert Olby, *The Path to the Double Helix* (New York: Dover, 1994 [1974]).

9. de Chadarevian, *Designs for Life*, chap. 4.

10. Ibid., chap. 4.

11. Hamilton, "Revolution in Crystallography," 136.

12. Ibid., 137.

13. Ibid., 138.

14. Douglas M. Collins et al., "Protein Crystal Structures: Quicker, Cheaper Approaches," *Science* 190, no. 4219 (Dec 12 1975): 1047–53.

15. de Chadarevian, *Designs for Life*, chap. 4.

16. Ibid., chap. 4.

17. November, *Biomedical Computing*, chap. 2; Martin Campbell-Kelly and William Aspray, *Computer: A History of the Information Machine* (Boulder: Westview Press, 2004), chap. 6.

18. Frederic M. Richards, "The Matching of Physical Models to Three-Dimensional Electron-Density Maps: A Simple Optical Device," *Journal of Molecular Biology* 37, no. 1 (Oct 14 1968): 225–30.

19. de Chadarevian, *Designs for Life*, chap. 4.

20. For these three arguments about the importance of computers for crystallography, see Soraya de Chadarevian, "Models and the Making of Molecular Biology," in *Models: The Third Dimension of Science*, ed. de Chadarevian and Nick Hopwood (Stanford: Stanford University Press, 2004), 348.

21. Eric Francoeur and Jérôme Segal, "From Model Kits to Interactive Graphics," in de Chadarevian and Hopwood, *Models*, 402–29.

22. Ibid., 334.

23. On Cyrus Levinthal's development of molecular graphics, see Francoeur and Segal, "From Model Kits to Interactive Graphics"; Eric Francoeur, "Cyrus Levinthal, the Kluge and the Origins of Interactive Molecular Graphics," *Endeavour* 26, no. 4 (Dec 2002): 127–31.

24. Cyrus Levinthal, "Molecular Model-Building by Computer," *Scientific American* 214 (1966): 41–52.

25. Ibid., 50.

26. Ibid., 52.

27. Francoeur and Segal, "From Model Kits to Interactive Graphics": over 15 million dollars for 2018, when adjusted for the Consumer Price Index.

28. These included C. David Barry at Oxford, Robert Langridge at Princeton, and Edgar F. Meyer Jr. at Texas A&M University.

29. David Ophir et al., "BRAD: Brookhaven Raster Display," *Communications of the ACM* 11, no. 6 (1968): 415–16.

30. Edgar F. Meyer Jr. "Three-Dimensional Graphical Models of Molecules and a Time-Slicing Computer," *Journal of Applied Crystallography* 3 (Oct 1 1970): 392.

31. Ibid., 85.

32. Olga Kennard, William G. Town, and David G. Watson, "Cambridge Crystallographic Data Center, 1. Bibliographic File," *Journal of Chemical Documentation* 12, no. 1 (1972): 14–19; R. M. S. Hall, "Development of the United Kingdom Data Program," ibid., 7 (1967): 18.

33. Edgar F. Meyer Jr. "Towards an Automatic, Three Dimensional Display of Structural Data," *Journal of Chemical Documentation* 10, no. 2 (1970): 85.

34. At the same time, C. David Barry at Oxford was developing a similar system also relying on small computers. C. David Barry and Anthony C. T. North, "The Use of a Computer-Controlled Display System in the Study of Molecular Conformations," *Cold Spring Harbor Symposia on Quantitative Biology* 36 (1972): 577–84.

35. Ophir et al., "BRAD," 415.

36. Meyer, "Three-Dimensional Graphical Models," 394.

37. Ibid., 393.

38. Edgar F. Meyer Jr. to Helen M. Berman, Sep 8 1970, Protein Data Bank Archives, Rutgers University, New Brunswick (PDB Archives hereafter). This letter mentions the "Protein Data Project." Ray A. Young also remembers this meeting as the origin of the PDB: "I first heard intimations of it [the PDB] in discussions with Helen [Berman] and Ed [Meyer] at Ottawa," Ray A. Young to Thomas Koetzle, Feb 9 1974. In 1966, Berman visited a friend at MIT, where she met Edgar F. Meyer Jr., who was working in Levinthal's group. Interview with Helen Berman, New Brunswick, NJ, Oct 30 2009.

39. Strasser, "World."

40. Moore, Spackman, and Stein, "Automatic Recording Apparatus"; Edman and Begg, "A Protein."

41. Berman, "A proposed crystallographic study of peptides contained in globular proteins," 1967, pp. 1–2, PDB Archives.

42. Ibid, p. 17.

43. Ibid, p. 18.

44. See for example Kendrew's descriptions of hemoglobin. John C. Kendrew et al., "Structure of Myoglobin: A Three-Dimensional Fourier Synthesis at 2 A. Resolution," *Nature* 185, no. 4711 (Feb 13 1960): 422–27.

45. Interview with Helen Berman, New Brunswick, NJ, Jul 17 2009.

46. Helen M. Berman, George A. Jeffrey, and Robert D. Rosenstein, "Crystal Structures of α' and β Forms of D-Mannitol," *Acta Crystallographica Section B—Structural Crystallography and Crystal Chemistry* B 24 (1968): 442.

47. Peter F. Stevens, "Haüy and A.-P. De Candolle: Crystallography, Botanical Systematics, and Comparative Morphology, 1780–1840," *Journal of the History of Biology* 17, no. 1 (1984): 49–82.

48. Bryan Craven, "George A. Jeffrey (1915–2000)," *Acta Crystallographica Section B—Structural Science* 56 (Aug 2000): 545–46.

49. Alfred D. French, "George Alan Jeffrey," *Advances in Carbohydrate Chemistry and Biochemistry* 57 (2001): 1–9.

50. George A. Jeffrey, Edith Mootz, and Dietrich Mootz, "A Knowledge Availability Survey of the Crystal Structural Data for Pyrimidine Derivatives," *Acta Crystallographica* 19, no. 4 (Oct 10 1965): 691–92.

51. James G. Williams, "Governance of Special Information-Centers: Knowledge Availability Systems Center at University of Pittsburgh," *Library Trends* 26, no. 2 (1977): 241–54.

52. William Aspray, "Command and Control, Documentation, and Library Science: The Origins of Information Science at the University of Pittsburgh," *IEEE Annals of the History of Computing* 21, no. 4 (Oct–Dec 1999): 4–20.

53. Cited in ibid., 6.

54. Edgar F. Meyer Jr., "Interactive Computer Display for the Three Dimensional Study of Macromolecular Structures," *Nature* 232, no. 5308 (Jul 23 1971): 255–57. The article was submitted March 1, 1971. See Edgar F. Meyer Jr. to J. Drenth, Jun 9 1971, PDB Archives. "I am assembling a library of protein structures." However, Meyer mentions

that the coordinates will be used only for teaching and research in Brookhaven, not for distribution.

55. Interview with Helen Berman, New Brunswick, NJ, Jul 17 2009; interview with Tom Koetzle, Long Island, NY, Nov 3 2009.

56. Edgar F. Meyer Jr. to Geoffrey C. Ford, Sep 15 1971, PDB Archives.

57. Edgar F. Meyer Jr. to Helen M. Berman, Sep 8 1970, PDB Archives.

58. Jenny Glusker, Minutes, Feb 1971, PDB Archives.

59. Ibid., p. 2.

60. Ibid., p. 4.

61. Interview with Helen Berman, New Brunswick, NJ, Jul 17 2009.

62. On the role of "hippies" in science, see David Kaiser and Patrick McCray, *Groovy Science: Knowledge, Innovation, and American Counterculture* (Chicago: University of Chicago Press, 2016); David Kaiser, *How the Hippies Saved Physics: Science, Counterculture, and the Quantum Revival* (New York: W. W. Norton, 2011).

63. David C. Phillips, "Protein Crystallography 1971: Coming of Age," *Cold Spring Harbor Symposia on Quantitative Biology* 36 (1972): 589–92.

64. Philip J. Pauly, "Summer Resort and Scientific Discipline: Woods Hole and the Structure of American Biology," in *The American Development of Biology*, ed. Ronald Rainger, Keith R. Benson, and Jane Maienschein (Philadelphia: University of Pennsylvania Press, 1988), 121–50.

65. Robert P. Crease, *Making Physics. A Biography of Brookhaven National Laboratory, 1946–1972* (Chicago: University of Chicago Press, 1999).

66. Mogens S. Lehmann, Thomas F. Koetzle, and Walter C Hamilton, "Precision Neutron Diffraction Structure Determination of Protein and Nucleic Acid Components, I. The Crystal and Molecular Structure of the Amino Acid L-Alanine," *Journal of the American Chemical Society* 94, no. 8 (Apr 19 1972): 2657–60.

67. Walter C. Hamilton et al., "Stereoscopic Atlas of Amino-Acid Structures," *Materials Research Bulletin* 7, no. 11 (1972): 1225.

68. Hamilton, "Revolution in Crystallography," 140.

69. Ibid.

70. Walter Hamilton to Olga Kennard, Jun 18 1971, PDB Archives.

71. Phillips, "Protein Crystallography," 592.

72. Walter Hamilton to Olga Kennard, Jun 18 1971, PDB Archives.

73. Interview with Helen Berman, New Brunswick, NJ, Jul 17 2009.

74. "Protein Data Bank," *Nature New Biology* 233, no. 42 (Oct 20 1971): 223.

75. James A. Ibers, "Walter Clark Hamilton, 1931–1973," *Acta Crystallographica Section A—Crystal Physics, Diffraction, Theoretical and General Crystallography* 29, no. 4 (1973): 483–84.

76. Edgar F. Meyer Jr. et al., "CRYSNET, a Crystallographic Computing Network with Interactive Graphics Display," *Federation Proceedings* 33, no. 12 (Dec 1974): 2402–5; Thomas F. Koetzle et al., "CRYSNET: A Network for Crystallographic Computing," in *Computer Networking and Chemistry*, ed. Peter Lykos (Washington, DC: American Chemical Society, 1975), 1–8 ; Herbert J. Bernstein et al., "Second Annual AEC Scientific Computer Information Exchange Meeting," *Proceedings of the Technical Program* 148 (1974): 158.

77. Edgar F. Meyer Jr. to Helen Berman, Aug 3 1971, PDB Archives.

78. Walter Hamilton to Eloise Clark, Feb 24 1972, PDB Archives.

79. Walter Hamilton to Eloise Clark, Feb 24 1972, PDB Archives.

80. Ibers, "Walter Clark Hamilton."

81. Thomas Koetzle to Frederic M. Richards, Mar 30 1973, PDB Archives.

82. Wayne Hendrickson to Thomas Koetzle, Apr 6 1973, PDB Archives.

83. Thomas Koetzle to Max F. Perutz, May 22, 1973, PDB Archives.

84. Thomas Koetzle to Max F. Perutz, May 22 1973, PDB Archives.

85. Thomas Koetzle to Wayne Hendrickson, May 4 1973, PDB Archives; "Crystallography Protein Data Bank," *Journal of Molecular Biology* 78 (1971): 587.

86. *PDB Annual Report*, 1974, 1975, 1976, PDB Archives.

87. *Protein Data Bank Newsletter* 3, Nov 1976, p. 2, PDB Archives.

88. *Protein Data Bank Newsletter* 5, Nov 1977, PDB Archives; Margaret O. Dayhoff, *Atlas of Protein Sequence and Structure, vol. 5, suppl.* 2 (Washington DC: National Biomedical Research Foundation, 1976).

89. Jürgen Bajorath, Ronald Stenkamp, and Alejandro Aruffo, "Knowledge-Based Model-Building of Proteins: Concepts and Examples," *Protein Science* 2, no. 11 (Nov 1993): 1798–1810.

90. Wynne J. Browne et al., "A Possible Three-Dimensional Structure of Bovine Lactalbumin Based on That of Hen's Egg-White Lysozyme," *Journal of Molecular Biology* 42, no. 1 (May 28 1969): 65–86.

91. K. Ravi Acharya et al., "A Critical Evaluation of the Predicted and X-Ray Structures of Alpha-Lactalbumin," *Journal of Protein Chemistry* 9, no. 5 (Oct 1990): 549–63.

92. As Joe November points out, the kind of real-time interactive graphical programs proposed by Meyers were a radical departure from the batch processing programs used by the crystallographers and might have been considered a less "noble" form of computing. Personal communication from Joseph A. November, 2012.

93. *Protein Data Bank Newsletter* 6, Nov 1978.

94. Cyrus Chothia and Arthur M. Lesk, "The Relation between the Divergence of Sequence and Structure in Proteins," *EMBO Journal* 5, no. 4 (Apr 1986): 823–26.

95. Cyrus Chothia, "Structural Invariants in Protein Folding," *Nature* 254, no. 5498 (Mar 27 1975): 304–8.

96. Michael Levitt and Cyrus Chothia, "Structural Patterns in Globular Proteins," *Nature* 261, no. 5561 (Jun 17 1976): 552–58.

97. Cyrus Chothia and Arthur M. Lesk, "Canonical Structures for the Hypervariable Regions of Immunoglobulins," *Journal of Molecular Biology* 196, no. 4 (Aug 20 1987): 901–17.

98. Jane S. Richardson, "β-Sheet Topology and the Relatedness of Proteins," *Nature* 268, no. 5620 (Aug 11 1977): 495–500.

99. Cited in Sonya Bahar, "Ribbon Diagrams and Protein Taxonomy: A Profile of Jane S. Richardson," *Biological Physicist* 4, no. 3 (2004): 5–8.

100. Michael Levitt and Jonathan Greer, "Automatic Identification of Secondary Structure in Globular Proteins," *Journal of Molecular Biology* 114, no. 2 (Aug 5 1977): 181–239.

101. Ibid., 182.

102. Ibid.

103. Daston and Galison, *Objectivity*.

104. Gordon M. Crippen, "The Tree Structural Organization of Proteins," *Journal of Molecular Biology* 126, no. 3 (Dec 15 1978): 315–32; Shoshana J. Wodak and Joel Janin, "Location of Structural Domains in Protein," *Biochemistry* 20, no. 23 (Nov 10 1981): 6544–52.

105. Liisa Holm and Chris Sander, "Parser for Protein Folding Units," *Proteins* 19, no. 3 (Jul 1994): 256–68.

106. Neena L. Summers and Martin Karplus, "Construction of Side-Chains in Homology Modelling. Application to the C-Terminal Lobe of Rhizopuspepsin," *Journal of Molecular Biology* 210, no. 4 (Dec 20 1989): 785–811; Christine A. Orengo et al., "Identification and Classification of Protein Fold Families," *Protein Engineering* 6, no. 5 (Jul 1993): 485–500; Liisa Holm and Chris Sander, "The FSSP Database of Structurally Aligned Protein Fold Families," *Nucleic Acids Research* 22, no. 17 (Sep 1994): 3600–3609.

107. Alexey G. Murzin et al., "SCOP: A Structural Classification of Proteins Database for the Investigation of Sequences and Structures," *Journal of Molecular Biology* 247, no. 4 (Apr 7 1995): 536–40.

108. Christine A. Orengo et al., "CATH: A Hierarchic Classification of Protein Domain Structures," *Structure* 5, no. 8 (Aug 15 1997): 1093–1108; Orengo et al., "Identification and Classification."

109. Orengo et al., "CATH," 1104.

110. Caroline Hadley and David T. Jones, "A Systematic Comparison of Protein Structure Classifications: SCOP, CATH and FSSP," *Structure* 7, no. 9 (Sep 15 1999): 1099–1112.

111. Malcolm W. MacArthur and Janet M. Thornton, "Influence of Proline Residues on Protein Conformation," *Journal of Molecular Biology* 218, no. 2 (Mar 20 1991): 397–412.

112. Cyrus Chothia, "One Thousand Families for the Molecular Biologist," *Nature* 357, no. 6379 (Jun 18 1992): 543–44.

113. *Protein Data Bank Newsletter* 6, May 1978, PDB Archives. For an overview of the deposition policies, see Helen M. Berman et al., "How Community Has Shaped the Protein Data Bank," *Structure* 21, no. 9 (Sep 3 2013): 1485–91.

114. *Protein Data Bank Newsletter* 7, December 1978, PDB Archives.

115. *Protein Data Bank Newsletter* 12, April 1980, PDB Archives.

116. *Protein Data Bank Newsletter* 20, April 1982, PDB Archives.

117. de Chadarevian, "Models," 349.

118. *Protein Data Bank Newsletter* 23, Jan 1983, PDB Archives.

119. Commission on Journals, "Deposition of Macromolecular Atomic Coordinates and Structure Factors with the Protein Data Bank," *Acta Crystallographica* B37 (1981): 1161–62.

120. "Deposition of Macromolecular Atomic Coordinates and Structure Factors with the Protein Data Bank: Modified Policy," *Acta Crystallographica* B38 (1982): 1050.

121. Jenny P. Glusker to Allen J. Bard, Jan 24 1984, PDB Archives.

122. Allen J. Bard to Jenny P. Glusker, Jan 31 1984, PDB Archives.

123. Jenny Glusker, "Lost Data," *Accounts of Chemical Research* 18, no. 4 (1985): 95.

124. Helen M. Berman to Eleano Butz, Aug 14 1985; Helen M. Berman to Daniel Koshland, Jan 16 1986, PDB Archives.

125. R. Schekman, "The Nine Lives of Daniel E. Koshland, Jr. (1920–2007)," *Proceedings of the National Academy of Sciences* 104, no. 37 (Sep 11 2007): 14551–52.

126. John C. Norvell to Helen M. Berman, Jun 10 1987, PDB Archives.

127. Richard E. Dickerson to Charles E. Bugg, Jul 27 1987, PDB Archives.

128. Ibid.

129. Robert K. Merton argued that science was governed by four universal norms, communism, universalism, disinterestedness, and organized skepticism; Steven Shapin, "Understanding the Merton Thesis," *Isis* 79 (1988): 594–605.

130. Ibid.

131. Ibid.

132. Ibid.

133. Charles E. Bugg to Jan Drenth, Aug 24 1987, PDB Archives.

134. Brian W. Matthews to Guy Dodson, Nov 8 1988, PDB Archives.

135. Charles E. Bugg to Jan Drenth, Aug 24 1987, PDB Archives.

136. Wendell A. Lim, "Frederic M. Richards 1925–2009 Obituary," *Nature Structural & Molecular Biology* 16, no. 3 (Mar 2009): 230–32.

137. Frederic M. Richards to Charles E. Bugg, Jan 20 1988, PDB Archives.

138. Marcia Barinaga, "The Missing Crystallography Data," *Science* 245, no. 4923 (Sep 15 1989): 1179–81.

139. John Maddox, "Making Good Databanks Better," *Nature* 341, no. 6240 (Sep 28 1989): 277.

140. Richard J. Roberts, "Benefits of Databases," *Nature* 342, no. 6246 (Nov 9 1989): 114.

141. Thomas F. Koetzle,"Benefits of Databases," *Nature* 342, no. 6246 (Nov 9 1989): 114.

142. Melinda Clare Baldwin, *Making Nature: The History of a Scientific Journal* (Chicago: University of Chicago Press, 2015), and specifically on peer review, see Baldwin, "Credibility, Peer Review, and Nature, 1945–1990," *Notes and Records of the Royal Society of London* 69, no. 3 (Sep 20 2015): 337–52.

143. Barinaga, "Missing Crystallography Data," 1180.

144. On the academic culture at Genentech, see Sally Smith Hughes, *Genentech: The Beginnings of Biotech* (Chicago: University of Chicago Press, 2011).

145. Barinaga, "Missing Crystallography Data."

146. Steven Epstein, *Impure Science: AIDS, Activism, and the Politics of Knowledge* (Berkeley: University of California Press, 1996).

147. John C. Norvell to Helen M. Berman, Jun 10 1987, PDB Archives.

148. Jim Cassatt to Helen M. Berman, Jan 4 1990, PDB Archives.

149. "Resolved," 1990, PDB Archives.

150. Joel L. Sussman, "Protein Data Bank Deposits," *Science* 282, no. 5396 (Dec 11 1998): 1993; Alexander Wlodawer, "Deposition of Macromolecular Coordinates Resulting from Crystallographic and NMR Studies," *Nature Structural Biology* 4, no. 3 (Mar 1997): 173–74.

151. Alexander Wlodawer et al., "Immediate Release of Crystallographic Data: A Proposal," *Science* 279, no. 5349 (Jan 16 1998): 306–7.

152. Floyd E. Bloom, "Policy Change," *Science* 281, no. 5374: 175; Philip Campbell, "New Policy for Structure Data," *Nature* 394, no. 6689 (Jul 9 1998): 105.

153. Alexander Wlodawer, "Deposition of Structural Data Redux," *Acta Crystallographica Section D—Biological Crystallography* 63, no. 3 (Mar 2007): 421–23.

154. Eliot Marshall, "Transfer of Protein Data Bank Sparks Concern," *Science* 281, no. 5383 (Sep 11 1998): 1584–85.

155. de Chadarevian, *Designs for Life*, chap. 7.

156. Edgar F. Meyer Jr., "The First Years of the Protein Data Bank," *Protein Science* 6, no. 7 (Jul 1997): 1591–97.

157. Chris P. Ponting and Robert R. Russell, "The Natural History of Protein Domains," *Annual Review of Biophysics and Biomolecular Structure* 31 (2002): 45–71.

158. Ann Blair, *The Theater of Nature: Jean Bodin and Renaissance Science* (Princeton: Princeton University Press, 1997); Brian W. Ogilvie, *The Science of Describing: Natural History in Renaissance Europe* (Chicago: University of Chicago Press, 2006); Staffan Müller-Wille, "Carl Von Linnés Herbarschrank. Zur epistemischen Funktion eines Sammlungsmöbels," in te Heesen and Spary, *Sammeln als Wissen*, and I. Charmantier and S. Muller-Wille, "Carl Linnaeus's Botanical Paper Slips (1767–1773)," *Intellectual History Review* 24, no. 2 (Apr 3 2014): 215–38.

159. For an ethnographic study of crystallographers' interactions with their virtual models, see Natasha Myers, "Molecular Embodiments and the Body-Work of Modeling in Protein Crystallography," *Social Studies of Science* 38, no. 2 (Apr 2008): 163–99.

160. On the idea of chain of transformation, see Latour, *Pandora's Hope*, chap. 2.

161. Jocelyn Kaiser, "Chemists Want NIH to Curtail Database," *Science* 308, no. 5723 (May 6 2005): 774.

162. Steven Shapin, *The Scientific Life: A Moral History of a Late Modern Vocation* (Chicago. University of Chicago Press, 2008).

163. Kelly Moore, *Disrupting Science: Social Movements, American Scientists, and the Politics of the Military, 1945–1975* (Princeton: Princeton University Press, 2008); Eric J. Vettel, *Biotech: The Countercultural Origins of an Industry* (Philadelphia: University of Pennsylvania Press, 2006).

Chapter Five

1. NIH press release, "Public Collections of DNA and RNA Sequence Reach 100 Gigabases," Aug 22 2005, available from https://www.nlm.nih.gov/archive//20120510/news/press_releases/dna_rna_100_gig.html (accessed Dec 5 2009). For a participant's overview of the rise of genetic databases, including GenBank, see Temple E. Smith, "The History of the Genetic Sequence Databases," *Genomics* 6 (1990): 701–7.

2. Peter Galison and Lorraine Daston, "Scientific Coordination as Ethos and Epistemology," in *Instruments in Art and Science: On the Architectonics of Cultural Boundaries in the 17th Century*, ed. Helmar Schramm, Ludger Schwarte, and Jan Lazardzig (Berlin: Walter de Gruyter, 2008), 296–333.

3. G. I. Bell and W. Goad to R. Ewald, Dec 4 1980, Water Goad Papers, American Philosophical Society, Philadelphia, Ms. Coll. 114, series III (APS Archives hereafter).

4. Stevens, *Life out of Sequence*, 139.

5. From an ENTREZ search on www.ncbi.nlm.nih.gov (accessed Dec 5 2015), using "1990:2017[date] AND species[rank] AND cellular organisms[subtree] NOT unspecified[prop] NOT uncultured[prop]." I thank Dennis Benson for proving this method to estimate the number of species in GenBank.

6. In his criticism of my previous work, Stevens claims that biological databases "cannot be thought of as just *collections*" but should thought of as "tools." But that was precisely my argument all along: "Databases, like earlier natural history collections, are not mere repositories; they are tools for producing knowledge." Bruno J. Strasser, "The Experimenter's Museum: GenBank, Natural History, and the Moral Economies of Biomedicine," *Isis* 102, no. 1 (2011): 63; Stevens, *Life out of Sequence*, 137.

7. On the early modern "information overload," see Daniel Rosenberg, "Early Modern Information Overload," *Journal of the History of Ideas* 64, no. 1 (2003): 1–9; Brian W. Ogilvie, "The Many Books of Nature: Renaissance Naturalists and Information Overload," ibid., 29–40; Ogilvie, *Science of Describing*.

8. Collections were key tools not only for systematics, but also for studies in anatomy and evolution, for example. The case of comparative anatomy is particularly illuminating; see for example Richard W. Burkhardt Jr., "The Leopard in the Garden: Life in Close Quarters at the Museum D'Histoire Naturelle," *Isis* 98, no. 4 (Dec 2007): 675–94, and Toby A. Appel, *The Cuvier-Geoffroy Debate: French Biology in the Decades before Darwin* (New York: Oxford University Press, 1987).

9. On early modern collections, see Findlen, *Possessing Nature*.

10. M. O. Dayhoff to C. Berkley, Feb 27 1967, Archives of the National Biomedical Research Foundation, Georgetown, Washington, DC (NBRF Archives hereafter). The archives are currently unsorted and being processed at the National Library of Medicine; no further location information can be provided.

11. Margaret O. Dayhoff, *Atlas of Protein Sequence and Structure* (Silver Spring, MD: National Biomedical Research Foundation, 1969).

12. RNA sequences could first be determined experimentally in 1965, although the process was slow. For the first RNA sequence, see Robert W. Holley et al., "Structure of a Ribonucleic Acid," *Science* 147 (Mar 19 1965): 1462–65.

13. Frederick Sanger and Alan R. Coulson, "A Rapid Method for Determining Sequences in DNA by Primed Synthesis with DNA Polymerase," *Journal of Molecular Biology* 94, no. 3 (May 25 1975): 441–48; Frederick Sanger, Steve Nicklen, and Alan R. Coulson, "DNA Sequencing with Chain-Terminating Inhibitors," *Proceedings of the National Academy of Sciences* 74, no. 12 (Dec 1977): 5463–67, and Allan M. Maxam and Walter Gilbert, "A New Method for Sequencing DNA," ibid., no. 2 (Feb 1977): 560–64. On Sanger's sequencing methods, see Miguel Garcia-Sancho, "A New Insight into Sanger's Development of Sequencing: From Proteins to DNA, 1943–1977," *Journal of the History of Biology* 43 (2010): 265–323.

14. At the meeting, the "increasing rate at which nucleic acid sequence information is becoming available" was cited as the first reason for the need to create a nucleic acid data bank; C. W. Anderson to H. Lewis, Nov 14 1980, Appendix II, NBRF Archives.

15. "Sequences Add Up," *Nature* 297 (1982): 96.

16. M. O. Dayhoff to D. de Solla Price, Sep 17 1980; D. de Solla Price to M. O. Dayhoff, Sep 11 1980, NBRF Archives.

17. Theodore Friedmann, Russell F. Doolittle, and Gernot Walter, "Amino-Acid Sequence Homology between Polyoma and SV40 Tumor Antigens Deduced from Nucleotide-Sequences," *Nature* 274, no. 5668 (1978): 291–93. Through the DNA-hybridization method, DNA sequences could indirectly be compared before sequencing techniques were available.

18. Morange, *History of Molecular Biology*, chap. 17.

19. Thomas R. Gingeras and Richard J. Roberts, "Steps toward Computer Analysis of Nucleotide Sequences," *Science* 209, no. 4463 (Sep 19 1980): 1322–28.

20. C. W. Anderson to M. O. Dayhoff, Jan 9 1979, NBRF Archives. The meeting also had a more local agenda, namely, to assess the possibility of establishing a "centralized computer facility" at the Rockefeller University to collect and analyze nucleic acid sequences. See "Report to the National Science Foundation," attached to C. W. Anderson to H. Lewis, Nov 14 1980, NBRF Archives; interview with Norton Zinder, Rockefeller University, Feb 10 2006.

21. C. W. Anderson to M. O. Dayhoff, Jan 9 1979, NBRF Archives. The terms "data base," "data bank," and "data library" were often used interchangeably by the historical actors.

22. C. W. Anderson, "Report to the National Science Foundation," Nov 14 1980, Appendix II, NBRF Archives. Participants included C. W. Anderson, H. Bilofsky, F. Blattner, M. Billeter, M. O. Dayhoff, G. Edelman, B. Erickson, R. J. Feldmann, W. Fitch, P. Friedland, T. Gingeras, J. Hahn, J. S. Haemer, C. Hutchinson, E. Kabat, L. Kedes, O. Kennard, L. Korn, J. Lederberg, C. Levinthal, H. Lewis, J. Maizel, A. M. Maxam, J. Milazzo, J. Pasta, G. Pieczenik, C. Queen, R. J. Roberts, T. Smith, R. Sommer, C. Squires, R. Staden, J. Vournakis, M. Waterman, and S. M. Weissman.

23. Paul A. Castleman et al., "The Implementation of the PROPHET System," *AFIPS Conference Proceedings* 43 (Jan–Mar 1974): 457–68.

24. Michael S. Waterman, *Skiing the Sun* (2007). Their most notable contribution would be an algorithm for local sequence alignment: Temple F. Smith and Michael S. Waterman, "Identification of Common Molecular Subsequences," *Journal of Molecular Biology* 147, no. 1 (1981): 195–97. For their collaborations prior to the Rockefeller meeting, see Michael S. Waterman et al., "Additive Evolutionary Trees," *Journal of Theoretical Biology* 64, no. 2 (Jan 21 1977): 199–213, and Michael S. Waterman and Temple F. Smith, "On the Similarity of Dendrograms," *Journal of Theoretical Biology* 73, no. 4 (Aug 21 1978): 789–800. In 1981, the laboratory was renamed Los Alamos National Laboratory (LANL). On sequence comparisons, see Hallam Stevens, "Coding Sequences: A History of Sequence Comparison Algorithms as a Scientific Instrument," *Perspectives on Science* 19, no. 3 (2011): 263–99.

25. C. W. Anderson, "Report to the National Science Foundation," Nov 14 1980, p. 2, NBRF Archives.

26. Ibid., p. 3.

27. Ibid., p. 2.

28. "Report to the National Science Foundation," attached to C. W. Anderson to H. Lewis, Nov 14 1980, p. 2, NBRF Archives.

29. Ibid., p. 2 and Appendix II, p. 5.

30. Olga Kennard, "Notes on the Preliminary Report and Recommendations of the Workshop on Computer Facilities for the Analysis of Protein and Nucleic Acid Sequence Information," p. 1, attached to C. W. Anderson to H. Lewis, Nov 14 1980, NBRF Archives.

31. Ibid., p. 1.

32. The minutes of the Rockefeller meeting were not released until November 1980, i.e., almost two years after the meeting took place. Temple F. Smith has argued that this delay prevented the NIH from knowing about the conclusions of the Rockefeller meeting, and perhaps delayed the development of the database project within the NIH. Smith, "History of the Genetic Sequence Databases." However, it seems unlikely that the NIH, and the NIGMS in particular, were unaware of the conclusions of the meeting before the minutes were released. Indeed, a member of the NIH's DRR was present, and a number of participants, including Richard J. Roberts, were in close contact with the directorship of the NIGMS and most likely made the conclusions known. Interview with Ruth L. Kirschstein, Bethesda, Feb 22 2006.

33. In 1981, the laboratory was renamed Los Alamos National Laboratory (LANL). Stanford University was considered an even more promising candidate than Los Alamos. However, for the sake of brevity, it will not be discussed here.

34. M. O. Dayhoff to C. W. Anderson, Jan 25 1979, NBRF Archives.

35. Ibid.

36. W. Goad to D. Kerr, Sep 19 1979, APS Archives.

37. Report, "Group T-10: Theoretical Biology and Biophysics," 1977, APS Archives.

38. Peter J. Westwick, *The National Labs: Science in an American System, 1947–1974* (Cambridge: Harvard University Press, 2003).

39. On the postwar "biophysics bubble," see Nicolas Rasmussen, "The Mid-Century Biophysics Bubble: Hiroshima and the Biological Revolution in America, Revisited," *History of Science* 35 (1997): 245–93.

40. On Goad, see Hallam Stevens, "From Bomb to Bank: Walter Goad and the Introduction of Computers into Biology," in *Outsider Scientists: Routes to Innovation in Biology*, ed. Oren Harmon and Michael Dietrich (Chicago: University of Chicago Press, 2013), 128–44.

41. Report, "Group T-10: Theoretical Biology and Biophysics," 1977, APS Archives.

42. Walter B. Goad's résumé is attached to BBN, "Establishment of a Nucleic Acid Sequence Data Bank," March 1982, NBRF Archives.

43. Walter B. Goad, "GenBank," *Los Alamos Science* 9 (1983): 52–63.

44. Wolfe, "Germs in Space"; James E. Strick, "Creating a Cosmic Discipline: The Crystallization and Consolidation of Exobiology, 1957–1973," *Journal of the History of Biology* 37, no. 1 (2004): 131–80.

45. Timothy Lenoir, "Shaping Biomedicine as an Information Science," in *Proceedings of the 1998 Conference on the History and Heritage of Science Information Systems*, ed. Mary Ellen Bowden, Trudi Bellardo Hahn, and Robert V. Williams (Medford, NJ: Information Today, 1999), 27–45.

46. Robert K. Lindsay, *Applications of Artificial Intelligence for Organic Chemistry: The Dendral Project* (New York: McGraw-Hill, 1980); November, *"Digitizing Life,"* chap. 7.

47. For a contemporary overview of computer networks and their uses for science, see Allen Newell and Robert F. Sproull, "Computer-Networks: Prospects for Scientists," *Science* 215, no. 4534 (1982): 843–52.

48. Lenoir, "Shaping Biomedicine," 34.

49. Peter Friedland and Laurence H. Kedes, "Discovering the Secrets of DNA," *Computer* 18, no. 11 (1985): 49–69.

50. P. Friedland to T. Smith, Sep 7 1989, Temple Smith personal papers.

51. Douglas L. Brutlag et al., "SEQ: A Nucleotide Sequence Analysis and Recombination System," *Nucleic Acids Research* 10, no. 1 (Jan 11 1982): 279–94. The program was eventually known as SEQ.

52. Lenoir, "Shaping Biomedicine," 34.

53. Edward A. Feigenbaum and Edward H. Shortliffe, "Sumex Annual Report—Year 09" (1982), available from Profiles in Science (profiles.nlm.nih.gov, accessed Jun 1 2008), p. 70.

54. Castleman et al., "Implementation of the PROPHET System." On BBN's role in the development of biomedical informatics, see Paul Castleman, "Medical Application of Computers at BBN," *IEEE Annals of the History of Computing* 28, no. 1 (Jan–Mar 2006): 6–16.

55. The PROPHET system required more sophisticated computer terminals than SUMEX, and thus reached a much smaller community than the Stanford computer resource. In 1974, the PROPHET system had only approximately thirty users. Castleman et al., "Implementation of the Prophet System," 457, whereas SUMEX had several hundred.

56. Carl W. Anderson, "Report to the National Science Foundation," Nov 14 1980, Appendix II, NBRF Archives.

57. W. Goad to R. Roberts, Jan 14 1981, APS Archives.

58. G. I. Bell to W. Goad, Jan 21 1980, APS Archives.

59. "EMBL Workshop on Computing and DNA Sequences, 24th and 25th April 1980," Archives of the European Molecular Biology Institute, Hinxton, UK (EBI Archives hereafter).

60. F. Murray to M. O. Dayhoff, Jan 19 1980, NBRF Archives.

61. F. R. Blattner, "Report on EMBL Workshop on Computing & DNA Sequences," Jun 24 1980, EBI Archives.

62. "Banking DNA Sequences," *Nature* 285 (1980).

63. The editorial overlooked the existence of Dayhoff's database, and she wrote a letter to *Nature* to correct this omission. Margaret O. Dayhoff et al., "Banking DNA Sequences," *Nature* 286 (1980): 326.

64. K. Murray to W. Goad, Jun 12 1980, APS Archives.

65. R. Grantham, "EMBL Workshop on Computing and Nucleic Acid Sequences, 27–28 April 1981," EBI Archives, folder 3.

66. John Krige, "The Birth of EMBO and the Difficult Road to EMBL," *Studies in the History and Philosophy of Biological and Biomedical Sciences* 33 (2002): 547–64.

67. A second EMBL workshop took place in Schönau, on "Computing and Nucleic Acid Sequences," Apr 27–28 1981. For a report, see John Fox, "Second EMBL Workshop,"

Apr 1981, EBI Archives, folder 3. On the career of the early EMBL database managers, see Miguel Garcia-Sancho, "From Metaphor to Practices: The Introduction of 'Information Engineers' into the First DNA Sequence Database," *History and Philosophy of the Life Sciences* 33, no. 1 (2011): 71–104.

68. Ken Murray, "EMBL Nucleotide Sequence Data Library," *Nucleotide Sequence Data Library News* 1, March/April 1982, EBI Archives, folder 6.

69. Greg Hamm, "EMBL Sequence Data Library: Status and Plan, Memo 17," Dec 6 1982, EBI Archives, folder 6.

70. A. M. Maxam to G. Hamm, May 4 1982, EBI Archives, folder 6.

71. M. O. Dayhoff to E. Jordan, Aug 13 1980, NBRF Archives.

72. M. O. Dayhoff et al., "Now available over 150 Kilobases," Aug 15 1980, NBRF Archives.

73. M. O. Dayhoff to D. DeVincenzi, Aug 20 1980, NBRF Archives.

74. M. O. Dayhoff to various pharmaceutical and biotech companies, Aug-Dec 1980, NBRF Archives.

75. Margaret O. Dayhoff et al., "Nucleic Acid Sequence Bank," *Science* 209, no. 4462 (Sep 12 1980): 1182.

76. R. Roberts to K. Stüber, Dec 10 1980, APS Archives.

77. M. O. Dayhoff to E. Jordan, Oct 23 1980, NBRF Archives.

78. T. Smith to NBRF, Apr 30 1982, NBRF Archives.

79. M. O. Dayhoff, "Progress report 08710; 1.2.1981–31.7.1982," Sep 15 1980.

80. Ibid.

81. On IntelliGenetics, see Lenoir, "Shaping Biomedicine."

82. L. Kedes to M. O. Dayhoff, Sep 18 1980, NBRF Archives.

83. Ibid. A similar arrangement was being made with a Japanese partner; see M. O. Dayhoff to K. Koike, Nov 7 1980, NBRF Archives.

84. M. O. Dayhoff to D. M. Moore, Sep 25 1981, NBRF Archives.

85. M. O. Dayhoff to D. A. Jackson (Genex), Nov 20 1981, NBRF Archives.

86. M. O. Dayhoff to L. H. Kedes, Jun 12 1981, NBRF Archives.

87. W. Barker to H. Aaslestad, Feb 23 1981, and M. O. Dayhoff to G. Milne, May 18 1981, NBRF Archives.

88. R. L. Kirschstein to M. O. Dayhoff, Jul 15 1981, NBRF Archives.

89. M. O. Dayhoff to R. L. Kirschstein, Aug 7 1981, NBRF Archives.

90. G. Hamm to K. Murray, Nov 20 1980, EBI Archives.

91. T. T. Puck to C. A. Thomas, Jul 16 1979, APS Archives.

92. At the second general meeting on the development of a DNA database in June 1980, Goad began to present "the big computers at Los Alamos" as a "solution." F. R. Blattner, "Report on EMBL Workshop on Computing & DNA," Jun 24 1980, EBI Archives.

93. *Los Alamos Sequence Library*, Mar 1982, p. 3. I thank Christian Burks for sharing this document with me.

94. S. Simon to T10, May 9 1980, APS Archives.

95. M. O. Dayhoff to D. DeVincenzi, Aug 20 1980, NBRF Archives.

96. T. Smith to M. O. Dayhoff, Sep 17 1980, NBRF Archives.

97. W. Goad to M. O. Dayhoff, Jan 9 1981, NBRF Archives.

98. M. O. Dayhoff to W. Goad, Aug 7 1981, NBRF Archives.

99. Goad obtained two important European collections: those of Kurt Stüber in Cologne and Richard Grantham in Lyon. W. Goad to K. Stüber, Jan 15 1981; W. Goad to R. Roberts, Jan 14 1981, APS Archives.

100. G. Hamm to M. O. Dayhoff, Mar 17 1981, NBRF Archives.

101. Schiebinger, *Plants and Empire*, 57.

102. "Minutes Workshop on the Need for a Nucleic Acid Sequence Data Bank, Jul 14–15 1980, Bethesda," APS Archives.

103. Possibly reflecting the growing competition between the American and the European projects, only American scientists were invited to the NIH meeting, whereas representatives from both continents had attended the previous EMBL meeting. EMBL explicitly asked to be informed of the NIH meeting's conclusions. K. Murray to E. Jordan, Jul 3 1980, NBRF Archives.

104. E. Jordan to M. O. Dayhoff, Jun 17 1980, NBRF Archives.

105. L. H. Kedes, collective email, Jul 15 1980, APS Archives.

106. M. Cassman and E. Jordan, "Minutes, Workshop on the Need for a Nucleic Acid Sequence Data Bank," Jul 22 1980, NBRF Archives.

107. "Minutes Workshop on the Need for a Nucleic Acid Sequence Data Bank, July 14–15 1980, Bethesda", APS Archives.

108. L. H. Kedes, collective email, Jul 15 1980, APS Archives.

109. Ibid.

110. M. Cassman and E. Jordan, Jul 21 1980 Minutes, Workshop on the Need for a Nucleic Acid Sequence Data Bank, NBRF Archives.

111. L. H. Kedes, collective email, Jul 15 1980, APS Archives.

112. E. Jordan to W. Raub, Jul 25 1980, GenBank Archives at the National Center for Biotechnology Information Archives, Bethesda, MD (NCBI Archives hereafter).

113. Westwick, *National Labs*, chap. 8.

114. S. W. Thornton, "RFP," Dec 1 1981, NBRF Archives.

115. Ibid.

116. Smith, "History of the Genetic Sequence Databases."

117. Interview with Ruth L. Kirschstein, Bethesda, Feb 22 2006.

118. Electronic message posted on SUMEX-AIM, Sep10 1980, NCBI Archives. Elke Jordan circulated a copy of this message within the NIH, most likely to gather support. E. Jordan to W. F. Raub et al., Sep 11 1980.

119. Sanger, "Sequences," 1.

120. M. O. Dayhoff, "Technical Proposal. Establishment of a nucleic acid sequence data bank," Mar 1 1982, NBRF Archives, p. 12.

121. Ibid., p. 18. Dayhoff planned to explore the remaining 2 percent by manually searching through bibliographic indexes.

122. Wise, *Values of Precision*.

123. BBN, "Establishment of a Nucleic Acid Sequence Data Bank," Mar 1982, NBRF Archives, p. 16.

124. Ibid., pp. 26, 87.

125. Ibid., p. 26.

126. NIGMS, "Second Staff Evaluation of Proposal from BBN," Jun 21 1982, NBRF Archives, p. 1.

127. Ibid., p. 1 When the NIH reviewers asked Dayhoff if she planned to collaborate with journal editors, she replied that she supported the mandatory submission scheme but that she would not make it a priority.

128. K. Murray, "Summary of telephone conversation with Margaret Dayhoff," Jun 8 1982, Jun 9 1982, EBI Archives, folder 6.

129. Hagstrom, "Gift Giving."

130. Lewis Wolpert, *The Unnatural Nature of Science* (London: Faber and Faber, 1993).

131. Eck and Dayhoff, *Atlas (1966)*, xiv.

132. Commission on Journals, "Deposition."

133. Roger Lewin, "Long-Awaited Decision on DNA Database," *Science* 217 (1982): 817–18. On how these concerns led researchers to withhold crystallographic data from the Protein Data Bank, see Marcia Barinaga, "The Missing Crystallography Data," ibid., 245, no. 4923 (Sep 15 1989): 1179–81.

134. Richard Rhodes, *Dark Sun: The Making of the Hydrogen Bomb* (New York: Simon & Schuster, 1995).

135. "Report, Group T-10: Theoretic Biology and Biophysics," 1977, APS Archives.

136. Stuart W. Leslie, *The Cold War and American Science: The Military-Industrial-Academic Complex at MIT and Stanford* (New York: Columbia University Press, 1993), and Moore, *Disrupting Science*.

137. Lewin, "Long-Awaited Decision," 818.

138. Ibid.

139. Interview with Richard J. Roberts, Jul 2 2008. On the Asilomar conference, see Wright, *Molecular Politics*.

140. BBN, "Establishment of a Nucleic Acid Sequence Data Bank," Mar 1982, NBRF Archives.

141. BBN to NIH, May 7 1982, APS Archives.

142. Ibid. However, calculations could be made on the Crays and then transferred to an accessible computer by Los Alamos personnel, as Bilofsky and Goad tried to explain.

143. W. Goad to P. Carruthers and M. Slaughter, Nov 3 1981, APS Archives.

144. Ibid.

145. Interview with Richard J. Roberts, Jul 2 2008.

146. BBN, "Establishment of a Nucleic Acid Sequence Data Bank," Mar 1982, NBRF Archives, p. 24.

147. Using the Needleman-Wunsch algorithms. See table 1 in Rodney A. Jue, Neal W. Woodbury, and Russell F. Doolittle, "Sequence Homologies among E. Coli Ribosomal Proteins: Evidence for Evolutionarily Related Groupings and Internal Duplications," *Journal of Molecular Evolution* 15, no. 2 (1980): 129–48.

148. Daniel J. Kevles, "*Diamond v. Chakrabarty* and Beyond: The Political Economy of Patenting Life," in *Private Science: Biotechnology and the Rise of the Molecular Sciences*, ed. Arnold Thackray (Philadelphia: University of Pennsylvania Press, 1998), 65–79 ; Sally Smith Hughes, "Making Dollars out of DNA: The First Major Patent in Biotechnology and

the Commercialization of Molecular Biology, 1974–1980," *Isis* 92, no. 3 (2001): 541–75; Kara Swanson, "Biotech in Court: A Legal Lesson on the Unity of Science," *Social Studies of Science* 37, no. 3 (2007): 357–84.

149. See for example David Dickson, "Stanford Ready to Fight for Patent," *Nature* 292, no. 5824 (Aug 13 1981): 573.

150. BBN to NIH, May 7 1982, APS Archives.

151. R. Ledley and M. O. Dayhoff to S. Thornton, May 7 1982, NBRF Archives.

152. J. Seeber to R. Ledley, Mar 5 1982, NBRF Archives.

153. BBN to NIH, May 7 1982, APS Archives.

154. Ibid.

155. Ibid.

156. R. Ledley and M. O. Dayhoff to S. Thornton, May 7 1982, NBRF Archives.

157. M. O. Dayhoff, "Replies to information requested by May 10, 1982," n.d. [1982], NBRF Archives, p. 1.

158. BBN to NIH, May 7 1982, APS Archives.

159. W. Goad to P. Carruthers and M. Slaughter, Nov 3 1981, APS Archives.

160. NIGMS, "Second Staff Evaluation of Proposal from NBRF," Jun 21 1982, NBRF Archives, p. 1. Elsewhere, R. Roberts reported that there was "widespread concern about the level of access that has been provided" by Dayhoff to the data she collected. John Fox, "Second EMBL Workshop," Apr 1981, EBI Archives, folder 3.

161. M. O. Dayhoff, "Technical Proposal. Establishment of a nucleic acid sequence data bank," Mar 1 1982, NBRF Archives. This computer was perhaps still "modern" but had already been introduced four years earlier.

162. BBN, "Establishment of a Nucleic Acid Sequence Data Bank," Mar 1982, NBRF Archives, p. 117. The extension of ARPANET ca. 1982 is visualized in Newell and Sproull, "Computer-Networks."

163. BBN, "Establishment of a Nucleic Acid Sequence Data Bank," Mar 1982, NBRF Archives, p. 24.

164. Ibid., p. 6.

165. Ibid., p. 27.

166. R. Ledley and M. O. Dayhoff to S. Thornton, May 7 1982, NBRF Archives.

167. Douglas Brutlag, "Meeting at DOE," Jan 9 1982, NCBI Archives.

168. R. Ledley and M. O. Dayhoff to S. Thornton, May 7 1982, NBRF Archives.

169. Lewin, "Long-Awaited Decision," 817.

170. Joshua Lederberg, "Digital Communications and the Conduct of Science; New Literacy," *Proceedings of the IEEE* 66, no. 11 (1978): 1314–19; see also Elliott C. Levinthal et al., "When Computers 'Talk' to Computers," *Industrial Research* 17, no. 12 (1975): 35–42, and Newell and Sproull, "Computer-Networks."

171. Lederberg, "Digital Communications," 1314.

172. Robert S. Ledley, "Chronology of NBRF/BBN Protest," Jan 14 1983; Department of Health and Human Services, "Negotiated Contract," Jun 30 1982, NBRF Archives.

173. Lewin, "Long-Awaited Decision"; Elke Jordan, "DNA Database," *Science* 218 (1982): 108.

174. Jordan, "DNA Database."

175. J. Thornton to C. Fretts, Aug 16 1982, NBRF Archives.

176. Ibid.

177. Interview with Winona Barker, Georgetown, Sep 1, 2005.

178. Stevens, *Life out of Sequence*, 138.

179. Leonelli, *Data-centric Biology*.

180. For an excellent example of the importance of collections for systematic work, see Johnson, *Ordering Life*.

181. Robert Teitelman, *Gene Dreams: Wall Street, Academia, and the Rise of Biotechnology* (New York: Basic Books, 1989); Vettel, *Biotech*; Kaushik Sunder Rajan, *Biocapital: The Constitution of Postgenomic Life* (Durham: Duke University Press, 2006); Hughes, *Genentech*; Rasmussen, *Gene Jockeys*; and specifically on the Chakrabarty case, Kevles, "*Diamond v. Chakrabarty* and Beyond."

182. Interview with Ruth L. Kirschstein, Bethesda, Feb 22 2006.

183. Richard J. Roberts, "The Early Days of Bioinformatics Publishing," *Bioinformatics* 16, no. 1 (Jan 2000): 2–4.

184. Bevin P. Engelward and Richard J. Roberts, "Open Access to Research Is in the Public Interest," *PLOS Biology* 5, no. 2 (2007): e48.

185. Turner, *From Counterculture to Cyberculture*.

186. A number of authors have connected open source software and data sharing, including Janet Hope, *Biobazaar: The Open Source Revolution and Biotechnology* (Cambridge: Harvard University Press, 2008); Alessandro Delfanti, *Biohackers: The Politics of Open Science* (London: Pluto Press, 2013); and Hallam Stevens, "Networking Biology: The Origins of Sequence-Sharing Practices in Genomics," *Technology and Culture* 56, no. 4 (Oct 2015): 839–67 (on the role of computer networking technologies and practices in the 1970s for sequence data sharing in the 1980s) and Stevens, "The Politics of Sequence: Data Sharing and the Open Source Software Movement," *Information & Culture: A Journal of History* 50, no. 4 (2015): 465–503 (on the open source movement and sequence data sharing in the genomics projects of the 1990s).

Chapter Six

1. Christine Carrico, "Dr. Allan Maxam," Oct 13 1982, MIT Archives; and Elvin Kabat's later assessment, "Part of the problem [with data collection] lies in the choice of Los Alamos as a site for GenBank," Elvin A. Kabat, "The Problem with GenBank," in *Biomolecular Data: A Resource in Transition*, ed. Rita Colwell (Oxford: Oxford University Press, 1989), 127–28.

2. Philip L. Altman et al., *Guidelines for Development of Biology Data Banks* (Bethesda, MD: Life Sciences Research Office, Federation of American Societies for Experimental Biology, 1981).

3. M. O. Dayhoff to Philip Abelson, May 13 1982, NBRF Archives.

4. GenBank advisors meeting, minutes, Nov 6 1987, EBI Archives.

5. Goad, "GenBank," 57.

6. James Lauterberger, "Report," Jan 4 1983, MIT Archives.

7. Wayne P. Rindone et al., "GenBank Tm—the Genetic Sequence Data-Bank," *DNA: A Journal of Molecular and Cellular Biology* 2, no. 2 (1983): 173–73.

8. GenBank advisors meeting, minutes, Oct 1 1985, EBI Archives.

9. "GenBank advisory meeting," Oct 31 1985, MIT Archives. See also Christian Burks, James Fickett, and Walter Goad, "GenBank," *Science* 235 (1987): 267–68.

10. "GenBank advisory meeting," Oct 31 1985, MIT Archives.

11. Roger Lewin, "Proposal to Sequence the Human Genome Stirs Debate," *Science* 232, no. 4758 (Jun 27 1986): 1598–1600.

12. Douglas Brutlag, "Meeting at DOE," Jan 9 1982, NCBI Archives.

13. A. Maxam in "GenBank Advisors Meeting," Nov 6 1987; "GenBank advisory meeting," Oct 31 1985, MIT Archives.

14. Lewin, "Proposal," 1599.

15. IntelliGenetics advertisement in *Science* 229, no. 4719 (Sep 20 1985): 1297, or Beckman advertisement in ibid., 229, no. 4714 (Aug 16 1985): 590.

16. GenBank Advisory Meeting, Oct 31 1985, MIT Archives.

17. Elvin A. Kabat, in "GenBank Advisory Meeting," Oct 31 1985, p. 5, MIT Archives.

18. R. J. Roberts to J. Cassatt, Jan 5 1987, MIT Archives.

19. James Lauterberger, "Report," Jan 4 1983, MIT Archives.

20. "Meeting of GenBank Advisors," Jan 25 1983, MIT Archives.

21. "GenBank Advisors Meeting," Nov 6 1987, MIT Archives.

22. Ibid.

23. Greg Hamm and Ken Murray to Journal Editors, Mar 18 1982, EBI Archives.

24. M. Bruber to K. Murray and G. Hamm, Aug 25 1982, EBI Archives.

25. R. Grantham to Giorgio Bernardi, May 26 1982, EBI Archives.

26. Jack Franklin to Patricia Kahn, Sep 15 1987, EBI Archives.

27. Graham Cameron, "EMBL Nucleotide Sequence Data Library," Sep 1984, EBI Archives, folder 6.

28. L. Philipson to EMBL staff, Feb 1 1985, EBI Archives, folder 3.

29. The Editors, "Instruction to Authors," *Nucleic Acids Research* 11, no. 24 (1983).

30. Interview with Graham Cameron, Jul 30 2007.

31. P. Kahn to members of the Data Library Advisory Board, Oct 29 1986, EBI Archives, folder 3.

32. P. Kahn to J. Franklin, Sep 4 1987, EBI Archives, folder 5.

33. Ibid.

34. Ibid.

35. R. J. Roberts to J. Cassatt, Jan 5 1987, MIT Archives.

36. Richard T. Walker, "A Method for the Rapid and Accurate Deposition of Nucleic Acid Sequence Data in an Acceptably-Annotated Form," in *Biomolecular Data: A Resource in Transition*, ed. Rita Colwell (Oxford: Oxford University Press, 1989), 45–51.

37. P. Kahn to J. Franklin, Sep 4 1987, EBI Archives, folder 5.

38. The Editors, "Deposition of Nucleotide Sequence Data in the Data Banks," *Nucleic Acids Research* 15, no. 18 (1987).

39. Patricia Kahn and David Hazledine, "NAR's New Requirement for Data Submission to the EMBL Data Library: Information for Authors," *Nucleic Acids Research* 16, no. 10 (May 25 1988): I–IV.

40. "GenBank Advisory Meeting," Apr 11–12 1988, National Library of Medicine Archives, Bethesda, MD (hereafter NLM Archives).

41. Walker, "Method," 48.

42. Igor B. Dawid, "Editorial Submission of Sequences," *Proceedings of the National Academy of Sciences* 86 (1989): 407.

43. John Maddox, "Making Authors Toe the Line," *Nature* 342 (1989): 855.

44. Baldwin, *Making Nature*, chap. 8 and conclusion.

45. "GenBank advisory meeting," Nov 15–16 1990, NLM Archives.

46. Walker, "Method."

47. Ibid., 49.

48. Christine McGourty, "Human Genome Project: Dealing with the Data," *Nature* 342, no. 6246 (Nov 9 1989): 108.

49. L. Philipson to K. Bauer, Nov 27 1986, folder 5, EBI Archives.

50. Michael J. Cinkosky et al., "Electronic Data Publishing and GenBank," *Science* 252, no. 5010 (May 31 1991): 1273–77.

51. Ibid.

52. Philip Bourne, "Will a Biological Database Be Different from a Biological Journal?," *PLoS Computational Biology* 1, no. 3 (Aug 2005): 179–81; Graham Cameron, "The Impact of Electronic Publishing on the Academic Community," in *Electronic Databases and the Scientific Record* (London: Portland, 1997).

53. James W. Fickett, "Recognition of Protein Coding Regions in DNA Sequences," *Nucleic Acids Research* 10, no. 17 (Sep 11 1982): 5303–18.

54. For an overview, see Laurence Jay Korn and Cary Queen, "Analysis of Biological Sequences on Small Computers," *DNA* 3, no. 6 (1984): 421–36.

55. For a good overview, see David W. Mount, "Computer Analysis of Sequence, Structure and Function of Biological Macromolecules," *BioTechniques* 3, no. 2 (1985): 102–12; Gunnar von Heijne, *Sequence Analysis in Molecular Biology: Treasure Trove or Trivial Pursuit* (San Diego: Academic Press, 1987).

56. Christian Burks, "Biotechnology and the Human Genome Innovations and Impact," in *Basic Life Sciences 46*, ed. Avril D. Woodhead and Benjamin J. Barnhart (Boston: Springer, 1988), 51–56.

57. On the origins of the term, see Joel B. Hagen, "The Origins of Bioinformatics," *Nature Reviews* 1 (2000): 231–36.

58. Christian Burks et al., "The GenBank Nucleic Acid Sequence Database," *Computer Applications in the Biosciences* 1, no. 4 (Dec 1985): 225–33.

59. Morange, *History of Molecular Biology*, chap. 17; "GenBank Advisory Meeting," Jan 13 1984, NLM Archives. For an early overview of the envisioned uses of sequences databases, see Geoff G. Kneale and Martin J. Bishop, "Nucleic Acid and Protein Sequence Databases," *Computer Applications in the Biosciences* 1, no. 1 (1985): 11–17, and Russell F. Doolittle, *Of Urfs and Orfs: A Primer on How to Analyze Derived Amino Acid Sequences* (Mill Valley, CA: University Science Books, 1986).

60. On the origins of sequence alignment algorithms, see Stevens, "Coding Sequences."

61. Temple F. Smith and Christian Burks, "Searchng for Sequence Similarities," *Nature* 301 (1983): 194.

62. Stevens, "Coding Sequences."

63. Philip E. Doggett and Frederick R. Blattner, "Personal Access to Sequence Databases on Personal Computers," *Nucleic Acids Research* 14, no. 1 (Jan 10 1986): 611–19.

64. Michael Fortun, "Projecting Speed Genomics," in *The Practices of Human Genetics*, ed. M. Fortun and E. Mendelsohn (Dordrecht: Kluwer, 1999), 25–87.

65. IntelliGenetics, "GenBank Quarterly Report 1 October to 31 December 1991," MIT Archives.

66. Anne Harding, "Blast: How 90.000 Lines of Code Helped Spark the Bioinfomatics Explosion," *Scientist* 19, no. 16 (2005): 21–26.

67. Theodosius Dobzhansky, "Biology, Molecular and Organismic," *American Zoologist* 4 (Nov 1964): 449.

68. Aldhous, "Managing the Genome Data Deluge."

69. Jackey S. Andersen et al., ed., *Nucleotide Sequences, 1984: A Compilation from the GenBank and EMBL Data Libraries* (Oxford: IRL Press, 1984); Edwin J. Atencio, ed., *Nucleotide Sequences, 1986/1987: A Compilation from the GenBank and EMBL Data Libraries* (Orlando: Academic Press, 1987).

70. Michael D. Gordin, "Beilstein Unbound: The Pedagogical Unraveling of a Man and His *Handbuch*," in *Pedagogy and the Practice of Science: Historical and Contemporary Perspectives*, ed. David Kaiser (Cambridge: MIT Press, 2005), 11–39.

71. "GenBank advisory meeting," Jan 13 1984, NIH Archives.

72. Ibid.

73. Anthony H. Slocum, "Sequences On-Line," *Nature* 299 (1982): 482.

74. Howard S. Bilofsky et al., "The GenBank Genetic Sequence Databank," *Nucleic Acids Research* 14, no. 1 (Jan 10 1986): 1–4.

75. Roger Lewin, "National Networks for Molecular Biologists," *Science* 223, no. 4643 (Mar 30 1984): 1379–80.

76. Ibid., 1379.

77. Richard J. Roberts and Dieter Söll, Preface, *Nucleic Acids Research* 14, no. 1 (1986).

78. "BIONET NAC meeting," Mar 23 1987, MIT Archives.

79. James C. Cassatt and Jane L. Peterson, "GenBank Information," *Science* 238, no. 4831 (Nov 27 1987): 1215.

80. Richard J. Roberts and Dieter Söll, Preface, *Nucleic Acids Research* 12, no. 1 (1984).

81. Richard J. Roberts and Dieter Söll, Preface, *Nucleic Acids Research* 14, no. 1 (1986).

82. Cassatt and Peterson, "GenBank Information." On the appreciation of BIONET, see Roger Lewin, "National Networks for Molecular Biologists," *Science* 223, no. 4643 (Mar 30 1984): 1379–80. However, it was discontinued in 1989; see Jean L. Marx, "Bionet Bites the Dust," ibid., 245, no. 4914 (Jul 14 1989): 126.

83. Dennis Benson, David J. Lipman, and James Ostell, "GenBank," *Nucleic Acids Research* 21, no. 13 (Jul 1 1993): 2963–65.

84. Dennis Benson et al., "The National Center for Biotechnology Information," *Genomics* 6, no. 3 (Mar 1990): 389–91; Leslie Roberts, "NIH, DOE Battle for Custody of DNA Sequence Data," *Science* 262, no. 5133 (Oct 22 1993): 504–5.

85. Daniel J. Kevles, "Big Science and Big Politics in the United States: Reflections on the Death of the SSC and the Life of the Human Genome Project," *Historical Studies in the*

Physical and Biological Sciences 27 (1997): 269–97; Rasmussen, "Mid-Century Biophysics Bubble."

86. Stephen F. Altschul et al., "Basic Local Alignment Search Tool," *Journal of Molecular Biology* 215, no. 3 (Oct 5 1990): 403–10; Harding, "Blast."

87. M. Mitchell Waldrop, "On-Line Archives Let Biologists Interrogate the Genome," *Science* 269, no. 5229 (Sep 8 1995): 1356–58.

88. Nigel Williams, "Europe Opens Institute to Deal with Gene Data Deluge," *Science*, no. 5224 (Aug 4 1995): 630.

89. Ibid.

90. Alison Abbott, "Bioinformatics Institute Plans Public Database for Gene Expression Data," *Nature* 398, no. 6729 (Apr 22 1999): 646.

91. Alvis Brazma et al., "Arrayexpress: A Public Repository for Microarray Gene Expression Data at the EBI," *Nucleic Acids Research* 31, no. 1 (Jan 1 2003): 68–71.

92. Catherine A. Ball et al., "Submission of Microarray Data to Public Repositories," *PLoS Biology* 2, no. 9 (Sep 2004): E317.

93. Stephen T. Sherry, Minghong Ward, and Karl Sirotkin, "dbSNP-Database for Single Nucleotide Polymorphisms and Other Classes of Minor Genetic Variation," *Genome Research* 9, no. 8 (Aug 1999): 677–79.

94. Garland R. Marshall, Jeremy G. Vinter, and Hans-Dieter Holtje, "The Veil of Commercialism," *Journal of Computer-Aided Molecular Design* 2, no. 1 (Apr 1988): 1–2.

95. Jon Cohen, "The Culture of Credit," *Science* 268, no. 5218 (Jun 23 1995): 1706–11.

96. See for example John Sulston and Georgina Ferry, *The Common Thread: A Story of Science, Politics, Ethics and the Human Genome* (London: Bantam Press, 2002). On the worm community, see de Chadarevian, "Of Worms and Programmes," and for a critical view of "data sharing," Leonelli, *Data-centric Biology* , chaps. 1–2, and her notion of the "affective" dimension of data, p. 64. Hallam Stevens focuses on the heritage of the open source movement and computer network technologies, "Politics of Sequence," "Networking Biology."

97. Cohen, "Culture of Credit," 1706. Not that he was entirely opposed to the appropriation of scientific knowledge, as he had obtained the first patent on a genetically modified animal, to become the famed OncoMouse™. See Daniel J. Kevles, "Of Mice and Money: The Story of the World's First Animal Patent," *Daedalus* 131, no. 2 (Spr 2002): 78–88.

98. D. David Blumenthal et al., "Data Withholding in Genetics and the Other Life Sciences: Prevalences and Predictors," *Academic Medicine* 81, no. 2 (Feb 2006): 137–45.

99. Robert A. Weinberg, "Reflections on the Current State of Data and Reagent Exchange among Biomedical Researchers," in *Responsible Science: Ensuring the Integrity of the Research Process*, ed. Committee on Science Engineering and Public Policy (US), Panel on Scientific Responsibility and the Conduct of Research (Washington, DC: Nationl Academy Press, 1993), 66–78.

100. Marvin Zelen, "Review: Sharing Research Data," *Journal of the American Statistical Association* 83, no. 398 (1987): 685–86.

101. Maddox, "Making Good Databanks Better."

102. Robert A. Weinberg, "The Case against Gene Sequencing," *Scientist* 1, no. 25 (Nov 16 1987): 11.

103. Salvador E. Luria, "Human Genome Program," *Science* 246 (1989): 873–74.

104. McGourty, "Human Genome Project"; Leslie Roberts, "Watson versus Japan," *Science* 246, no. 4930 (Nov 3 1989): 576, 578.

105. McGourty, "Human Genome Project."

106. Patricia Kahn, "Human Genome Project: Sequencers Split over Data Release," *Science* 271, no. 5257 (Mar 29 1996): 1798–99.

107. Ibid. For the genesis of the Bermuda meeting and a detailed analysis, see Jorge L. Contreras, "Bermuda's Legacy: Policy, Patents, and the Design of the Genome Commons," *Minnesota Journal of Law, Science and Technology* 12, no. 1 (2011): 61–125.

108. Eliot Marshall, "Sequencers Call for Faster Data Release," *Science* 276, no. 5316 (May 23 1997): 1189–90.

109. Eliot Marshall and Elizabeth Pennisi, "NIH Launches the Final Push to Sequence the Genome," *Science* 272, no. 5259 (Apr 12 1996): 188–89.

110. Eliot Marshall, "Genome Researchers Take the Pledge," *Science*, no. 5261 (Apr 26): 477–78.

111. On the public consortium's view of the competition, see Sulston and Ferry, *Common Thread*, chap. 5; on Venter's view, see J. Craig Venter, *A Life Decoded: My Genome, My Life* (New York: Viking, 2007), chap. 11.

112. Francis Collins, "In the Crossfire: Collins on Genomes, Patents, and 'Rivalry.' Interview by Eliot Marshall, Elizabeth Pennisi, and Leslie Roberts," *Science* 287, no. 5462 (Mar 31 2000): 2396–98. The same argument was made for the mouse genome; see Eliot Marshall, "Genome Sequencing: Claim and Counterclaim on the Human Genome," ibid., 288, no. 5464 (Apr 14): 242–43.

113. Eliot Marshall, "Genome Sequencing: Clinton and Blair Back Rapid Release of Data," *Science* 287, no. 5460 (Mar 17 2000): 1903.

114. Eliot Marshall, "Human Genome: Rival Genome Sequencers Celebrate a Milestone Together," *Science* 288, no. 5475 (Jun 30 2000): 2294–95.

115. Eliot Marshall, "The Human Genome: Sharing the Glory, Not the Credit," *Science* 291, no. 5507 (Feb 16 2001): 1189–93.

116. Ibid.

117. Ibid., 1189.

118. Eliot Marshall, "Human Genome: Storm Erupts over Terms for Publishing Celera's Sequence," *Science* 290, no. 5499 (Dec 15 2000): 2042–43.

119. Eliot Marshall, "Data Sharing: DNA Sequencer Protests Being Scooped with His Own Data," *Science* 295, no. 5558 (Feb 15 2002): 1206–7.

120. Ibid., 1206. For a more recent example showing that the resentment against "research parasites" is still vivid, see Dan L. Longo and Jeffrey M. Drazen, "Data Sharing," *New England Journal of Medicine* 374, no. 3 (Jan 21 2016): 276–77.

121. Leslie Roberts, "Genome Research: A Tussle over the Rules for DNA Data Sharing," *Science* 298, no. 5597 (Nov 15 2002): 1312–13.

122. Søren Brunak et al., "Nucleotide Sequence Database Policies," *Science* 298, no. 5597 (Nov 15 2002): 1333.

123. Leslie Roberts, "Bioinformatics: Private Pact Ends the DNA Data War," *Science* 299, no. 5606 (Jan 24 2003): 487–89.

124. http://grants.nih.gov/grants/policy/data_sharing/data_sharing_guidance.htm (posted Mar 5 2003; accessed Apr 13 2015).

125. For a broad overview of open access, see Peter Suber, *Open Access* (Cambridge: MIT Press, 2012).

126. Glenn S. McGuigan, "Publishing Perils in Academe," *Journal of Business and Finance Librarianship* 10, no. 1 (2004): 13–26.

127. Harold Varmus, *The Art and Politics of Science* (New York: W. W. Norton, 2009).

128. Eliot Marshall, "Varmus Defends E-Biomed Proposal, Prepares to Push Ahead," *Science* 284, no. 5423 (Jun 25 1999): 2062–63; Marshall, "Varmus Circulates Proposal for NIH-Backed Online Venture," *Science* 284, no. 5415 (Apr 30 1999): 718. A preprint service for the biomedical literature was briefly available for the NIH in the 1960s, before being shot down by journal publishers; see Matthew Cobb, "The Prehistory of Biology Preprints: A Forgotten Experiment from the 1960s," *PeerJ Preprints* 5:e3174v1 (2017).

129. Richard J. Roberts, "Pubmed Central: The GenBank of the Published Literature," *Proceedings of the National Academy of Sciences* 98, no. 2 (Jan 16 2001): 381–82.

130. Richard J. Roberts et al., "Building a 'GenBank' of the Published Literature," *Science* 291, no. 5512 (Mar 23 2001): 2318–19.

131. The Editors, "Is a Government Archive the Best Option?," *Science* 291, no. 5512 (Mar 23 2001).

132. E. Sequeira, J. McEntyre, and D. Lipman, "Pubmed Central Decentralized," *Nature* 410, no. 6830 (Apr 12 2001): 740.

133. Simon Hodson, "Data-Sharing Culture Has Changed," *Research Information,* Dec 2009–Jan 2010, available at https://www.researchinformation.info/feature/data-sharing-culture-has-changed (accessed Jul 4 2017).

134. David Malakoff, "Scientific Publishing: Opening the Books on Open Access," *Science* 302, no. 5645 (Oct 24 2003): 550–54.

135. http://grants.nih.gov/grants/guide/notice-files/NOT-OD-05-022.html (accessed Apr 13 2015).

136. http://grants.nih.gov/grants/guide/notice-files/NOT-OD-08-033.html (accessed Aug 13 2015).

137. Jim Giles, "PR's 'Pit Bull' Takes on Open Access," *Nature* 445, no. 7126 (Jan 25 2007): 347; Owen Dyer, "Publishers Hire PR Heavyweight to Defend Themselves against Open Access," *British Medical Journal* 334, no. 7587 (Feb 3 2007): 227.

138. Interview with Patricia Kahn, Nyack, NY, Jul 22 2010.

139. On the specific function of model organism databases, see Leonelli, *Data-centric Biology*.

140. R. J. Roberts to Dr. [Tamara] Namaroff, undated [before Jun 1 2005], available from https://mx2.arl.org/Lists/SPARC-OAForum/Message/1977.html (accessed Nov 30 2015).

Conclusion

1. For a similarly broad perspective, see Lorraine Daston, *Science in the Archives: Pasts, Presents, Futures* (Chicago: University of Chicago Press, 2017).

2. David Sepkoski, "Towards 'a Natural History of Data': Evolving Practices and Epistemologies of Data in Paleontology, 1800–2000," *Journal of the History of Biology* 46, no. 3 (Aug 2013): 401–44.

3. I thank Rob Kohler for suggesting this useful comparison.

4. Stevens, *Life out of Sequence*, 148.

5. Endersby, *Guinea Pig's History*; and for a most insightful discussion about the epistemic role of model organisms, see Ankeny and Leonelli, "What's So Special."

6. http://www.nih.gov/science/models/ (accessed Aug 19 2015). On the effect of this list on model organism research publications, see M. R. Dietrich, R. A. Ankeny, and P. M. Chen, "Publication Trends in Model Organism Research," *Genetics* 198, no. 3 (Nov 2014): 787–94.

7. Jonathan M. W. Slack. "Emerging Market Organisms," *Science* 323 (2009): 1674–75.

8. On the history of the derogatory expression "stamp collecting," see Johnson, "Natural History as Stamp Collecting."

9. Mayr, *Growth of Biological Thought*, 30.

10. On Mayr's institutional agenda, see Milam, "Equally Wonderful Field."

11. Edward Osborne Wilson, *Naturalist* (Washington, DC: Island Books, 1994).

12. Ibid., 219. For the broader picture, see Rainger, Benson, and Maienschein, *American Development of Biology*; Keith R. Benson, Ronald Rainger, and Jane Maienschein, eds., *The Expansion of American Biology* (New Brunswick: Rutgers University Press, 1987); Allen, *Life Science*; Hagen, "Naturalists, Molecular Biology and the Challenge of Molecular Evolution"; Michael R Dietrich, "Paradox and Persuasion: Negotiating the Place of Molecular Evolution within Evolutionary Biology," *Journal of the History of Biology* 31 (1998): 85–111; Suárez-Díaz and Anaya-Muñoz, "History, Objectivity, and the Construction of Molecular Phylogenies"; Edna Suárez-Díaz, "Molecular Evolution: Concepts and the Origin of Disciplines," *Studies in History and Philosophy of Science Part C* 40, no. 1 (Mar 2009): 43–53.

13. Wilson, *Naturalist*, 219.

14. Auguste Comte, *Cours de philosophie positive* (Paris: Bachelier, 1830), 40th lesson.

15. Strasser and de Chadarevian, "The Comparative and the Exemplary."

16. The total figure results from adding the species numbers from the different departments, the vast majority of species coming from the entomology departments, which contain half a million species. For the British Museum (Natural History), http://www.nhm.ac.uk (accessed Oct 21 2015), and for the American Museum of Natural History, http://research.amnh.org/ (accessed Oct 21 2015).

17. Number using Taxonomy ENTREZ; see chap. 5.

18. Howard Dene, Morris Goodman, and Alejo E. Romero-Herrera, "The Amino Acid Sequence of Elephant (*Elephas Maximus*) Myoglobin and the Phylogeny of Proboscidea," *Proceedings of the Royal Society B—Biological Sciences* 207, no. 1166 (Feb 13 1980): 111–27.

19. Daniel A. Bisig et al., "Crystal-Structure of Asian Elephant (*Elephas Maximus*) Cyano-Metmyoglobin at 1.78-Å Resolution," *Journal of Biological Chemistry* 270, no. 35 (Sep 1 1995): 20754–62.

20. Ibid., 20762.

21. Denis Duboule et al., "A New Homeo-Box Is Present in Overlapping Cosmid Clones Which Define the Mouse Hox-1 Locus," *EMBO Journal* 5, no. 8 (Aug 1986): 1973–80.

22. Denis Duboule, "Temporal Colinearity and the Phylotypic Progression: A Basis for the Stability of a Vertebrate Bauplan and the Evolution of Morphologies through Heterochrony," *Development Supplement* (1994): 135–42.

23. Denis Duboule, "The Rise and Fall of Hox Gene Clusters," *Development* 134, no. 14 (Jul 2007): 2549–60.

24. Nicilas Di-Poi et al., "Changes in Hox Genes' Structure and Function during the Evolution of the Squamate Body Plan," *Nature* 464, no. 7285 (Mar 4 2010): 99–103.

25. Isabel Guerreiro et al., "Role of a Polymorphism in a Hox/Pax-Responsive Enhancer in the Evolution of the Vertebrate Spine," *Proceedings of the National Academy of Sciences* 110, no. 26 (Jun 25 2013): 10682–86.

26. Denis Duboule, "The Hox Complex: An Interview with Denis Duboule. Interviewed by Michael K. Richardson," *International Journal of Developmental Biology* 53, nos. 5–6 (2009): 717–23. More generally on the rise of Evo-Devo, see Manfred Dietrich Laubichler and Jane Maienschein, *From Embryology to Evo-Devo: A History of Developmental Evolution* (Cambridge: MIT Press, 2007).

27. Comparative perspectives existed in experimental research before the end of the twentieth century, for example in embryology, comparative physiology, and comparative biochemistry. However, except for embryology, where this tradition remained perhaps the strongest, it was always a minor strand in fields dominated by model organisms and model systems.

28. Coleman, *Biology in the Nineteenth Century*; Allen, *Life Science*; Lenoir, "Shaping Biomedicine."

29. Johnson, *Ordering Life*.

30. Mayr, Linsley, and Usinger, *Methods and Principles*, 282. On the type specimen, see Daston, "Type Specimens."

31. David Weinberger, *Too Big to Know: Rethinking Knowledge Now That the Facts Aren't the Facts, Experts Are Everywhere, and the Smartest Person in the Room Is the Room* (New York: Basic Books, 2012).

32. Kohler, *Lords of the Fly*.

33. Suber, *Open Access*; Bruno J. Strasser and Paul Edwards, *Open Access: Publishing, Commerce, and the Scientific Ethos* (Bern: Swiss Science and Innovation Council, 2016).

34. Christopher M. Kelty, *Two Bits: The Cultural Significance of Free Software* (Durham: Duke University Press, 2008); Christopher J. Tozzi, *For Fun and Profit: A History of the Free and Open Source Software Revolution* (Cambridge: MIT Press, 2017).

35. Richard Van Noorden, "Funders Punish Open-Access Dodgers," *Nature* 508, no. 7495 (Apr 10 2014): 161.

36. David Tyfield, "Transition to Science 2.0: 'Remoralizing' the Economy of Science," *Spontaneous Generations: A Journal for the History and Philosophy of Science* 7, no. 1 (2013): 29–48; Michael Hagner, "Open Access, Data Capitalism and Academic Publishing," *Swiss Medical Weekly* 148 (2018): w14600, https://doi.org/10.4414/smw.2018.14600; Philip

Mirowski, "The Future(s) of Open Science," *Social Studies of Science* 48, no. 2 (2018): 171–203.

37. J. Hodgson, "A Certain Lack of Coordination," *Trends in Biotechnology* 5, no. 3 (Mar 1987): 59–60.

38. Protein Data Bank, Policies & References, available at http://www.rcsb.org/pdb/static.do?p=general_information/about_pdb/policies_references.html (accessed Jul 10 2014).

39. "More Bang for Your Byte," *Scientific Data* 1 (2014): 140010.

40. Björn Brembs, Katherine Button, and Marcus Munafò, "Deep Impact: Unintended Consequences of Journal Rank," *Frontiers in Human Neuroscience* 7, no. 291 (2013): 1–12.

41. Group on Data Citation Standards and Practices CODATA-ICSTI, "Out of Cite, out of Mind: The Current State of Practice, Policy, and Technology for the Citation of Data," *Data Science Journal* 12 (2013): CIDCR1-CIDCR75; see also Paul F. Uhlir, *For Attribution: Developing Data Attribution and Citation Practices and Standards: Summary of an International Workshop* (Washington, DC: National Academies Press, 2012).

42. Thomson Reuters, "Thomson Reuters launches data citation index for discovering global data sets" (Apr 2 2013, accessed Nov 17 2014), http://thomsonreuters.com/content/press_room/science/730914.

43. Université de Liège, Extrait du procès verbal de la séance du Conseil d'administration du 23 mai 2007, available at http://orbi.ulg.ac.be/files/extrait_moniteur_CA.pdf (accessed Aug 28 2017). On the different kinds of open-access mandates, see Suber, *Open Access*, chap. 4.

44. Van Noorden, "Funders Punish Open-Access Dodgers." On the mix of "carrots" and "sticks," see Sabina Leonelli, Daniel Spichtinger, and Barbara Prainsack, "Sticks and Carrots: Encouraging Open Science at Its Source," *Geo* 2, no. 1 (Jun 30 2015): 12–16.

45. Michael Nielsen, *Reinventing Discovery: The New Era of Networked Science* (Princeton: Princeton University Press, 2011).

46. Varmus, *Art and Politics of Science.*

47. Mark J. Costello, "Motivating Online Publication of Data," *Bioscience* 59, no. 5 (May 2009): 418–27; Leonelli, Spichtinger, and Prainsack, "Sticks and Carrots."

48. Especially since the 1990s, argues Stevens, *Life out of Sequence*, chap. 1. Although the professionalization of bioinformatics (and computational biology) clearly takes place at this time, I find it more useful to take a broader historical view.

49. ENCODE Project Consortium, "An Integrated Encyclopedia of DNA Elements in the Human Genome," *Nature* 489, no. 7414 (Sep 6 2012): 57–74.

50. Stevens, *Life out of Sequence.*

51. Mark B. Gerstein et al., "Architecture of the Human Regulatory Network Derived from ENCODE Data," *Nature* 489, no. 7414 (Sep 6 2012): 91–100.

52. David Landsman et al., "Database: A New Forum for Biological Databases and Curation," *Database* (2009): bap002.

53. Kyle Burkhardt, Bohdan Schneider, and Jeramia Ory, "A Biocurator Perspective: Annotation at the Research Collaboratory for Structural Bioinformatics Protein Data Bank," *PLoS Computational Biology* 2, no. 10 (Oct 2006): 1186. For a survey of the professional expectation of curators, see Sarah Burge et al., "Biocurators and Biocuration: Surveying the 21st Century Challenges," *Database (Oxford)* (2012): bar059.

54. Kai Wang, "Gene-Function Wiki Would Let Biologists Pool Worldwide Resources," *Nature* 439, no. 7076 (Feb 2 2006): 534; Jim Giles, "Internet Encyclopaedias Go Head to Head," ibid., 438, no. 7070 (Dec 15 2005): 900–901. For similar suggestions, see Steven L. Salzberg, "Genome Re-annotation: A Wiki Solution?," *Genome Biology* 8, no. 1 (2007): 102.

55. Martin I. Bidartondo, "Preserving Accuracy in GenBank," *Science* 319, no. 5870 (Mar 21 2008): 1616.

56. Elizabeth Pennisi, "Proposal to 'Wikify' GenBank Meets Stiff Resistance," *Science* 319, no. 5870 (Mar 21 2008): 1598–99.

57. Jon W. Huss III et al., "A Gene Wiki for Community Annotation of Gene Function," *PLoS Biology* 6, no. 7 (Jul 8 2008): e175. For other similar efforts, see M. Mitchell Waldrop, "Big Data: Wikiomics," *Nature* 455, no. 7209 (Sep 4 2008): 22–25, and Raja Mazumder et al., "Community Annotation in Biology," *Biology Direct* 5 (2010): 12.

58. Robert Hoffmann, "A Wiki for the Life Sciences Where Authorship Matters," *Nature Genetics* 40, no. 9 (Sep 2008): 1047–51.

59. Andrew I. Su, Benjamin M. Good, and Abdre J. van Wijnen, "Gene Wiki Reviews: Marrying Crowdsourcing with Traditional Peer Review," *Gene* 531, no. 2 (Dec 1 2013): 125.

60. Ibid.

61. Strasser et al., "'Citizen Science'?"; Caren Cooper, *Citizen Science: How Ordinary People Are Changing the Face of Discovery* (New York: Overlook Press, 2016).

62. Jinseop S. Kim et al., "Space-Time Wiring Specificity Supports Direction Selectivity in the Retina," *Nature* 509, no. 7500 (May 15 2014): 331–36.

63. On the role of amateurs in natural history, see the introduction, and specifically on amateurs conducting their own intellectual agendas, and not just those of the professional scientists, see Anne Secord, "Science in the Pub: Artisan Botanists in Early Nineteenth-Century Lancashire," *History of Science* 32 (1994): 269–315, and Etienne Benson, "A Centrifuge of Calculation: Spinning Off Data and Enthusiasts in the US Bird Banding Program, 1920–1940," *Osiris* 32 (2017): 286–306.

64. Good examples can be found on platforms such as Zooniverse (www.zooniverse.org) or SciStarter (www.scistarter.com).

65. This was one of the central claims of Steven Shapin and Simon Schaffer's Leviathan and the Air Pump, developed by Shapin in subsequent work: *Social History of Truth*; *The Scientific Revolution* (Chicago: University of Chicago Press, 1996); *Scientific Life*.

66. "Democracy and the Recognition of Science," *Popular Science*, Mar 1902, 477–78.

Index

Note: References to figures are denoted by an "f" in italics following the page number.